International Conference on Algebra, Analysis and Quantum Probability 2015

Journal of Physics: Conference Series Volume 697

Tashkent, Uzbekistan
10 – 12 September 2015

Editors:

Shavkat Ayupov
Vladimir Chilin
Nasir Ganikhodjaev
Farrukh Mukhamedov
Isamiddin Rakhimov

ISBN: 978-1-5108-2238-2
ISSN: 1742-6588

Printed from e-media with permission by:

Curran Associates, Inc.
57 Morehouse Lane
Red Hook, NY 12571

Some format issues inherent in the e-media version may also appear in this print version.

Printed by Curran Associates, Inc. (2016)

For permission requests, please contact the Institute of Physics
at the address below.

Institute of Physics
Dirac House, Temple Back
Bristol BS1 6BE UK

Phone: 44 1 17 929 7481
Fax: 44 1 17 920 0979

techtracking@iop.org

Additional copies of this publication are available from:

Curran Associates, Inc.
57 Morehouse Lane
Red Hook, NY 12571 USA
Phone: 845-758-0400
Fax: 845-758-2633
Email: curran@proceedings.com
Web: www.proceedings.com

Table of contents

Volume 697

Algebra, Analysis and Quantum Probability10–12 September 2015, Tashkent, Uzbekistan

Accepted papers received: 26 February 2016Published online: 24 March 2016 Preface

Papers

Operator algebras

Structure theory of non-associative algebras

Quadratic operators and processses

Applications of operator theory to statistical physics

Foreword

The present volume of the *Journal of Physics: Conference Series* represents contributions from participants of the International Conference "Algebra, Analysis and Quantum Probability" (Tashkent, 10-12 September 2015) organized by the Institute of Mathematics and the Faculty of Mechanics and Mathematics of the National University of Uzbekistan (NUUz) in collaboration with University Putra Malaysia (UPM) and International Islamic University Malaysia (IIUM).

The Conference is dedicated to the 100th anniversary of one of the outstanding scientists of Uzbekistan, the founder of the Tashkent scientific school of functional analysis, who has initiated the investigations on operator algebras and quantum probability theory in Uzbekistan - Professor Tashmukhamed Alievich Sarymsakov (10 Sept. 1915 - 19 Dec. 1995). Among the mathematical community Professor T. A. Sarymsakov is widely known for his research in the fields of probability theory, functional analysis, general topology and their applications. A gifted teacher and skilful organizer he had a beneficial effect on the development of many new mathematicians in Uzbekistan. Professor T.A. Sarymsakov, an outstanding organizer of science in Uzbekistan, was one of the founders of the Uzbekistan Academy of Sciences, where from 1943 he wass a member and Vice President, and from 1946 to 1952 president of the Academy of Sciences. Professor Sarymsakov successfully combined his fruitful scientific research with teaching and social work. During 1943-1944, 1952-1958 and 1971-1983 he was the rector of Tashkent State University (now the National University of Uzbekistan). He has made a significant contribution to the development of higher education in Uzbekistan, serving from 1959 to 1960 as the Chairman of the State Committee, and from 1960 to 1971 as the Minister of Higher and Secondary Special Education of Uzbekistan.

The main objective of the scientific conference was to facilitate communication and collaboration between mathematicians from different countries and it was focused on contemporary issues in the theory of operator algebras and non-commutative integration, structure theory of non-associative algebras and their applications in the theory of dynamical systems and quantum probability.

The Conference venue was the Mechanics and Mathematics Faculty of the National University of Uzbekistan and Institute of Mathematics.

The program of conference included plenary and invited lectures (40 minutes), short communications (15 minutes) given by more than 100 participants, including invited speakers from Australia, Brazil, France, Germany, Italy, Malaysia, Russia, Spain, USA and Uzbekistan in the following main sections of the conference:

1. Operator algebras and non-commutative integration;
2. Structure theory of non-associative algebras;
3. Quadratic operators and quantum probability;
4. Theory of dynamical systems and statistical physics.

We would like to thank the Invited Speakers for their significant contributions to the conference. We would also like to thank the members of the International Scientific Committee and the members of the Organizing Committee. We cannot end without expressing our many thanks to our sponsors for their financial support.

Guest Editors

Shavkat Ayupov, Vladimir Chilin, Nasir Ganikhodjaev, Farrukh Mukhamedov, Isamiddin Rakhimov

Journal of Physics: Conference Series **697** (2016) 011002 doi:10.1088/1742-6596/697/1/011002

Peer review statement

All papers published in this volume of *Journal of Physics: Conference Series* have been peer reviewed through processes administered by the proceedings Editors. Reviews were conducted by expert referees to the professional and scientific standards expected of a proceedings journal published by IOP Publishing.

2-Local derivations on matrix algebras over semi-prime Banach algebras and on AW^*-algebras

Shavkat Ayupov[1], **Karimbergen Kudaybergenov**[2]

[1] Dormon yoli 29, Institute of Mathematics, National University of Uzbekistan, 100125 Tashkent Uzbekistan

[2]Ch. Abdirov 1, Department of Mathematics Karakalpak state university Nukus 230113, Uzbekistam

E-mail: sh_ayupov@mail.ru, karim2006@mail.ru

Abstract. The paper is devoted to 2-local derivations on matrix algebras over unital semi-prime Banach algebras. For a unital semi-prime Banach algebra $\mathcal{A}$ with the inner derivation property we prove that any 2-local derivation on the algebra $M_{2^n}(\mathcal{A})$, $n \geq 2$, is a derivation. We apply this result to AW^*-algebras and show that any 2-local derivation on an arbitrary AW^*-algebra is a derivation.

1. Introduction

Given an algebra $\mathcal{A}$, a linear operator $D : \mathcal{A} \to \mathcal{A}$ is called a *derivation*, if $D(xy) = D(x)y+xD(y)$ for all $x, y \in \mathcal{A}$ (the Leibniz rule). Each element $a \in \mathcal{A}$ implements a derivation D_a on $\mathcal{A}$ defined as $D_a(x) = \mathrm{ad}(a)(x) = ax - xa$, $x \in \mathcal{A}$. Such derivations D_a are said to be *inner derivations*.

In 1997, P. Semrl [10] introduced the concepts of 2-local derivations and 2-local automorphisms. Recall that a map $\Delta : \mathcal{A} \to \mathcal{A}$ (not linear in general) is called a *2-local derivation* if for every $x, y \in \mathcal{A}$, there exists a derivation $D_{x,y} : \mathcal{A} \to \mathcal{A}$ such that $\Delta(x) = D_{x,y}(x)$ and $\Delta(y) = D_{x,y}(y)$. In particular, he has described 2-local derivations on the algebra $B(H)$ of all bounded linear operators on the infinite-dimensional separable Hilbert space H. In [8] S.O.Kim and J.S.Kim gave a short proof that every 2-local derivation on a finite-dimensional complex matrix algebras is a derivation. The methods of the proofs above mentioned results from [8] and [10] are essentially based on the fact that the algebra $B(H)$ is generated by two elements for separable Hilbert space H. Later J. H. Zhang and H. X. Li [11] have extended the above mentioned result of [8] for arbitrary symmetric digraph matrix algebras and constructed an example of 2-local derivation which is not a derivation on the algebra of all upper triangular complex 2×2-matrices.

As it was mentioned above, the proofs of the papers [8] and [10] are essentially based on the fact that the algebra $B(H)$ is generated by two elements for separable Hilbert space H. Since the algebra $B(H)$ is not generated by two elements for non separable H, one cannot directly apply the methods of the above papers in this case. In [2] the authors suggested a new technique and have generalized the above mentioned results of [8] and [10] for arbitrary Hilbert spaces. Namely, we considered 2-local derivations on the algebra $B(H)$ of all bounded linear operators on an arbitrary (no separability is assumed) Hilbert space H and proved that every 2-local derivation on $B(H)$ is a derivation. A similar result for 2-local derivations on finite von

Neumann algebras was obtained in [5]. In [1] the authors extended all above results and give a short proof of this result for arbitrary semi-finite von Neumann algebras. Finally, in [3], using the analogue of Gleason Theorem for signed measures, we have extended this result to type III von Neumann algebras. This implies that on arbitrary von Neumann algebra each 2-local derivation is a derivation.

In the present paper we consider 2-local derivations on matrix algebras over unital semi-prime Banach algebras. Let $\mathcal{A}$ be a unital semi-prime Banach algebra with the inner derivation property. We prove that any 2-local derivation on the algebra $M_{2^n}(\mathcal{A})$, $n \geq 2$, is a derivation. We also apply this result to AW^*-algebras and prove that any 2-local derivation on an arbitrary AW^*-algebra is a derivation.

2. 2-local derivations on matrix algebras

If $\Delta : \mathcal{A} \to \mathcal{A}$ is a 2-local derivation, then from the definition it easily follows that Δ is homogenous. At the same time,

$$\Delta(x^2) = \Delta(x)x + x\Delta(x) \tag{1}$$

for each $x \in \mathcal{A}$.

In [6] it is proved that any Jordan derivation (i.e. a linear map satisfying the above equation) on a semi-prime algebra is a derivation. So, in the case semi-prime algebras in order to prove that a 2-local derivation $\Delta : \mathcal{A} \to \mathcal{A}$ is a derivation it is sufficient to prove that $\Delta : \mathcal{A} \to \mathcal{A}$ is additive.

We say that an algebra $\mathcal{A}$ has the *inner derivation property* if every derivation on $\mathcal{A}$ is inner. Recall that an algebra $\mathcal{A}$ is said to be *semi-prime* if $a\mathcal{A}a = 0$ implies that $a = 0$.

The following theorem is the main result of this section.

Theorem 2.1 *Let $\mathcal{A}$ be a unital semi-prime Banach algebra with the inner derivation property and let $M_{2^n}(\mathcal{A})$ be the algebra of $2^n \times 2^n$-matrices over $\mathcal{A}$. Then any 2-local derivation Δ on $M_{2^n}(\mathcal{A})$ is a derivation.*

The proof of Theorem 2.1 consists of two steps. In the first step we shall show additivity of Δ on the the subalgebra of diagonal matrices from $M_{2^n}(\mathcal{A})$.

Let $M_n(\mathcal{A})$ be the algebra of $n \times n$-matrices over $\mathcal{A}$ and let $\{e_{i,j}\}_{i,j=1}^n$ be the system of matrix units in $M_n(\mathcal{A})$. For $x \in M_n(\mathcal{A})$ by $x_{i,j}$ we denote the (i,j)-entry of x, i.e. $x_{i,j} = e_{i,i}xe_{j,j}$, where $1 \leq i,j \leq n$. We shall, when necessary, identify this element with the matrix from $M_n(\mathcal{A})$ whose (i,j)-entry is $x_{i,j}$, other entries are zero.

Further in Lemmata 2.2–2.6 we assume that $n \geq 2$.

Lemma 2.2 *Let $\mathcal{A}$ be a unital Banach algebra with the inner derivation property. Then the algebra $M_n(\mathcal{A})$ also has the inner derivation property.*

Proof. Let D be a derivation on $M_n(\mathcal{A})$. Set

$$a = \sum_{i=1}^n D(e_{i,1})e_{1,i}.$$

We have

$$\sum_{i=1}^n D(e_{i,1})e_{1,i} + \sum_{i=1}^n e_{i,1}D(e_{1,i}) = \sum_{i=1}^n D(e_{i,1}e_{1,i}) = \sum_{i=1}^n D(e_{i,i}) = D\left(\sum_{i=1}^n e_{i,i}\right) = D(\mathbf{1}) = 0,$$

where $\mathbf{1}$ is the unit matrix. Therefore $a = -\sum\limits_{i=1}^{n} e_{i,1}D(e_{1,i})$.

From the above equalities by direct calculations we obtain that

$$ae_{k,s} - e_{k,s}a = D(e_{k,1})e_{1,s} + e_{k,1}D(e_{1,s}) = D(e_{k,s}),$$

i.e. $D(e_{k,s}) = [a, e_{k,s}]$ for all $k, s \in \overline{1,n}$.

Put $D_0 = D - \mathrm{ad}(a)$. Since $D_0(e_{1,1}) = 0$, it follows that D_0 maps $e_{1,1}M_n(\mathcal{A})e_{1,1} \equiv \mathcal{A}$ into itself. Therefore the restriction $D_0|_{\mathcal{A}}$ of D_0 onto $\mathcal{A}$ is a derivation. Since $\mathcal{A}$ has the inner derivation property there exists an element $a_{1,1} \in \mathcal{A}$ such that $D_0(x) = [a_{1,1}, x]$ for all $x \in \mathcal{A}$.

Set $b = \sum\limits_{i=1}^{n} e_{i,1}a_{1,1}e_{1,i}$. Let x be a matrix such that $x = e_{k,k}xe_{s,s}$. Then

$$bx - xb = e_{k,1}a_{1,1}e_{1,k}x - xe_{s,1}a_{1,1}e_{1,s} = e_{k,1}[a_{1,1}, e_{1,k}xe_{s,1}]e_{1,s}$$

and

$$\begin{aligned} D_0(x) &= D_0(e_{k,1}e_{1,k}xe_{s,1}e_{1,s}) = D_0(e_{k,1})e_{1,k}xe_{s,1}e_{1,s} + \\ &+ e_{k,1}D_0(e_{1,k}xe_{s,1})e_{1,s} + e_{k,1}e_{1,k}xe_{s,1}D_0(e_{1,s}) = \\ &= 0 + e_{k,1}D_0(e_{1,k}xe_{s,1})e_{1,s} + 0 = e_{k,1}[a_{1,1}, e_{1,k}xe_{s,1}]e_{1,s}, \end{aligned}$$

i.e. $D_0(x) = [b, x]$ for all x of the form $x = e_{k,k}xe_{s,s}$. By linearity of D_0 we have that $D_0 = \mathrm{ad}(b)$. So, $D = \mathrm{ad}(a + b)$. The proof is complete. □

Consider the following two matrices:

$$u = \sum_{i=1}^{n} \frac{1}{2^i} e_{i,i}, \ v = \sum_{i=2}^{n} e_{i-1,i}. \tag{2}$$

It is easy to see that an element $x \in M_n(\mathcal{A})$ commutes with u if and only if it is diagonal, and if an element a commutes with v, then a is of the form

$$a = \begin{pmatrix} a_1 & a_2 & a_3 & \dots & a_n \\ 0 & a_1 & a_2 & \dots & a_{n-1} \\ 0 & 0 & a_1 & \dots & a_{n-2} \\ \vdots & \vdots & \vdots & \vdots & \vdots \\ 0 & 0 & \dots & a_1 & a_2 \\ 0 & 0 & \dots & 0 & a_1 \end{pmatrix}. \tag{3}$$

A result, similar to the following one, was proved in [4, Lemma 4.4] for matrix algebras over commutative regular algebras.

Lemma 2.3 *For every 2-local derivation Δ on $M_n(\mathcal{A})$, there exists a derivation D such that $\Delta|_{M_n(Z(\mathcal{A}))} = D|_{M_n(Z(\mathcal{A}))}$, where $Z(\mathcal{A})$ is the center of the algebra $\mathcal{A}$. In particular, $\Delta|_{sp\{e_{i,j}\}_{i,j=1}^n} = D|_{sp\{e_{i,j}\}_{i,j=1}^n}$, where $sp\{e_{i,j}\}_{i,j=1}^n$ is the linear span of the set $\{e_{i,j}\}_{i,j=1}^n$.*

Proof. By Lemma 2.2 there exists an element a in $M_n(\mathcal{A})$ such that

$$\Delta(u) = [a, u], \ \Delta(v) = [a, v],$$

where u, v are the elements from (2). Replacing Δ by $\Delta - \mathrm{ad}(a)$, if necessary, we can assume that $\Delta(u) = \Delta(v) = 0$.

Let $i, j \in \overline{1, n}$. Take a matrix h such that

$$\Delta(e_{i,j}) = [h, e_{i,j}],\ \Delta(u) = [h, u].$$

Since $\Delta(u) = 0$, it follows that h has diagonal form, i.e. $h = \sum_{i=1}^{n} h_{i,i}$. So we have

$$\Delta(e_{i,j}) = he_{i,j} - e_{i,j}h.$$

In the same way, but starting with the element v instead of u, we obtain

$$\Delta(e_{i,j}) = be_{i,j} - e_{i,j}b,$$

where b has the form (3), depending on $e_{i,j}$. So

$$\Delta(e_{i,j}) = he_{i,j} - e_{i,j}h = be_{i,j} - e_{i,j}b.$$

Since

$$he_{i,j} - e_{i,j}h = (h_{i,i} - h_{j,j})e_{i,j}$$

and

$$[be_{i,j} - e_{i,j}b]_{i,j} = 0,$$

it follows that $\Delta(e_{i,j}) = 0$.

Now let us take a matrix $x = \sum_{i,j=1}^{n} f_{i,j}e_{i,j} \in M_n(Z(\mathcal{A}))$. Then

$$\begin{aligned} e_{i,j}\Delta(x)e_{i,j} &= e_{i,j}D_{e_{i,j},x}(x)e_{i,j} = \\ &= D_{e_{i,j},x}(e_{i,j}xe_{i,j}) - D_{e_{i,j},x}(e_{i,j})xe_{i,j} - e_{i,j}xD_{e_{i,j},x}(e_{i,j}) = \\ &= D_{e_{i,j},x}(f_{j,i}e_{i,j}) - 0 - 0 = f_{j,i}D_{e_{i,j},x}(e_{i,j}) = 0, \end{aligned}$$

i.e. $e_{i,j}\Delta(x)e_{i,j} = 0$ for all $i, j \in \overline{1, n}$. This means that $\Delta(x) = 0$. The proof is complete. □

Further in Lemmata 2.4–2.9 we assume that Δ is a 2-local derivation on the algebra $M_n(\mathcal{A})$, such that $\Delta|_{\mathrm{sp}\{e_{i,j}\}_{i,j=1}^{n}} = 0$.

Let $\Delta_{i,j}$ be the restriction of Δ onto $\mathcal{M}_{i,j} = e_{i,i}M_n(\mathcal{A})e_{j,j}$, where $1 \le i, j \le n$.

Lemma 2.4 *$\Delta_{i,j}$ maps $\mathcal{M}_{i,j}$ into itself.*

Proof. First, let us show that

$$\Delta_{i,j}(x) = e_{i,i}\Delta(x)e_{j,j} \tag{4}$$

for all $x \in \mathcal{M}_{i,j}$.

Let $x = x_{i,j} \in \mathcal{M}_{i,j}$. Take a derivation D on $M_n(\mathcal{A})$ such that

$$\Delta(x) = D(x),\ \Delta(e_{i,j}) = D(e_{i,j}).$$

It is suffices to consider the following two cases.

Case 1. Let $i = j$. Then

$$\begin{aligned} \Delta_{i,i}(x) &= \Delta(x) = D(x) = D(e_{i,i}xe_{i,i}) = \\ &= D(e_{i,i})xe_{i,i} + e_{i,i}D(x)e_{i,i} + e_{i,i}xD(e_{i,i}) = \\ &= 0 + e_{i,i}\Delta(x)e_{i,i} + 0 = e_{i,i}\Delta(x)e_{i,i}. \end{aligned}$$

Case 2. Let $i \neq j$. Denote by $\mathbf{1}$ the unit matrix. Since $e_{i,i}x(\mathbf{1} - e_{i,i}) = x$ and $(\mathbf{1} - e_{j,j})xe_{j,j} = x$, we obtain that

$$\begin{aligned} e_{i,i}\Delta(x)(\mathbf{1} - e_{i,i}) &= e_{i,i}D_{e_{i,i},x}(x)(\mathbf{1} - e_{i,i}) = \\ &= D_{e_{i,i},x}(e_{i,i}x(\mathbf{1} - e_{i,i})) - D_{e_{i,i},x}(e_{i,i})x(\mathbf{1} - e_{i,i}) - e_{i,i}xD_{e_{i,i},x}(\mathbf{1} - e_{i,i}) = \\ &= D_{e_{i,i},x}(x) - 0 - 0 = \Delta(x) \end{aligned}$$

and

$$\begin{aligned} (\mathbf{1} - e_{j,j})\Delta(x)e_{j,j} &= (\mathbf{1} - e_{j,j})D_{e_{j,j},x}(x)e_{j,j} = \\ &= D_{e_{j,j},x}((\mathbf{1} - e_{j,j})xe_{j,j}) - D_{e_{j,j},x}(\mathbf{1} - e_{j,j})xe_{j,j} - (\mathbf{1} - e_{j,j})xD_{e_{j,j},x}(e_{j,j}) = \\ &= D_{e_{j,j},x}(x) - 0 - 0 = \Delta(x). \end{aligned}$$

Hence

$$e_{i,i}\Delta(x)e_{j,j} = (\mathbf{1} - e_{j,j})e_{i,i}\Delta(x)(\mathbf{1} - e_{i,i})e_{j,j} = (\mathbf{1} - e_{j,j})\Delta(x)e_{j,j} = \Delta(x).$$

The proof is complete. □

Lemma 2.5 *Let* $x = \sum_{i=1}^{n} x_{i,i}$ *be a diagonal matrix. Then*

$$e_{k,k}\Delta(x)e_{k,k} = \Delta(x_{k,k}) \tag{5}$$

for all $k \in \overline{1,n}$.

Proof. Take an element a from $M_n(\mathcal{A})$ such that

$$\Delta(x) = [a, x],\ \Delta(x_{k,k}) = [a, x_{k,k}].$$

Since x is a diagonal matrix, the equality (4) implies that

$$\Delta(x_{k,k}) = e_{k,k}\Delta(x_{k,k})e_{k,k} = e_{k,k}[a, x_{k,k}]e_{k,k} = [a_{k,k}, x_{k,k}]$$

and

$$e_{k,k}\Delta(x)e_{k,k} = e_{k,k}[a, x]e_{k,k} = [a_{k,k}, x_{k,k}].$$

Thus $e_{k,k}\Delta(x)e_{k,k} = \Delta(x_{k,k})$. The proof is complete. □

Lemma 2.6 *Let* $x = x_{i,i} \in \mathcal{M}_{i,i}$. *Then*

$$e_{j,i}\Delta_{i,i}(x)e_{i,j} = \Delta_{j,j}(e_{j,i}xe_{i,j}). \tag{6}$$

Proof. For $i = j$ we have already proved (see Lemma 2.5).

Suppose that $i \neq j$. For arbitrary element $x = x_{i,i} \in \mathcal{M}_{i,i}$, consider $y = x + e_{j,i}xe_{i,j} \in \mathcal{M}_{i,i} + \mathcal{M}_{j,j}$. Take an element $a \in \mathcal{A}$ such that

$$\Delta(x) = [a, y] \text{ and } \Delta(v) = [a, v],$$

where v is the element from (2). Since $\Delta(v) = 0$, it follows that a has the form (3). By Lemma 2.5 we obtain that

$$e_{j,i}\Delta_{i,i}(x)e_{i,j} = e_{j,i}e_{i,i}\Delta(y)e_{i,i}e_{i,j} = e_{j,i}[a, y]e_{i,j} = e_{j,i}[a_1, x]e_{i,j}$$

and

$$\Delta_{j,j}(e_{j,i}xe_{i,j}) = e_{j,j}\Delta(y)e_{j,j} = e_{j,j}[a, y]e_{j,j} = e_{j,j}[a, x + e_{j,i}xe_{i,j}]e_{j,j} = e_{j,i}[a_1, x]e_{i,j}.$$

The proof is complete. □

Further in Lemmata 2.7–2.9 we assume that $n \geq 3$.

Lemma 2.7 *$\Delta_{i,i}$ is additive for all $i \in \overline{1, n}$.*

Proof. Let $i \in \overline{1, n}$. Since $n \geq 3$, we can take different numbers k, s such that $k \neq i, s \neq i$.

For arbitrary $x, y \in \mathcal{M}_{i,i}$ consider $z = x + y + e_{k,i}xe_{i,k} + e_{s,i}ye_{i,s} \in \mathcal{M}_{i,i} + \mathcal{M}_{k,k} + \mathcal{M}_{s,s}$. Take a matrix $a \in M_n(\mathcal{A})$ such that

$$\Delta(z) = [a, z] \text{ and } \Delta(v) = [a, v],$$

where v is the element from (2). Since $\Delta(v) = 0$, it follows that a has the form (3). Using Lemma 2.5 we obtain that

$$\begin{aligned}
\Delta_{i,i}(x + y) &= e_{i,i}\Delta(z)e_{i,i} = e_{i,i}[a, z]e_{i,i} = [a_1, x + y],\\
\Delta_{i,i}(x) &= e_{i,k}\Delta(e_{k,i}xe_{i,k})e_{k,i} = e_{i,k}e_{k,k}\Delta(z)e_{k,k}e_{k,i} = e_{i,k}[a, z]e_{k,i} = [a_1, x],\\
\Delta_{i,i}(y) &= e_{i,s}\Delta(e_{s,i}ye_{i,s})e_{s,i} = e_{i,s}e_{s,s}\Delta(z)e_{s,s}e_{s,i} = e_{i,s}[a, z]e_{s,i} = [a_1, y].
\end{aligned}$$

Hence

$$\Delta_{i,i}(x + y) = \Delta_{i,i}(x) + \Delta_{i,i}(y).$$

The proof is complete. □

As it was mentioned in the beginning of the section any additive 2-local derivation on a semi-prime algebra is a derivation. Since $\mathcal{M}_{i,i} \equiv \mathcal{A}$ is semi-prime, Lemma 2.7 implies the following result.

Lemma 2.8 *$\Delta_{i,i}$ is a derivation for all $i \in \overline{1, n}$.*

Since $\Delta_{1,1}$ is a derivation on $e_{1,1}M_n(\mathcal{A})e_{1,1} \equiv \mathcal{A}$ and $\mathcal{A}$ has the inner derivation property, it follows that there exists an element $a_{1,1}$ in $\mathcal{A}$ such that $\Delta_{1,1} = \mathrm{ad}(a_{1,1})$. Set $\tilde{a} = \sum\limits_{i=1}^{n} e_{i,1}a_{1,1}e_{1,i}$.

Denote by $\mathcal{D}_n$ the set of all diagonal matrices from $M_n(\mathcal{A})$, i.e. the set of all matrices of the following form

$$x = \begin{pmatrix} x_1 & 0 & 0 & \dots & 0 \\ 0 & x_2 & 0 & \dots & 0 \\ \vdots & \vdots & \vdots & \vdots & \vdots \\ 0 & 0 & \dots & x_{n-1} & 0 \\ 0 & 0 & \dots & 0 & x_n \end{pmatrix}.$$

Let $x \in \mathcal{D}_n$. Then

$$\begin{aligned}
[\tilde{a}, x]_{i,i} &= e_{i,i}\tilde{a}xe_{i,i} - e_{i,i}x\tilde{a}e_{i,i} = e_{i,1}a_{1,1}e_{1,i}x_{i,i} - x_{i,i}e_{1,i}a_{1,1}e_{1,i} =\\
&= e_{i,1}[a_{1,1}, e_{1,i}x_{i,i}e_{i,1}]e_{1,i} = e_{i,1}\Delta_{1,1}(e_{1,i}x_{i,i}e_{i,1})e_{1,i} \overset{(6)}{=}\\
&= \Delta_{i,i}(x_{i,i}) \overset{(5)}{=} e_{i,i}\Delta(x)e_{i,i} = \Delta(x)_{i,i},
\end{aligned}$$

i.e. $\operatorname{ad}(\tilde{a})|_{\mathcal{D}_n} = \Delta|_{\mathcal{D}_n}$.

Further,

$$\begin{aligned}[\tilde{a}, e_{i,j}] &= \tilde{a}e_{i,j} - e_{i,j}\tilde{a} = e_{i,1}a_{1,1}e_{1,i}e_{i,j} - e_{i,j}e_{j,1}a_{1,1}e_{1,j} = \\ &= e_{i,1}a_{1,1}e_{1,j} - e_{i,1}a_{1,1}e_{1,j} = 0,\end{aligned}$$

i.e. $\operatorname{ad}(\tilde{a})|_{\operatorname{sp}\{e_{i,j}\}_{i,j=1}^n} \equiv 0$, where $\operatorname{sp}\{e_{i,j}\}_{i,j=1}^n$ is the linear span of the set $\{e_{i,j}\}_{i,j=1}^n$.

So, we have proved the following result.

Lemma 2.9 $\Delta|_{\mathcal{D}_n} = ad(\tilde{a})|_{\mathcal{D}_n}$ *and* $ad(\tilde{a})|_{sp\{e_{i,j}\}_{i,j=1}^n} = 0$.

Replacing, if necessary, Δ by $\Delta - \operatorname{ad}(\tilde{a})$, below in this section we assume that

$$\Delta|_{\mathcal{D}_n} \equiv 0,\ \Delta|_{\operatorname{sp}\{e_{i,j}\}_{i,j=1}^n} \equiv 0.$$

Now we are in position to pass to the second step of our proof. As the second step let us show that if a 2-local derivation Δ on a matrix algebra equals to zero on all diagonal matrices and on the linear span of matrix units, then it is identically zero on the whole algebra. In order to prove this we first consider the 2×2-matrix algebras case.

2.1. *The case of 2×2-matrices*

In this subsection we shall assume that $\mathcal{B}$ is a unital Banach algebra with the inner derivation property and Δ is a 2-local derivation on $M_2(\mathcal{B})$, such that

$$\Delta|_{\operatorname{sp}\{e_{i,j}\}_{i,j=1}^2} \equiv 0 \text{ and } \Delta|_{\mathcal{D}_2} \equiv 0.$$

We also denote by e the unit of the algebra $\mathcal{B}$.

Lemma 2.10 $\Delta(x)_{1,1} = \Delta(x)_{2,2} = 0$ *for all* $x \in M_2(\mathcal{B})$.

Proof. Let $x = \begin{pmatrix} x_{1,1} & x_{1,2} \\ x_{2,1} & x_{2,2} \end{pmatrix}$. Since Δ is homogeneous, we can assume that $\|x_{1,1}\| < 1$, where $\|\cdot\|$ is the norm on $\mathcal{B}$. Set $y = \begin{pmatrix} e + x_{1,1} & 0 \\ 0 & 0 \end{pmatrix}$. Since $\|x_{1,1}\| < 1$, it follows that $e + x_{1,1}$ is invertible in $\mathcal{B}$. Take an element $a \in M_2(\mathcal{B})$ such that

$$\Delta(x) = [a, x],\ \Delta(y) = [a, y].$$

Since $y \in \mathcal{D}_2$ we have that $0 = \Delta(y) = [a, y]$, and therefore

$$0 = \Delta(y)_{1,1} = a_{1,1}(e + x_{1,1}) - (e + x_{1,1})a_{1,1} = 0,$$

$$0 = \Delta(y)_{2,1} = a_{2,1}(e + x_{1,1}) = 0,$$

and

$$0 = \Delta(y)_{1,2} = -(e + x_{1,1})a_{1,2} = 0.$$

Thus

$$a_{1,1}x_{1,1} - x_{1,1}a_{1,1} = 0$$

and

$$a_{2,1} = a_{1,2} = 0.$$

The above equalities imply that

$$\Delta(x)_{1,1} = a_{1,1}x_{1,1} - x_{1,1}a_{1,1} = 0.$$

In a similar way we can show that $\Delta(x)_{2,2} = 0$. The proof is complete. □

Lemma 2.11 *Let x be a matrix with $x_{k,s} = \lambda e$, where $\lambda \in \mathbb{C}$. Then $\Delta(x)_{k,s} = 0$.*

Proof.

$$\begin{aligned} e_{s,k}\Delta(x)e_{s,k} &= e_{s,k}D_{e_{s,k},x}(x)e_{s,k} = \\ &= D_{e_{s,k},x}(e_{s,k}xe_{s,k}) - D_{e_{s,k},x}(e_{s,k})xe_{s,k} - e_{s,k}xD_{e_{s,k},x}(e_{s,k}) = \\ &= \lambda D_{e_{s,k},x}(e_{s,k}) - 0 - 0 = 0. \end{aligned}$$

Thus

$$\Delta(x)_{k,s} = e_{k,k}\Delta(x)e_{s,s} = e_{k,s}e_{s,k}\Delta(x)e_{s,k}e_{k,s} = 0.$$

The proof is complete. □

Lemma 2.12 *Let $x = \begin{pmatrix} x_{1,1} & x_{1,2} \\ x_{2,1} & x_{2,2} \end{pmatrix}$ and $y = \begin{pmatrix} x_{1,1} & x_{1,2} \\ 0 & x_{2,2} \end{pmatrix}$. Then $\Delta(x)_{1,2} = \Delta(y)_{1,2}$.*

Proof. Take a matrix $a \in M_2(\mathcal{B})$ such that

$$\Delta(x) = [a, x] \text{ and } \Delta(y) = [a, y].$$

Then

$$\Delta(x)_{1,2} = a_{1,1}x_{1,2} + a_{1,2}x_{2,2} - x_{1,1}a_{1,2} - x_{1,2}a_{2,2} = \Delta(y)_{1,2}.$$

The proof is complete. □

Lemma 2.13 *Let $x = \begin{pmatrix} e + x_{1,1} & x_{1,1} \\ \lambda e & 0 \end{pmatrix}$, where $x_{1,1} \in \mathcal{B}$, $\|x_{1,1}\| < 1$, $\lambda \in \mathbb{C}$. Then $\Delta(x) = 0$.*

Proof. From Lemmata 2.10 and 2.11, it follows that

$$\Delta(x)_{1,1} = \Delta(x)_{2,2} = \Delta(x)_{2,1} = 0.$$

Let us to show that $\Delta(x)_{1,2} = 0$.

Case 1. Let $\lambda = 0$. Take a matrix $a \in M_2(\mathcal{B})$ such that

$$\Delta(x) = [a, x] \text{ and } \Delta(e_{2,1}) = [a, e_{2,1}].$$

Since the element a commutes with $e_{2,1}$, it follows that a is of the form $a = \begin{pmatrix} a_1 & 0 \\ a_2 & a_1 \end{pmatrix}$. Then

$$0 = \Delta(x)_{2,1} = a_2(e + x_{1,1}) = 0.$$

Since $e + x_{1,1}$ is invertible in $\mathcal{B}$, it follows that $a_2 = 0$. From the last equality we obtain that

$$0 = \Delta(x)_{1,1} = a_1x_{1,1} - x_{1,1}a_1 = \Delta(x)_{1,2}.$$

i.e. $\Delta(x)_{1,2} = 0$. Therefore, $\Delta(x) = 0$.

Case 2. Let $\lambda \neq 0$. Set $y = \begin{pmatrix} e + x_{1,1} & x_{1,1} \\ 0 & 0 \end{pmatrix}$. By Case 1, $\Delta(y) = 0$. Applying Lemma 2.12 we obtain that

$$0 = \Delta(y)_{1,2} = \Delta(x)_{1,2}.$$

i.e. $\Delta(x)_{1,2} = 0$. Thus $\Delta(x) = 0$. The proof is complete. □

Lemma 2.14 *If $x = \begin{pmatrix} x_{1,1} & x_{1,2} \\ \lambda e & 0 \end{pmatrix}$, where $x_{1,1}$ is an invertible element in $\mathcal{B}$, then $\Delta(x) = 0$.*

Proof. As in the proof of Lemma 2.13 it is suffices to show that $\Delta(x)_{1,2} = 0$. Since Δ is homogeneous, we can assume that $\|x_{1,2}\| < 1$.

Case 1. Let $\lambda = 0$. Set $y = \begin{pmatrix} e + x_{1,2} & x_{1,2} \\ e & 0 \end{pmatrix}$. Take a matrix $a \in M_2(\mathcal{B})$ such that

$$\Delta(x) = [a, x] \text{ and } \Delta(y) = [a, y].$$

By Lemma 2.13 we have that $\Delta(y) = 0$. Since

$$0 = \Delta(x)_{2,1} = a_{2,1}x_{1,1},$$

and $x_{1,1}$ is invertible in $\mathcal{B}$, it follows that $a_{2,1} = 0$. From

$$0 = \Delta(y)_{2,2} = a_{2,1}x_{1,2} - a_{1,2},$$

it follows that $a_{1,2} = 0$. So, a is a diagonal matrix. This implies that

$$0 = \Delta(y)_{1,2} = a_{1,1}x_{1,2} - x_{1,2}a_{2,2} = \Delta(x)_{1,2},$$

i.e. $\Delta(x)_{1,2} = 0$. Therefore $\Delta(x) = 0$.

Case 2. Let $\lambda \neq 0$. Set $y = \begin{pmatrix} x_{1,1} & x_{1,2} \\ 0 & 0 \end{pmatrix}$. By Case 1, $\Delta(y) = 0$. From Lemma 2.12 we obtain that

$$0 = \Delta(y)_{1,2} = \Delta(x)_{1,2}.$$

i.e. $\Delta(x)_{1,2} = 0$, and hence $\Delta(x) = 0$. The proof is complete. □

Lemma 2.15 *Let* $x = \begin{pmatrix} x_{1,1} & x_{1,2} \\ 0 & x_{2,2} \end{pmatrix}$, *where* $x_{1,1}$ *is an invertible element in* $\mathcal{B}$. *Then* $\Delta(x) = 0$.

Proof. Set $y = \begin{pmatrix} x_{1,1} & x_{1,2} \\ e & 0 \end{pmatrix}$. Take a matrix $a \in M_2(\mathcal{B})$ such that

$$\Delta(x) = [a, x] \text{ and } \Delta(y) = [a, y].$$

By Lemma 2.14 we have that $\Delta(y) = 0$. Since

$$0 = \Delta(y)_{1,1} - \Delta(x)_{1,1} = a_{1,2},$$

it follows that $a_{1,2} = 0$. From the last equality we obtain that

$$0 = \Delta(y)_{1,2} = a_{1,1}x_{1,2} - x_{1,2}a_{2,2} = \Delta(x)_{1,2},$$

i.e. $\Delta(x)_{1,2} = 0$. Therefore $\Delta(x) = 0$. The proof is complete. □

Lemma 2.16 *Let* Δ *be a 2-local derivation on* $M_2(\mathcal{B})$ *such that*

$$\Delta|_{sp\{e_{i,j}\}_{i,j=1}^2} \equiv 0, \ \Delta|_{\mathcal{D}_2} \equiv 0.$$

Then $\Delta \equiv 0$.

Proof. Let $x = \begin{pmatrix} x_{1,1} & x_{1,2} \\ x_{2,1} & x_{2,2} \end{pmatrix}$. By Lemma 2.10, $\Delta(x)_{1,1} = \Delta(x)_{2,2} = 0$.

Let us to show that $\Delta(x)_{1,2} = 0$. Since Δ is homogeneous we can assume that $\|x_{1,1}\| < 1$. Further, since $\Delta(x) = \Delta(\mathbf{1} + x)$, replacing, if necessary, x by $\mathbf{1} + x$, we may assume that $x_{1,1}$ is invertible in $\mathcal{B}$.

Put $y = \begin{pmatrix} x_{1,1} & x_{1,2} \\ 0 & x_{2,2} \end{pmatrix}$. Lemma 2.15 implies that $\Delta(y) = 0$. Now from Lemma 2.12 we obtain that

$$\Delta(x)_{1,2} = \Delta(y)_{1,2} = 0.$$

In a similar way we can show that $\Delta(x)_{2,1} = 0$. The proof is complete. □

Journal of Physics: Conference Series **697** (2016) 012001 doi:10.1088/1742-6596/697/1/012001

2.2. The general case

Now we are in position to prove Theorem 2.1.

Proof of Theorem 2.1. Let Δ be a 2-local derivation on $M_{2^n}(\mathcal{A})$, where $n \geq 2$. By Lemma 2.3 there exists a derivation D on $M_{2^n}(\mathcal{A})$ such that $\Delta|_{\mathrm{sp}\{e_{i,j}\}_{i,j=1}^{2n}} = D|_{\mathrm{sp}\{e_{i,j}\}_{i,j=1}^{2n}}$. Replacing, if necessary, Δ by $\Delta - D$, we may assume that Δ is equal to zero on $\mathrm{sp}\{e_{i,j}\}_{i,j=1}^{2n}$. Further, by Lemma 2.9 there exists a diagonal element $\tilde{a}$ in $M_{2^n}(\mathcal{A})$ such that $\Delta|_{\mathcal{D}_{2n}} = \mathrm{ad}(\tilde{a})|_{\mathcal{D}_{2n}}$. Now replacing Δ by $\Delta - \mathrm{ad}(\tilde{a})$, we can assume that Δ is identically zero on $\mathcal{D}_{2^n}$. So, we can assume that

$$\Delta|_{\mathrm{sp}\{e_{i,j}\}_{i,j=1}^{2n}} \equiv 0 \text{ and } \Delta|_{\mathcal{D}_{2^n}} \equiv 0.$$

Let us to show that $\Delta \equiv 0$. We proceed by induction on n.

Let $n = 2$. We identify the algebra $M_4(\mathcal{A})$ with the algebra of 2×2-matrices $M_2(\mathcal{B})$, over $\mathcal{B} = M_2(\mathcal{A})$.

Let $\{e_{i,j}\}_{i,j=1}^{4}$ be a system of matrix units in $M_4(\mathcal{A})$. Then

$$p_{1,1} = e_{1,1} + e_{2,2},\ p_{2,2} = e_{3,3} + e_{4,4},\ p_{1,2} = e_{1,3} + e_{2,4},\ p_{2,1} = e_{3,1} + e_{4,2}$$

is the system of matrix units in $M_2(\mathcal{B})$. Since $\Delta|_{\mathrm{sp}\{e_{i,j}\}_{i,j=1}^{4}} \equiv 0$, it follows that $\Delta|_{\mathrm{sp}\{p_{i,j}\}_{i,j=1}^{2}} \equiv 0$.

Take an arbitrary element $x \in p_{1,1}M_2(\mathcal{B})p_{1,1} \equiv \mathcal{B}$. Choose a derivation D on $M_2(\mathcal{B})$ such that

$$\Delta(x) = D(x),\ \Delta(p_{1,1}) = D(p_{1,1}).$$

Since $\Delta(p_{1,1}) = 0$, we obtain that

$$p_{1,1}\Delta(x)p_{1,1} = p_{1,1}D(x)p_{1,1} = D(p_{1,1}xp_{1,1}) - D(p_{1,1})xp_{1,1} - p_{1,1}xD(p_{1,1}) = \Delta(x).$$

This means that the restriction $\Delta_{1,1}$ of Δ onto $p_{1,1}M_2(\mathcal{B})p_{1,1} \equiv \mathcal{B}$ maps $\mathcal{B} = M_2(\mathcal{A})$ into itself.

If $\mathcal{D}_4$ is the subalgebra of diagonal matrices from $M_4(\mathcal{A})$, then $p_{1,1}\mathcal{D}_4p_{1,1}$ is the subalgebra of diagonal matrices in the algebra $M_2(\mathcal{A})$. Since $\Delta|_{\mathcal{D}_4} \equiv 0$, it follows that $\Delta_{1,1}$ is identically zero on the diagonal matrices from $M_2(\mathcal{A})$. So,

$$\Delta_{1,1}|_{\mathrm{sp}\{e_{i,j}\}_{i,j=1}^{2}} \equiv 0 \text{ and } \Delta_{1,1}|_{p_{1,1}\mathcal{D}_4p_{1,1}} \equiv 0.$$

By Lemma 2.16 it follows that $\Delta_{1,1} \equiv 0$.

Let $\mathcal{D}_2$ be the set of diagonal matrices from $M_2(\mathcal{B})$. Since

$$\mathcal{D}_2 = \begin{pmatrix} \mathcal{B} & 0 \\ 0 & \mathcal{B} \end{pmatrix}$$

and $\Delta_{1,1} = 0$, Lemma 2.5 implies that $\Delta|_{\mathcal{D}_2} \equiv 0$. Hence, Δ is a 2-local derivation on $M_2(\mathcal{B})$ such that

$$\Delta|_{\mathrm{sp}\{p_{i,j}\}_{i,j=1}^{2}} \equiv 0 \text{ and } \Delta|_{\mathcal{D}_2} \equiv 0.$$

Again by Lemma 2.16 it follows that $\Delta \equiv 0$.

Now assume that the assertion of the Theorem is true for $n - 1$.

Considering the algebra $M_{2^n}(\mathcal{A})$ as the algebra of 2×2-matrices $M_2(\mathcal{B})$ over $\mathcal{B} = M_{2^{n-1}}(\mathcal{A})$ and repeating the above arguments we obtain that $\Delta \equiv 0$. The proof is complete. □

The condition on the algebra $\mathcal{A}$ to be a Banach algebra was applied only for the invertibility of elements of the forms $\mathbf{1} + x$, where $x \in \mathcal{A}$, $\|x\| < 1$. In this connection the following question naturally arises.

Problem 2.17 *Does Theorem 2.1 hold for arbitrary (not necessarily normed) algebra $\mathcal{A}$ with the inner derivation property?*

3. An application to AW^*-algebras

In this section we apply Theorem 2.1 to the description of 2-local derivations on AW^*-algebras.

Theorem 3.1 *Let $\mathcal{A}$ be an arbitrary AW^*-algebra. Then any 2-local derivation Δ on $\mathcal{A}$ is a derivation.*

Proof. Let us first note that any AW^*-algebra is semi-prime. It is also known [9] that AW^*-algebra has the inner derivation property.

Let z be a central projection in $\mathcal{A}$. Since $D(z) = 0$ for an arbitrary derivation D, it is clear that $\Delta(z) = 0$ for any 2-local derivation Δ on $\mathcal{A}$. Take $x \in \mathcal{A}$ and let D be a derivation on $\mathcal{A}$ such that $\Delta(zx) = D(zx), \Delta(x) = D(x)$. Then we have $\Delta(zx) = D(zx) = D(z)x + zD(x) = z\Delta(x)$. This means that every 2-local derivation Δ maps $z\mathcal{A}$ into $z\mathcal{A}$ for each central projection $z \in \mathcal{A}$. So, we may consider the restriction of Δ onto $e\mathcal{A}$. Since an arbitrary AW^*-algebra can be decomposed along a central projection into the direct sum of an abelian AW^*-algebra, and AW^*-algebras of type $\mathrm{I}_n, n \geq 2$, type I_∞, type II and type III, we may consider these cases separately.

Let $\mathcal{A}$ be an abelian AW^*-algebra. It is well-known that any derivation on a such algebra is identically zero. Therefore any 2-local derivation on an abelian AW^*-algebra is also identically zero.

If $\mathcal{A}$ is an AW^*-algebra of type I_n, $n \geq 2$, with the center $Z(\mathcal{A})$, then it is isomorphic to the algebra $M_n(Z(\mathcal{A}))$. By Lemma 2.3 there exists a derivation D on $\mathcal{A} \equiv M_n(Z(\mathcal{A}))$ such that $\Delta \equiv D$. So, Δ is a derivation.

Let the AW^*-algebra $\mathcal{A}$ have one of the types I_∞, II or III. Then the halving Lemma [7, Lemma 4.5] for type I_∞-algebras and [7, Lemma 4.12] for type II or III algebras, imply that the unit of the algebra $\mathcal{A}$ can be represented as a sum of mutually equivalent orthogonal projections e_1, e_2, e_3, e_4 from $\mathcal{A}$. Then the map $x \mapsto \sum\limits_{i,j=1}^{4} e_i x e_j$ defines an isomorphism between the algebra $\mathcal{A}$ and the matrix algebra $M_4(\mathcal{B})$, where $\mathcal{B} = e_{1,1}\mathcal{A}e_{1,1}$. Therefore Theorem 2.1 implies that any 2-local derivation on $\mathcal{A}$ is a derivation. The proof is complete. □

References

[1] Ayupov Sh A and Arzikulov F N 2014 *Glasgow Math. Jour.* **56** 9–12
[2] Ayupov Sh A and Kudaybergenov K K 2012 *J. Math. Anal. Appl.* **395** 15-18
[3] Ayupov Sh A and Kudaybergenov K K 2015 *Positivity,* **19** 445–455
[4] Ayupov Sh A, Kudaybergenov K K and Alauadinov A K 2013 *Linear Alg. Appl.* **439** 1294–1311
[5] Ayupov Sh A and Kudaybergenov K K, Nurjanov B O and Alauadinov A K 2014 *Mathematica Slovaca,* **64** 423–432
[6] Brešar M 1988 *Proc. Amer. Math. Soc.* **104** 1003-1006
[7] Kaplansky I 1951 *Ann. Math.* **53** 235-249
[8] Kim S O and Kim H S 2004 *Proc. Amer. Math. Soc.* **132** 1389–1392
[9] Olesen D 1974 *Pacific J. Math.* **53** 555-561
[10] Šemrl P 1997 *Proc. Amer. Math. Soc.* **125** 2677–2680
[11] Zhang J H and Li H X 2006 *Acta Math. Sinica, Chinese ser.* **49** 1401–1406

Strictly homogeneous laterally complete modules

V I Chilin [1], **J A Karimov**[2]

[1,2] Faculty of Mechanics and Mathematics, National University of Uzbekistan, 100174, Vuzgorodok, Tashkent, Uzbekistan

E-mail: vladimirchil@gmail.com, karimovja@mail.ru

Abstract. Let $\mathcal{A}$ be a laterally complete commutative regular algebra and X be a laterally complete $\mathcal{A}$-module. In this paper we introduce a notion of homogeneous and strictly homogeneous $\mathcal{A}$-modules. It is proved that any homogeneous $\mathcal{A}$-module is strictly homogeneous $\mathcal{A}$-module, if the Boolean algebra of all idempotents in $\mathcal{A}$ is multi-σ-finite.

1. Introduction

J. Kaplansky [5] introduced a class of AW^*-algebras to describe C^*-algebras, which is close to von Neumann algebras by their algebraic and order structure. The class of AW^*-algebras became a subject of many researches in the operator theory (see review in [2]). One of the important results in this direction is the realization of an arbitrary AW^*-algebra M of type I as a $*$-algebra of all linear bounded operators, which act in a special Banach module over the center $Z(M)$ of the algebra M [6]. The Banach $Z(M)$-valued norm in this module is generated by the scalar product with values in the commutative AW^*-algebra $Z(M)$. Later, these modules were called Kaplansky-Hilbert modules (KHM). Detailed exposition of many useful properties of KHM is given, for example, in ([10], 7.4). One of the important properties is a representation of an arbitrary Kaplansky-Hilbert module as a direct sum of homogeneous KHM ([7], [10], 7.4.7).

Development of the noncommutative integration theory stimulated an interest to the different classes of algebras of unbounded operators, in particular, to the $*$-algebras $LS(M)$ of locally measurable operators, affiliated with von Neumann algebras or AW^*-algebras M. If M is a von Neumann algebra, then the center $Z(LS(M))$ in the algebra $LS(M)$ identifies with the algebra $L^0(\Omega, \Sigma, \mu)$ of all classes of equal almost everywhere measurable complex functions, defined on some measurable space (Ω, Σ, μ) with a complete locally finite measure μ ([12], 2.1, 2.2). If M is an AW^*-algebra, then $Z(LS(M))$ is an extended f-algebra $C_\infty(Q)$, where Q is the Stone compact coresponding to the Boolean algebra of central projectors in M [2]. The problem (like the one in the work of J. Kaplansky [6] for AW^*-algebras) on possibility of realization of $*$-algebras $LS(M)$, in the case, when M has the type I, as $*$-algebras of linear $L^0(\Omega, \Sigma, \mu)$-bounded (respectively, $C_\infty(Q)$-bounded) operators, which act in corresponding KHM over the $L^0(\Omega, \Sigma, \mu)$ or over the $C_\infty(Q)$ naturally arises. In order to solve this problem it is necessary to construct corresponding theory of KHM over the algebras $L^0(\Omega, \Sigma, \mu)$ and $C_\infty(Q)$. In a particular case of KHM over the algebras $L^0(\Omega, \Sigma, \mu)$ this problem is solved in [8], where the decomposition of KHM over $L^0(\Omega, \Sigma, \mu)$ as a direct sum of homogeneous KHM is given. Similar decomposition as a direct sum of strictly γ-homogeneous modules is given in the paper [3] for arbitrary regular laterally complete modules over the algebra $C_\infty(Q)$ (the definitions see in the Section 3 below).

In this paper is studied laterally complete modules over laterally complete regular algebras $\mathcal{A}$ and it is proved that any homogeneous $\mathcal{A}$-module is strictly homogeneous if the Boolean algebra of all idempotents in $\mathcal{A}$ is multi-σ-finite.

2. Cyclic hull of set

Let $\mathcal{A}$ be a commutative algebra over the field K with the unity $\mathbf{1}$ and $\nabla = \{e \in \mathcal{A} : e^2 = e\}$ be a set of all idempotents in $\mathcal{A}$. For all $e, f \in \nabla$ we write $e \leq f$ if $ef = e$. It is well known (see, for example [11, Prop. 1.6]) that this binary relation is partial order in ∇ and ∇ is a Boolean algebra with respect to this order. Moreover, we have the following equalities: $e \vee f = e + f - ef$, $e \wedge f = ef$, $Ce = \mathbf{1} - e$ with respect to the lattice operations and the complement Ce in ∇.

The commutative unital algebra $\mathcal{A}$ is called regular if the following equivalent conditions hold [14, §2]:

1. For any $a \in \mathcal{A}$ there exists $b \in \mathcal{A}$ such that $a = a^2 b$;
2. For any $a \in \mathcal{A}$ there exists $e \in \nabla$ such that $a\mathcal{A} = e\mathcal{A}$.

A regular algebra $\mathcal{A}$ is a regular semigroup with respect to the multiplication operation [4, Ch. I, §1.9]. In this case $\mathcal{A}$ is a commutative inverse semigroup, i.e. for any $a \in \mathcal{A}$ there exists an unique element $i(a) \in \mathcal{A}$, which is an unique solution of the system: $a^2 x = a$, $ax^2 = x$ [4, Ch. I, §1.9]. The element $i(a)$ is called an inversion of the element a. Obviously, $ai(a) \in \nabla$ for any $a \in \mathcal{A}$. In this case the map $i : \mathcal{A} \to \mathcal{A}$ is a bijection and an automorphism (by multiplication) in semigroup $\mathcal{A}$. Moreover, $i(i(a)) = a$ and $i(g) = g$ for all $a \in \mathcal{A}$, $g \in \nabla$.

Idempotent $s(a) \in \nabla$ is called the support of an element $a \in \mathcal{A}$ if $s(a)a = a$ and $ga = a$, $g \in \nabla$ imply $s(a) \leq g$. It is clear that $s(a) = ai(a)$. In particular, $s(e) = ei(e) = e$ for any $e \in \nabla$.

It is easy to show that supports of elements in a commutative regular unital algebra $\mathcal{A}$ satisfy the following properties (compare with [13, Ch. II, §4]):

Proposition 2.1. *If $a, b \in \mathcal{A}$, $e, f \in \nabla$, $0 \neq \alpha \in K$, then*
(i). $s(i(a)) = s(a)$, $s(\alpha a) = s(a)$ and $s(a^n) = s(a)$ for any positive integer n;
(ii). $s(ab) = s(a)s(b)$;
(iii). $ab = 0 \Leftrightarrow s(a)s(b) = 0$;
(iv). If $ab = 0$, then $i(a + b) = i(a) + i(b)$, $s(a + b) = s(a) + s(b)$.

Two elements a and b in a commutative unital regular algebra $\mathcal{A}$ are called *disjoint elements*, if $ab = 0$, which equivalent to the equality $s(a)s(b) = 0$ (see Proposition 2.1, *(iii)*). If the Boolean algebra ∇ of all idempotents in $\mathcal{A}$ is complete, $a \in \mathcal{A}$ and $r(a) = \sup\{e \in \nabla : ae = 0\}$, then

$$s(a)r(a) = s(a) \wedge r(a) = s(a) \wedge (\sup\{e : ae = 0\}) =$$
$$= \sup\{s(a) \wedge e : ae = 0\} = \sup\{s(a)e : ae = 0\} = 0.$$

Hence $s(a) \leq \mathbf{1} - r(a)$. If $q = (\mathbf{1} - r(a) - s(a))$, then $aq = as(a)q = 0$, thus $q \leq r(a)$. This yields that $q = 0$, i.e. $s(a) = \mathbf{1} - r(a)$.

Proposition 2.2. *Let $\mathcal{A}$ be a commutative unital regular algebra and let ∇ be complete Boolean algebra of idempotents in $\mathcal{A}$. If $\{e_i\}_{i \in I}$ is a partition of unity in ∇, $a, b \in \mathcal{A}$ and $ae_i = be_i$ for all $i \in I$, then $a = b$.*

Proof. Since $(a - b)e_i = 0$ for any $i \in I$, then $\mathbf{1} = \sup\limits_{i \in I} e_i \leq r(a - b)$, i.e. $r(a - b) = \mathbf{1}$. Hence, $s(a - b) = 0$, i.e. $a = b$. □

Commutative unital regular algebra $\mathcal{A}$ is called *laterally complete* (*l-complete*) if the Boolean algebra of its idempotents is complete and for any set $\{a_i\}_{i\in I}$ of pairwise disjoint elements in $\mathcal{A}$ there exists an element $a \in \mathcal{A}$ such that $as(a_i) = a_i$ for all $i \in I$. The element $a \in \mathcal{A}$ such that $as(a_i) = a_i$, $i \in I$, in general, is not uniquely determined. However, by Proposition 2.2, it follows that the element a is unique in the case, when $\sup\limits_{i\in I} s(a_i) = \mathbf{1}$. In general case, due to the equality $as(a_i) = a_i = bs(a_i)$ for all $i \in I$ and $a, b \in \mathcal{A}$, it follows that $a \sup_{i\in I} s(a_i) = b \sup_{i\in I} s(a_i)$.

Let $\mathcal{A}$ be a laterally complete commutative regular algebra and let ∇ be a Boolean algebra of all idempotents in $\mathcal{A}$. Let X be a left $\mathcal{A}$-module with algebraic operations $x + y$ and ax, $x, y \in X$, $a \in \mathcal{A}$. Since the algebra $\mathcal{A}$ is commutative, then a left $\mathcal{A}$-module X becomes a right $\mathcal{A}$-module, if we put $xa := ax$, $x \in X$, $a \in \mathcal{A}$. Hence, we can assume, that X is a bimodule over $\mathcal{A}$, where the following equality $ax = xa$ holds for any $x \in X$, $a \in \mathcal{A}$. Next, an $\mathcal{A}$-bimodule X we shall call an $\mathcal{A}$-module.

An $\mathcal{A}$-module X is called faithful, if for any nonzero $e \in \nabla$ there exists $x \in X$ such that $ex \neq 0$. Clearly, for a faithful $\mathcal{A}$-module X the set $X_e := eX$ is a faithful $\mathcal{A}_e$-module for any $0 \neq e \in \nabla$, where $\mathcal{A}_e := e\mathcal{A}$.

An $\mathcal{A}$-module X is said to be a regular module, if for any $x \in \mathcal{A}$ the condition $ex = 0$, for all $e \in L \subset \nabla$, implies $(\sup L)x = 0$. In this case, for $x \in X$ the idempotent $s(x) = \mathbf{1} - \sup\{e \in \nabla : ex = 0\}$ is called the support of an element x. In case, when $X = \mathcal{A}$, the notions of support of an element in an $\mathcal{A}$-module X and of support of an element in $\mathcal{A}$ coincide. If X is a regular $\mathcal{A}$-module, then X_e is also a regular $\mathcal{A}_e$-module for any nonzero $e \in \nabla$.

Let X be a regular $\mathcal{A}$-module and $Y \subset X$. An idempotent $s(Y) = \sup\{s(y) : y \in Y\}$ we call the support of the subset Y. Clearly, that any regular $\mathcal{A}$-module X is a faithful module if and only if $s(X) = \mathbf{1}$.

We say that a regular $\mathcal{A}$-module X is laterally complete (l-complete), if for any set $\{x_i\}_{i\in I} \subset X$ and for any partition $\{e_i\}_{i\in I}$ of unity of the Boolean algebra ∇ there exists $x \in X$ such that $e_i x = e_i x_i$ for all $i \in I$. In this case, the element x is called mixing of the set $\{x_i\}_{i\in I}$ with respect to the partition of unity $\{e_i\}_{i\in I}$ and denote by $\operatorname{mix}\limits_{i\in I}(e_i x_i)$. Mixing $\operatorname{mix}\limits_{i\in I}(e_i x_i)$ is defined uniquely, whereas the equalities $e_i x = e_i x_i = e_i y$, $x, y \in X$, $i \in I$, implies $e_i(x - y) = 0$ for all $i \in I$, and, by regularity of the $\mathcal{A}$-module X, we obtain $x = y$.

Let $\{x_i\}_{i\in I} \subset E \subset X$ and let $\{e_i\}_{i\in I}$ be a partition of unity in ∇. The set of all mixings $\operatorname{mix}\limits_{i\in I}(e_i x_i)$ is called a cyclic hull of the set E in X and denotes by $\operatorname{mix}(E)$. Obviously, the inclusion $E \subset \operatorname{mix}(E)$ is always true. If $E = \operatorname{mix}(E)$, then E is called a cyclic set in X (compare with [9], 1.1.2).

Thus, a regular $\mathcal{A}$-module X is a l-complete $\mathcal{A}$-module if and only if X is a cyclic set. In particular, in any l-complete $\mathcal{A}$-module X its submodule X_e is also a l-complete $\mathcal{A}_e$-module for any nonzero idempotent e in $\mathcal{A}$.

We need the following properties of cyclic hulls of sets.

Proposition 2.3. *Let X be a l-complete $\mathcal{A}$-module and let E be a nonempty subset in X, $a \in \mathcal{A}$. Then*

(i). $\operatorname{mix}(\operatorname{mix}(E)) = \operatorname{mix}(E)$*;*

(ii). $\operatorname{mix}(aE) = a\operatorname{mix}(E)$*;*

(iii). If Y is an $\mathcal{A}$-submodule in X, then $\operatorname{mix}(Y)$ is a l-complete $\mathcal{A}$-submodule in X;

(iv). If U is an isomorphism from $\mathcal{A}$-module X onto $\mathcal{A}$-module Z, then Z is a l-complete $\mathcal{A}$-module and $\operatorname{mix}(U(E)) = U(\operatorname{mix}(E))$.

Proof. *(i)*. It is sufficient to show that $\operatorname{mix}(\operatorname{mix}(E)) \subset \operatorname{mix}(E)$. If $x \in \operatorname{mix}(\operatorname{mix}(E))$, then $x = \operatorname{mix}\limits_{i\in I}(e_i x_i)$, where $x_i \in \operatorname{mix}(E)$, $i \in I$. Since $x_i \in \operatorname{mix}(E)$, then $x_i = \operatorname{mix}\limits_{j\in J(i)}(e_j^{(i)} x_j^{(i)})$, where $x_j^{(i)} \in E$, $j \in J(i)$ and $\{e_j^{(i)}\}_{j\in J(i)}$ is a partition of unity in the Boolean algebra ∇ for all $i \in I$.

Fix $i \in I$ and put $g_j^{(i)} := e_i e_j^{(i)}$. It is clear that $\{g_j^{(i)}\}_{j \in J(i)}$ is a partition of the idempotent e_i. Hence, $\{g_j^{(i)}\}_{j \in J(i), i \in I}$ is a partition of unity $\mathbf{1}$. Besides, $g_j^{(i)} x = g_j^{(i)} e_i x = g_j^{(i)} e_i x_i = e_i e_j^{(i)} x_i = e_i e_j^{(i)} x_j^{(i)} = g_j^{(i)} x_j^{(i)}$. This yields that $x = \underset{j \in J(i), i \in I}{\text{mix}} (g_j^{(i)} x_j^{(i)}) \in \text{mix}(E)$.

(ii). If $x \in \text{mix}(aE)$, then $x = \underset{i \in I}{\text{mix}}(e_i a y_i)$, where $y_i \in E, i \in I$. Since X is a l-complete $\mathcal{A}$-module, then there exists $y = \underset{i \in I}{\text{mix}}(e_i y_i) \in \text{mix}(E)$ and $e_i x = a e_i y_i = e_i(ay)$ for all $i \in I$. Hence, $e_i(x - ay) = 0$, and regularity of the $\mathcal{A}$-module X implies the equality $x = ay$. Thus, $\text{mix}(aE) \subset a\text{mix}(E)$.

Conversely, if $x \in a\text{mix}(E)$, then $x = az$, where $z = \underset{i \in I}{\text{mix}}(e_i z_i)$, $z_i \in E, i \in I$. Since $az_i \in aE$ and $e_i x = e_i(az) = e_i a e_i z = e_i(az_i)$ for all $i \in I$, we have that $x = \underset{i \in I}{\text{mix}}(e_i(az_i)) \in \text{mix}(aE)$. Hence, $a\text{mix}(E) \subset \text{mix}(aE)$.

(iii). Let $x, y \in \text{mix}(Y)$, $x = \underset{i \in I}{\text{mix}}(e_i x_i)$, $y = \underset{j \in J}{\text{mix}}(g_j y_j)$, where $x_i, y_j \in Y$, $i \in I$, $j \in J$, $\{e_i\}_{i \in I}$, $\{g_j\}_{j \in J}$ are partitions of unity in ∇. Clearly, that $p_{ij} = e_i g_j$, $i \in I$, $j \in J$, is also a partition of unity in ∇ and $p_{ij}(x + y) = p_{ij}(x_i + y_j)$, where $x_i + y_j \in Y$ for all $i \in I$, $j \in J$. This means that $(x + y) \in \text{mix}(Y)$.

Since $aY \subset Y$, then by *(ii)* we have that $ax \in a\text{mix}(Y) = \text{mix}(aY) \subset \text{mix}(Y)$. Hence, $\text{mix}(Y)$ is an $\mathcal{A}$-submodule in X, and by regularity of the $\mathcal{A}$-module X, it is a regular $\mathcal{A}$-module. The equality $\text{mix}(Y) = \text{mix}(\text{mix}(Y))$ (see *(i)*) implies that $\text{mix}(Y)$ is a l-complete $\mathcal{A}$-module.

(iv). If $U(x) = y \in Z$, $x \in X$, $\emptyset \neq L \subset \nabla$ and $ey = 0$ for all $e \in L$, then $U(ex) = eU(x) = ey = 0$. Since U is a bijection, then $ex = 0$ for any $e \in L$. By regularity of the $\mathcal{A}$-module X, we have that $(\sup L)x = 0$, and, therefore, $(\sup L)y = U((\sup L)x) = 0$. Hence, Z is a regular $\mathcal{A}$-module. In the same way we show that Z is a l-complete $\mathcal{A}$-module and the equality $\text{mix}(U(E)) = U(\text{mix}(E))$ holds. □

3. Homogenous $\mathcal{A}$-modules

Let $\mathcal{A}$ be a laterally complete commutative regular algebra, let ∇ be a complete Boolean algebra of all idempotents in $\mathcal{A}$, let X be a faithful $\mathcal{A}$-module. The following $\mathcal{A}$-submodule in X is called $\mathcal{A}$-linear hull of a nonempty subset $Y \subset \mathcal{A}$

$$\text{Lin}(Y, \mathcal{A}) = \left\{ \sum_{i=1}^{n} a_i y_i : a_i \in \mathcal{A}, y_i \in Y, i = 1, \ldots, n, n \in \mathcal{N} \right\},$$

where $\mathcal{N}$ is the set of all natural numbers. If X is a l-complete $\mathcal{A}$-module, then by proposition 2.3 *(iii)*, $\text{mix}(\text{Lin}(Y, \mathcal{A}))$ is also a l-complete $\mathcal{A}$-submodule in X.

A set $\{x_i\}_{i \in I}$ in an $\mathcal{A}$-module X is called *$\mathcal{A}$-linearly independent*, if for any $a_1, \ldots, a_n \in \mathcal{A}$, $x_{i_1}, \ldots, x_{i_n} \in \{x_i\}_{i \in I}$, $n \in \mathcal{N}$, the equality $\sum_{k=1}^{n} a_k x_{i_k} = 0$ implies the equality $a_1 = \ldots = a_n = 0$.

Proposition 3.1. *If $Y = \{x_1, \ldots, x_k\}$ is a finite $\mathcal{A}$-linearly independent subset in a l-complete $\mathcal{A}$-module X, then* $\text{mix}(\text{Lin}(Y, \mathcal{A})) = \text{Lin}(Y, \mathcal{A})$.

Proof. It is sufficient to show the following inclusion $\text{mix}(\text{Lin}(Y, \mathcal{A})) \subset \text{Lin}(Y, \mathcal{A})$. Let $x \in \text{mix}(\text{Lin}(Y, \mathcal{A}))$, $\{e_i\}_{i \in I}$ be a partition of unity in the Boolean algebra ∇ and let $\{y_i\}_{i \in I} \subset \text{Lin}(Y, \mathcal{A})$ be such that $e_i x = e_i y_i$ for all $i \in I$. Since $e_i x = e_i y_i \in \text{Lin}(Y, \mathcal{A})$, then $e_i x = \sum_{j=1}^{k} a_j^{(i)} x_j$ for some $a_j^{(i)} \in \mathcal{A}$, $j = 1, \ldots, k$. Hence, $e_i x = e_i(e_i x) = \sum_{j=1}^{k} e_i a_j^{(i)} x_j$. Since $\mathcal{A}$ is a l-complete commutative regular algebra and $\{e_i\}_{i \in I}$ is a partition of unity in ∇, then there exists a unique element $\beta_j \in \mathcal{A}$ such that $e_i \beta_j = e_i a_j^{(i)}$ for all $i \in I$, where

$j \in \{1, \ldots, k\}$. Thus, $e_i x = \sum_{j=1}^{k} e_i \beta_j x_j = e_i \left(\sum_{j=1}^{k} \beta_j x_j \right)$ for any $i \in I$, and this implies the equality $x = \sum_{j=1}^{k} \beta_j x_j \in \mathrm{Lin}(Y, \mathcal{A})$. □

We say that an $\mathcal{A}$-linearly independent system $\{x_i\}_{i \in I}$ from a l-complete $\mathcal{A}$-module X is *$\mathcal{A}$-Hamel basis*, if

$$\mathrm{mix}(\mathrm{Lin}(\{x_i\}_{i \in I}, \mathcal{A})) = X.$$

In the case when an $\mathcal{A}$-Hamel basis is a finite set, we say that it is an $\mathcal{A}$-basis in X (see Proposition 3.1).

Theorem 3.2. *If $\{x_i\}_{i=1}^{n}$, $\{y_j\}_{j=1}^{k}$ are $\mathcal{A}$-basises in an $\mathcal{A}$-module X, then $n = k$.*

Proof. First we shall show the following $\mathcal{A}$-variant of one known fact from the linear algebra.

Lemma 3.3. *Let $\{z_i\}_{i=1}^{n} \subset X, \{y_j\}_{j=1}^{k} \subset X, \{ey_j\}_{j=1}^{k} \subset \mathrm{Lin}(\{ez_i\}_{i=1}^{n}, \mathcal{A}_e)$ for nonzero $e \in \nabla$. If the set $\{ey_1, \ldots, ey_k\}$ is $\mathcal{A}_e$-linearly independent, then $k \le n$.*

Proof. of lemma 3.3. We use the mathematical induction. Let us suppose that for $n = 1$, $k > 1$ the equalities $ey_1 = a_1 ez_1, \ldots, ey_k = a_k ez_1$ hold, where $a_i \in \mathcal{A}_e$, $i = 1, \ldots, k$. Since $a_2 ey_1 + (-a_1) ey_2 = 0$, then $ea_1 = ea_2 = 0$, i.e. $ey_1 = ey_2 = 0$, this contradicts to $\mathcal{A}_e$-linear independence of the elements ey_1 and ey_2. Hence, $k = 1$.

Now assume that the lemma holds for $n = l - 1$. Let $\{z_i\}_{i=1}^{l} \subset X$ and the following equalities hold

$$ey_j = \sum_{i=1}^{l} a_{ji} ez_i, a_{ji} \in \mathcal{A}_e, \quad j = 1, \ldots, k, i = 1, \ldots, l. \tag{1}$$

Let $a_{j_0 l} ex_l \neq 0$ for some $j_0 \in \{1, \ldots, k\}$. By reindexing $\{y_j\}_{j=1}^{k}$, we can assume that $a_{kl} ex_l \neq 0$, in particular $p = s(a_{kl} e) \neq 0$, wherein $p \le e$. Since the set $\{ey_j\}_{j=1}^{k}$ is $\mathcal{A}_e$-linearly independent, then the set $\{py_j\}_{j=1}^{k}$ is $\mathcal{A}_p$-linearly independent, wherein, by (1), we have

$$py_j = \sum_{i=1}^{l} a_{ji} pz_i, \quad j = 1, \ldots, k. \tag{2}$$

Since $\mathcal{A}$ is a regular algebra, then for the inversion $h = i(a_{kl}) \in \mathcal{A}$ the equality $ha_{kl} = s(a_{kl})$ holds. Therefore the following equality

$$py_k = \sum_{i=1}^{l-1} a_{ki} pz_i + a_{kl} pz_l$$

implies

$$pz_l = hpy_k - \sum_{i=1}^{l-1} a_{ki} hpz_i. \tag{3}$$

Substitute pz_l from (3) in the first $(k-1)$ equalities from (2) and collect similar terms, we obtain

$$py_j - ha_{jl} py_k = \sum_{i=1}^{l-1} \beta_{ji} pz_i \in \mathrm{Lin}(\{pz_i\}_{i=1}^{l-1}, \mathcal{A}_p)$$

for some $\beta_{ji} \in \mathcal{A}_p$, $i = 1, \ldots, l-1$, $j = 1, \ldots, k-1$.

Let us show that the elements $u_j = py_j - ha_{jl}py_k$, $j = 1, \ldots, k-1$, are $\mathcal{A}_p$-linearly independent. Let

$$\sum_{j=1}^{k-1} \gamma_j p y_j - \left(\sum_{j=1}^{k-1} \gamma_j h a_{jl}\right) p y_k = \sum_{j=1}^{k-1} \gamma_j u_j = 0,$$

where $\gamma_j \in \mathcal{A}_p$, $j = 1, \ldots, k-1$. Since $\{py_j\}_{j=1}^k$ is $\mathcal{A}_p$-linearly independent, then $p\gamma_1 = p\gamma_2 = \ldots = p\gamma_{k-1} = 0$, i.e. the set $\{u_j\}_{j=1}^{k-1}$ is $\mathcal{A}_p$-linearly independent in pX. By the assumption of the mathematical induction we have that $k-1 \le l-1$, and thus $k \le l$. The Lemma 3.3 is proved. □

Return to the proof of Theorem 3.2. As $\{x_i\}_{i=1}^n$ is an $\mathcal{A}$-basis in X, then by Proposition 3.1 we obtain that $X = \mathrm{Lin}(\{x_i\}_{i=1}^n, \mathcal{A})$. On the other hand, $\{y_j\}_{j=1}^k \subset X$ and $\{y_j\}_{j=1}^k$ is an $\mathcal{A}$-linearly independent set. Therefore, by Lemma 3.3 it follows that $k \le n$.

Similarly, we show that $n \le k$, and thus $n = k$. □

Fix some cardinal number γ. A faithful l-complete $\mathcal{A}$-module X is called *γ-homogeneous*, if there exists an $\mathcal{A}$-Hamel basis $\{x_i\}_{i\in I}$ in X with $\mathrm{card}\, I = \gamma$. We say that $\mathcal{A}$-module X *homogeneous*, if it is a γ-homogeneous $\mathcal{A}$-module for some cardinal number γ.

If X is a γ-homogeneous $\mathcal{A}$-module, then obviously, eX is also γ-homogeneous $\mathcal{A}_e$-module for any nonzero idempotent $e \in \mathcal{A}$. Besides, by Proposition 2.3 (*iv*) it follows that, if $\mathcal{A}$-module Y is isomorphic to a γ-homogeneous $\mathcal{A}$-module X, then Y is also a γ-homogeneous module.

By repeating the proof of Theorem 3.8 from [3], we establish the following proposition on isomorphisms of γ-homogeneous $\mathcal{A}$-modules.

Proposition 3.4. *If X and Y are γ-homogeneous $\mathcal{A}$-modules, then X and Y are isomorphic.*

Proof. Let $\{x_i\}_{i\in I}$ be an $\mathcal{A}$-Hamel bases in X, $\{y_i\}_{i\in I}$ be an $L^0(\mathcal{B})$-Hamel bases in Y, $\mathrm{card}\, I = \gamma$, $X_0 = \mathrm{Lin}\,(\{x_i\}_{i\in I}, \mathcal{A})$, $Y_0 = \mathrm{Lin}\,(\{y_i\}_{i\in I}, \mathcal{A})$. Any $x \in X_0$, $y \in Y_0$ uniquely representing as $x = \sum\limits_{k=1}^{n} \alpha_k x_{i_k}$, $y = \sum\limits_{l=1}^{m} \beta_l y_{i_l}$ for some $\alpha_k, \beta_l \in \mathcal{A}$, $k = 1, \ldots, n$, $l = 1, \ldots, m$, $n, m \in \mathcal{N}$. A map $U_0 : X_0 \to Y_0$ which is defined by equality $U_0\left(\sum\limits_{k=1}^{n} \alpha_k x_{i_k}\right) = \sum\limits_{k=1}^{n} \alpha_k y_{i_k}$ is $\mathcal{A}$-linear bijection from X_0 onto Y_0.

For any $a \in X$ there exist a set $\{a_j\}_{j\in J} \subset X_0$ and a partition $\{e_j\}_{j\in J}$ of identity in ∇, such that $e_j a = e_j a_j$ for all $j \in J$. By l-completeness of $\mathcal{A}$-module Y, there exists unique $b \in Y$, such that $e_j b = e_j U_0(a_j)$ for all $j \in J$. We put $U(a) = b$. Hence U is a $\mathcal{A}$-linear bijection from X onto Y. □

Let us give examples of γ-homogeneous $\mathcal{A}$-modules for an arbitrary cardinal number γ and for any l-complete commutative regular untaly algebra $\mathcal{A}$. Consider an arbitrary set of indexes I with $\mathrm{card}\, I = \gamma$. Since the algebra $\mathcal{A}$ is l-complete, then the Cartesian product

$$Y = \prod_{i\in I} \mathcal{A} = \{\hat{\alpha} = \{\alpha_i\}_{i\in I} : \alpha_i \in \mathcal{A}, i \in I\}$$

is a l-complete $\mathcal{A}$-module with coordinate-wise algebraic operations.

For any $j \in I$ consider an element $\hat{g}_j = \{g_i^{(j)}\}_{i\in I}$ from Y, where $g_i^{(j)} = 0$, $i \ne j$ and $g_i^{(i)} = \mathbf{1}$, $i \in I$. Clearly, that the set $\{\hat{g}_j\}_{j\in I}$ is $\mathcal{A}$-linearly independent, and, therefore, the $\mathcal{A}$-submodule $X = \mathrm{mix}\,(\mathrm{Lin}\,(\{\hat{g}_j\}_{j\in I}, \mathcal{A}))$ in Y is a γ-homogeneous $\mathcal{A}$-module.

If γ is a positive integer n, then for the faithful l-complete $\mathcal{A}$-module $Y = \prod\limits_{i=1}^{n} \mathcal{A} = \mathcal{A}^n$ and for $\hat{g}_j = \{g_i^{(j)}\}_{i=1}^n$, $j = 1, \ldots, n$ we have that $\mathrm{Lin}\,(\{\hat{g}_j\}_{j=1}^n, \mathcal{A}) = Y$, i.e. the set $\{\hat{g}_j\}_{j=1}^n$ is an $\mathcal{A}$-Hamel basis in Y. Thus, Proposition 3.4 implies the following

Corollary 3.5. *For any positive integer n there exists a unique, up to isomorphism, n-homogeneous $\mathcal{A}$-module, which is isomorphic to $\mathcal{A}^n$.*

Let X be a faithful l-complete $\mathcal{A}$-module, which is γ-homogeneous and λ-homogeneous simultaneously. There is a natural question, whether in this case the equality $\gamma = \lambda$ holds. Similar question was studied in classification of Kaplansky-Hilbert modules (KHM) X over a commutative AW^*-algebra $\mathcal{A}$ with the Boolean algebra of projections ∇ (see [7]). In the case, when ∇ is a *multi-σ-finite* Boolean algebra (definition see below) in [7] it is proved that for a KHM X the equality $\lambda = \gamma$ is always true. However, for an arbitrary complete Boolean algebra ∇ this equality cannot be established. Thereby, in ([10], 7.4.6) the notion of *strictly γ-homogeneous* KHM X is defined, and this gave an opportunity to classify KHM X over an arbitrary commutative AW^*-algebra $\mathcal{A}$. For the same reason, below we introduce the notion of strictly γ-homogeneous faithful l-complete modules over laterally complete algebras $\mathcal{A}$.

Let X be a faithful l-complete $\mathcal{A}$-module, $0 \neq e \in \nabla$. By $\varkappa(e) = \varkappa_X(e)$ we denote the smallest cardinal number γ such that the $\mathcal{A}_e$-module X_e is γ-homogeneous. If the $\mathcal{A}$-module X is homogeneous, then the cardinal number $\varkappa(e)$ is defined for all nonzero $e \in \nabla$. Further, by ([10], 7.4.7), we assume that $\varkappa(0) = 0$.

We say that an $\mathcal{A}$-module X is *strictly γ-homogeneous* (compare with [10], 7.4.6), if X is γ-homogeneous and $\gamma = \varkappa(e)$ for all nonzero $e \in \nabla$. If an $\mathcal{A}$-module X is strictly γ-homogeneous for some cardinal number γ, then such $\mathcal{A}$-module X is called *strictly homogeneous*.

Clearly, any strictly γ-homogeneous $\mathcal{A}$-module is a γ-homogeneous $\mathcal{A}$-module. By Theorem 3.2 it follows that every n-homogeneous $\mathcal{A}$-module X is a strictly n-homogeneous module. By Proposition 2.3 (iv) every $\mathcal{A}$-module Y, which is isomorphic to a strictly γ-homogeneous $\mathcal{A}$-module X, is also strictly γ-homogeneous.

Let ∇ be an arbitrary complete Boolean algebra. For any nonzero element $e \in \nabla$ we put $\nabla_e = \{q \in \nabla : q \leq e\}$. The set ∇_e is a Boolean algebra with the unity e with respect to partial order, induced from ∇.

We say that a set B in ∇ is a minorant subset for nonempty set $E \subset \nabla$, if for any nonzero $e \in E$ there exists nonzero $q \in B$ such that $q \leq e$. We need the following important property of complete Boolean algebras.

Theorem 3.6. ([10], 1.1.6) *If ∇ is a complete Boolean algebra, e is a nonzero element in ∇ and B is a minorant subset for ∇_e, then there exists a disjoint subset $L \subset B$ such that* $\sup L = e$.

We say that a Boolean algebra ∇ has *a countable type* or is *σ-finite*, if any nonfinite family of nonzero pairwise disjoint elements in ∇ is a countable set. A complete Boolean algebra ∇ is called *multi-σ-finite*, if for any nonzero element $g \in \nabla$ there exists $0 \neq e \in \nabla$ such that $e \leq g$ and the Boolaen algebra ∇_e has a countable type. By Theorem 3.6, a multi-σ-finite Boolean algebra ∇ always has a partition $\{e_i\}_{i \in I}$ of unity $\mathbf{1}$ such that the Boolean algebra ∇_{e_i} has a countable type for all $i \in I$.

The next theorem gives coincidence of properties of λ-homogeneous and γ-homogeneous $\mathcal{A}$-module in case when the Boolean algebra ∇ of all idempotents in $\mathcal{A}$ has a countable type.

Theorem 3.7. *Let λ and γ be infinite cardinal numbers and let the Boolean algebra ∇ of all idempotents in a l-complete commutative regular algebra $\mathcal{A}$ has countable type. If a faithful l-complete $\mathcal{A}$-module X is λ-homogeneous and γ-homogeneous simultaneously, then $\gamma = \lambda$.*

Proof. Let $\{x_i\}_{i \in I}$, $\{y_j\}_{j \in J}$ be two $\mathcal{A}$-Hamel basises in X, i. e.

$$\mathrm{mix}(\mathrm{Lin}(\{x_i\}_{i \in I}, \mathcal{A})) = X = \mathrm{mix}(\mathrm{Lin}(\{y_j\}_{j \in J}, \mathcal{A}))$$

and $\mathrm{card}\, I = \lambda$, $\mathrm{card}\, J = \gamma$. Since $y_j \in \mathrm{mix}(\mathrm{Lin}(\{x_i\}_{i \in I}, \mathcal{A}))$ and the Boolean algebra ∇ has a countable type, then there exist a countable partition $\{e_n\}_{n \in \mathcal{N}}$ of identity of the

Boolean algebra ∇ and the set $\{a_n\}_{n\in\mathcal{N}} \subset \mathrm{Lin}(\{x_i\}_{i\in I}, \mathcal{A})$, that $e_n y_j = e_n a_n$ for all $n \in \mathcal{N}$. Chose $x_{i_k(j,n)}$ and $\alpha_{i_k(j,n)} \in \mathcal{A}$, $k = 1, \dots, m(n)$ such that $a_n = \sum_{k=1}^{m(n)} \alpha_{i_k(j,n)} x_{i_k(j,n)}$. Put $I(j) = \{i_k(j,n) : k = 1, \dots, m(n),\, n \in \mathcal{N}\}$ and $I(J) = \bigcup_{j\in J} I(j)$. It is clear, that set $I(j)$ is countable or finite and $I(J) \subset I$. In this case $y_j \in \mathrm{mix}(\mathrm{Lin}(\{x_i\}_{i\in I(j)}, \mathcal{A}))$ for any $j \in J$.

We show that $I(J) = I$. Assume that $i_0 \in I \setminus I(J)$. Since $x_{i_0} \in \mathrm{mix}(\mathrm{Lin}(\{y_j\}_{j\in J}, \mathcal{A}))$ and the Boolean algebra ∇ has a countable type, then there exist a partition of identity $\{g_k\}_{k\in\mathcal{N}}$ and a set $\{b_k\}_{k\in\mathcal{N}} \subset \mathrm{Lin}(\{y_j\}_{j\in J}, \mathcal{A})$, such that $g_k x_{i_0} = g_k b_k = g_k \sum_{s=1}^{p(k)} \beta_{j_s(i_0,k)} y_{j_s(i_0,k)}$ for some $\beta_{j_s}(i_0, k) \in \mathcal{A}$, $y_{j_s(i_0,k)} \in \{y_j\}_{j\in J}$, $s = 1, \dots, p(k)$. Since $y_{j_s(i_0,k)} \in \mathrm{mix}(\mathrm{Lin}(\{x_i\}_{i\in I(j_s(i_0,k))}, \mathcal{A})) \subset \mathrm{mix}(\mathrm{Lin}(\{x_i\}_{i\in I(J)}, \mathcal{A}))$ then $g_k b_k \in \mathrm{mix}(\mathrm{Lin}(\{x_i\}_{i\in I(J)}, \mathcal{A}))$ for any $k \in \mathcal{N}$ and therefore $x_{i_0} \in \mathrm{mix}(\mathrm{Lin}(\{x_i\}_{i\in I(J)}, \mathcal{A}))$ (see Proposition 2.3 (i)). In particular, there exists $0 \neq e \in \nabla$, such that $0 \neq e x_{i_0} = \sum_{t=1}^{l} e \delta_{i_t} x_{i_t}$, here $i_t \in I(J)$, $\delta_{i_t} \in \mathcal{A}$, $t = 1, \dots, l$, which contradicts to the $\mathcal{A}$-linear independence of the set $\{x_i\}_{i\in I}$. Thus, $I = I(J)$.

As γ is an infinite cardinal number, then $\mathrm{card}\, I(j) \leqslant \aleph_0 \leqslant \gamma$ for all $j \in J$, here $\aleph_0$ is a countable cardinal number. By Theorem 24 ([1], Chapter 3, §6) it follows that

$$\lambda = \mathrm{card}\, I = \mathrm{card}\, I(J) = \mathrm{card}\left(\bigcup_{j\in J} I(j)\right) \leqslant \mathrm{card}\, J = \gamma.$$

In the same way we may show $\mathrm{card}\, J \leqslant \mathrm{card}\, I$. Therefore $\mathrm{card}\, I = \mathrm{card}\, J$, i.e. $\lambda = \gamma$. □

Using Theorem 3.7 to the $\mathcal{A}_e$-module X_e, we have, that Theorem 3.7 holds in the case, when in the Boolean algebra ∇ of idempotents in $\mathcal{A}$ there exists nonzero element e, which has a countable type. Thus, repeating the proof of Corollary 3.7 in [3], we obtain the following necessary and sufficient conditions for coincidence of strictly γ-homogeneous and γ-homogeneous notions for $\mathcal{A}$-modules.

Theorem 3.8. *Let a Boolean algebra ∇ of all idempotents on a l-complete commutative regular algebra $\mathcal{A}$ be multi-σ-finite. If γ is an infinite cardinal number and X is a γ-homogeneous $\mathcal{A}$-module, then the module X is strictly γ-homogeneous.*

References

[1] Aleksandrov P S 1977 *Introduction in the set theory and general topology* (Moscow: Nauka)
[2] Chilin V I 1988 *In book: Modern problems of Mathematics: The latest trends* (Moscow: VINITI) **27** 99-128
[3] Chilin V I and Karimov J A 2014 *Vladikavkazskii Matem. Zhurnal* **16** 69-78
[4] Clifford A N and Preston G B 1964 *The algebraic theory of semigroup* (New York: AMS)
[5] Kaplansky J 1951 *Ann. Math* **53** 235-49
[6] Kaplansky J 1952 *Ann. Math* **56** 450-72
[7] Kaplansky J 1953 *Amer. J. Math* **75** 839-858
[8] Karimov J A 2010 *Uzbek math. J.* **4** 74-81
[9] Kusraev A G 1985 *Vector duality and its applications* (Novosibirsk: Nauka)
[10] Kusraev A G 2000 *Dominated operators* (Netherlands: Springer)
[11] Maeda F 1958 *Kontinuierliche Geometrien* (Berlin)
[12] Muratov M A and Chilin V I 2007 *Algebra of measurable and locally measurbale operators* (Kiev, Math. Inst.)
[13] Sarimsakov T A, Ayupov Sh A, Khadjiev D and Chilin V I 1983 *Ordered algebras* (Tashkent: FAN)
[14] Skornyakov L A 1961 *Dedekind's lattices with complements and regular rings* (Moscow: Fizmatgiz)

Algebra, Analysis and Quantum Probability IOP Publishing
Journal of Physics: Conference Series **697** (2016) 012003 doi:10.1088/1742-6596/697/1/012003

On a genaralized uniform zero-two law for positive contractions of non-commutative L_1-spaces

Inomjon Ganiev[1], **Farrukh Mukhamedov**[2], **Dilmurod Bekbaev**[3]

[1] Department of Science in Engineering, Faculty of Engineering, International Islamic University Malaysia, P.O. Box 10, 50728 Kuala Lumpur, Malaysia

[2,3] Department of Computational & Theoretical Sciences, Faculty of Science, International Islamic University Malaysia, P.O. Box, 25200, Kuantan, Pahang, Malaysia

E-mail: inam@iium.edu.my, farrukh_m@iium.edu.my

Abstract. In this paper we prove a non-commutative version of the uniform "zero-two" law for positive contractions of L_1-spaces associated with von Neumann algebras.

1. Introduction

Let $(X, \mathcal{F}, \mu)$ be a measure space with a positive σ-additive measure μ and let $L_1(X, \mathcal{F}, \mu)$ be the usual associated real L_1-space. A linear operator $T : L_1(X, \mathcal{F}, \mu) \to L_1(X, \mathcal{F}, \mu)$ is called a *positive contraction* if $Tf \geq 0$ whenever $f \geq 0$ and $\|T\| \leq 1$.

In [14] it was proved that if T is a positive contraction of L_1-space then $\sup\limits_{\|f\|\leq 1} \lim\limits_{n\to\infty} \|T^{n+1} - T^n\|$ is 0 or 2, and this was called "zero-two" law. Using the method of [14], Foguel [4] proved that $\lim_{n\to\infty} \|T^{n+1} - T^n\|$ is 0 or 2. In [16] the last result was restated as follows:

Theorem 1.1. *[16] Let $T : L_1 \to L_1$ be a positive contraction. If there is some $m \in \mathbb{N}$ such that $\|T^{m+1} - T^m\| < 2$ then*

$$\lim_{n\to\infty} \|T^{n+1} - T^n\| = 0.$$

To prove this theorem Zahoropol [16] reduced it to the following theorem.

Theorem 1.2. *[16] Let $T : L_1 \to L_1$ be a positive contraction. Then for the following statements:*

(i) there is some $m \in \mathbb{N}$ such that $\|T^{m+1} - T^m\| < 2$;

(ii) there is some $m \in \mathbb{N}$ such that $\|T^{m+1} - (T^{m+1} \wedge T^m)\| < 1$;

(iii) one has

$$\lim_{n\to\infty} \|T^{n+1} - T^n\| = 0.$$

the implications hold: (i)⇒(ii)⇒ (iii).

To establish the implication (ii)⇒(iii) the following auxiliary fact is needed.

Theorem 1.3. *[16] Let $T, S : L_1 \to L_1$ be two positive contractions such that $T \leq S$. If $\|S - T\| < 1$ then $\|S^n - T^n\| < 1$ for all $n \in \mathbb{N}$.*

We note that a "zero-two" law for Markov processes was proved in [2], which allowed to study random walks on locally compact groups. Other extensions and generalizations of the formulated law have been investigated by many authors [3, 5, 8, 17]. The aim of this paper is to prove a non-commutative version of the "zero-two" law for positive contractions of L_1-spaces associated with von Neumann algebras. We emphasize that Theorem 1.1 is included in the main result as a particular case.

2. Preliminaries

Throughout the paper M would be a von Neumann algebra with the unit $\mathbb{1}$ and let τ be a faithful normal semi finite trace on M. Therefore we omit this condition from the formulation of theorems. Recall that an element $x \in M$ is called *self-adjoint* if $x = x^*$. The set of all self-adjoint elements is denoted by M_{sa}. By M_* we denote a pre-dual space to M (see for definitions [1],[15]).

Let $\mathfrak{N} = \{x \in M : \tau(|x|) < \infty\}$, here $|x|$ denotes the modules of an element x, i.e. $|x| = \sqrt{x^*x}$. Define the map $\|\cdot\|_1 : \mathfrak{N} \to [0, \infty)$ defined by the formula $\|x\|_1 = \tau(|x|)$ is a norm (see [12]). The completion of $\mathfrak{N}$ with respect to the norm $\|\cdot\|_1$ is denoted by $L_1(M,\tau)$. It is known [12] that the spaces $L_1(M,\tau)$ and M_* are isometrically isomorphic, therefore they can be identified. Further we will use this fact without noting.

Theorem 2.1. *[12] The space $L_1(M,\tau)$ coincides with the set*

$$L^1 = \{x = \int_{-\infty}^{\infty} \lambda de_\lambda \ : \int_{-\infty}^{\infty} |\lambda| d\tau(e_\lambda) < \infty\}.$$

Moreover,

$$\|x\|_1 = \int_{-\infty}^{\infty} |\lambda| d\tau(e_\lambda).$$

It is known [12] that the equality

$$L_1(M,\tau) = L_1(M_{sa},\tau) + iL_1(M_{sa},\tau) \tag{2.1}$$

is valid. Note that $L_1(M_{sa},\tau)$ is a pre-dual to M_{sa}.

Let $T : L_1(M,\tau) \to L_1(M,\tau)$ be any bounded linear operator, by $\tilde{T}$ we denote its restriction to $L_1(M_{sa},\tau)$. Then due to (2.1) we have $T(x+iy) = \tilde{T}(x) + i\tilde{T}(y)$, where $x, y \in L_1(M_{sa},\tau)$. This means that any linear bounded operator is uniquely defined by its restriction to $L_1(M_{sa},\tau)$. Therefore, in what follows, we only consider linear operators on $L_1(M_{sa},\tau)$ over real numbers.

Recall that a linear operator T is called *positive* if $Tx \geq 0$ whenever $x \geq 0$. A linear operator T is said to be a *contraction* if $\|T(x)\|_1 \leq \|x\|_1$ for all $x \in L_1(M_{sa},\tau)$. Denote

$$\|T\| = \sup\{\|Tx\|_1 : \ \|x\|_1 = 1, x \in L_1(M_{sa},\tau)\}.$$

Let $T, S : L_1 \to L_1$ be two positive contractions. In what follows, we write $T \leq S$ if $S - T$ is a positive operator.

The following auxiliary facts are well known (see for example [11]).

Lemma 2.2. *Let $T : L_1(M_{sa},\tau) \to L_1(M_{sa},\tau)$ be a positive operator. Then*

$$\|T\| = \sup_{\|x\|=1} \|Tx\| = \sup_{\|x\|=1, x\geq 0} \|Tx\|.$$

Lemma 2.3. *Let $T, S : L_1(M_{sa},\tau) \to L_1(M_{sa},\tau)$ be two positive contraction such that $T \leq S$. Then for every $x \in L_1(M_{sa},\tau)$, $x \geq 0$ the equality holds*

$$\|Sx - Tx\| = \|Sx\| - \|Tx\|.$$

3. Dominant operators

Using the methods of [11, 9] one can prove the following result which is related to dominant operators.

Theorem 3.1. *Let $Z, T, S : L_1(M_{sa}, \tau) \to L_1(M_{sa}, \tau)$ be positive contractions such that $T \le S$ and $ZS = SZ$. If there is an $n_0 \in \mathbb{N}$ such that $\|Z(S^{n_0} - T^{n_0})\| < 1$. Then $\|Z(S^n - T^n)\| < 1$ for every $n \ge n_0$.*

We note that the last theorem extends a main result of the paper [11]. Namely, we have the following corollary.

Corollary 3.2. *Let $T, S : L_1(M_{sa}, \tau) \to L_1(M_{sa}, \tau)$ be positive contractions such that $T \le S$. If there is an $n_0 \in \mathbb{N}$ such that $\|S^{n_0} - T^{n_0}\| < 1$. Then $\|S^n - T^n\| < 1$ for every $n \ge n_0$.*

The proof immediately follows if one takes $Z = Id$. Note that if $n_0 = 1$ and M is a commutative von Neumann algebra, then from Corollary 3.2 we immediately get the Zaharopol's result (see Theorem 1.3).

4. A generalized uniform zero-two law

In this section we are going to prove a non-commutative version of the uniform zero-two law for positive contractions on L^1.

Let $T : L_1(M_{sa}, \tau) \to L_1(M_{sa}, \tau)$ be a positive contraction. Then its conjugate T^* acts on M_{sa} and it is also positive and enjoys $T^*\mathbb{1} \le \mathbb{1}$. If one has $T^*\mathbb{1} = \mathbb{1}$, then T is called *unital positive contraction.*

Let us formulate our main result.

Theorem 4.1. *Let $Z : L_1(M_{sa}, \tau) \to L_1(M_{sa}, \tau)$ be a unital positive contraction. Assume that $T : L_1(M_{sa}, \tau) \to L_1(M_{sa}, \tau)$ be unital positive contractions such that $ZT = TZ$. If there are $m, k \in \mathbb{N}_0$ and a positive contraction $S : L_1(M_{sa}, \tau) \to L_1(M_{sa}, \tau)$ such that $SZ = ZS$ with*

$$ZT^{m+k} \ge ZS, \quad ZT^m \ge ZS \quad \text{with} \tag{4.1}$$

$$\|Z(T^{m+k} - S)\| < 1, \quad \|Z(T^m - S)\| < 1. \tag{4.2}$$

then for any $\varepsilon > 0$ there are $M \in \mathbb{N}$ and $n_0 \in \mathbb{N}_0$ such that

$$\|Z^M(T^{n+k} - T^n)\| < \varepsilon \quad \text{for all} \quad n \ge n_0.$$

Proof. First we note that for any positive contraction T on L_1-spaces [?, p. 310] there is $\gamma > 0$ such that

$$\left\| \left(\frac{I + T^k}{2}\right)^\ell - T^k \left(\frac{I + T^k}{2}\right)^\ell \right\| \le \frac{\gamma}{\sqrt{\ell}}. \tag{4.3}$$

Now take any $\varepsilon > 0$ and fix $\ell_\varepsilon \in \mathbb{N}$ such that $\gamma/\sqrt{\ell_\varepsilon} < \varepsilon/2$.

Define

$$Q_1 = \frac{1}{2}(T^{m+k} - S) + \frac{1}{2}T^k(T^m - S).$$

It then follows from (4.1),(4.2) that ZQ_1 is positive and $\|ZQ_1\| < 1$. Moreover, one has

$$T^{m+k} = \left(\frac{I + T^k}{2}\right)S + Q_1$$

where I stands for the identity mapping.

For each $\ell \in \mathbb{N}$ let us define

$$Q_{\ell+1} = \left(\frac{I+T^k}{2}\right)^{\ell} Q_1 S^{\ell} + T^{m+k} Q_{\ell}, \quad \ell \in \mathbb{N}.$$

Taking into account the positivity of S and Q_1, one can see that Q_ℓ is a positive operator on $L_1(M_{sa},\tau)$ and $ZQ_\ell = Q_\ell Z$. Moreover, one has

$$T^{\ell(m+k)} = \left(\frac{I+T^k}{2}\right)^{\ell} S^{\ell} + Q_{\ell}, \quad \ell \in \mathbb{N}. \tag{4.4}$$

Now let us put $V_\ell^{(1)} = S^\ell$, and

$$V_\ell^{(d+1)} = T^{\ell(m+k)} V_\ell^{(d)} + V_\ell^{(1)} Q_\ell^d, \quad d \in \mathbb{N}.$$

One can see that for every $d, \ell \in \mathbb{N}$ the operator $ZV_\ell^{(d)}$ is positive, since Z and S are commuting. Moreover, one has

$$T^{d\ell(m+k)} = \left(\frac{I+T^k}{2}\right)^{\ell} V_\ell^{(d)} + Q_\ell^d, \quad d, \ell \in \mathbb{N}. \tag{4.5}$$

From $Z^*(\mathbb{1}) = \mathbf{T}^*(\mathbb{1}) = \mathbb{1}$, it follows from (4.5) that

$$V_\ell^{(d)*}(\mathbb{1}) + Q_\ell^{*d}(\mathbb{1}) = \mathbb{1}.$$

Now the positivity of $ZV_\ell^{(d)}$ and ZQ_ℓ imply that $\|ZV_\ell^{(d)}\| \le 1$ and $\|ZQ_\ell\| \le 1$.

From (4.1) and (4.2), due to Theorem 3.1, one finds that $\|Z(T^{\ell m} - S^\ell)\| < 1$ for all $\ell \in \mathbb{N}$. Using this inequality with $T^*(\mathbb{1}) = \mathbb{1}$ and the positivity of $Z(T^{\ell m} - S^\ell)$ we find that

$$\|Z(T^{\ell m} - S^\ell)\| = \|((T^*)^{\ell m} - S^{*\ell})Z^*\| = \|\mathbb{1} - S^{*\ell}(\mathbb{1})\| < 1. \tag{4.6}$$

The equality (4.4) yields that

$$Q_\ell^*(\mathbb{1}) = \mathbb{1} - S^{*\ell}(\mathbb{1}).$$

Hence, from (4.6) with the positivity of ZQ_ℓ we obtain

$$\|ZQ_\ell\| = \|Q_\ell^*(\mathbb{1})\| = \|\mathbb{1} - S^{*\ell}(\mathbb{1})\| < 1$$

for all $\ell \in \mathbb{N}$.

Therefore, there is a number $d_\varepsilon \in \mathbb{N}$ such that $\|(ZQ_{\ell_\varepsilon})^{d_\varepsilon}\| < \frac{\varepsilon}{4}$. From the commutativity of Z and Q_ℓ one finds

$$\|Z^{d_\varepsilon} Q_{\ell_\varepsilon}^{d_\varepsilon}\| < \frac{\varepsilon}{4}. \tag{4.7}$$

Now putting $n_0 = d_\varepsilon \ell_\varepsilon (m+k)$, from (4.5) with (4.7) we obtain

$$\begin{aligned}
\|Z^{d_\varepsilon}(T^{n_0+k} - T^{n_0})\| &\le \left\| Z^{d_\varepsilon}\left(T^k\left(\frac{I+T^k}{2}\right)^{\ell_\varepsilon} - \left(\frac{I+T^k}{2}\right)^{\ell_\varepsilon}\right)V_{\ell_\varepsilon}^{(d_\varepsilon)}\right\| \\
&\quad + \|Z^{d_\varepsilon} Q_{\ell_\varepsilon}^{d_\varepsilon}(T^k - I)\| \\
&\le \left\| T^k\left(\frac{I+T^k}{2}\right)^{\ell_\varepsilon} - \left(\frac{I+T^k}{2}\right)^{\ell_\varepsilon}\right\| + 2\|Z^{d_\varepsilon} Q_{\ell_\varepsilon}^{d_\varepsilon}\| \\
&\le \frac{\gamma}{\sqrt{\ell_\varepsilon}} + 2 \cdot \frac{\varepsilon}{4} < \varepsilon.
\end{aligned}$$

Take any $n \geq n_0$ then from the last inequality one gets

$$\|Z^{d_\varepsilon}(T^{n+k} - T^n)\| = \|Z^{d_\varepsilon}T^{n-n_0}(T^{n_0+k} - T^{n_0})\| \leq \|T^{n_0+k} - T^{n_0}\| < \varepsilon$$

which completes the proof. □

Remark 4.2. *We note the followings:*

(a) *If $Z = I$ then a similar kind of result has been proved in [13] for positive contractions of C^*-algebras.*

(b) *If the algebra M is commutative, then Theorem 4.1 covers the main result of [9, 10]. Note that vector-valued versions of the zero-two law have been proved in [6, 7].*

(c) *Since the dual of $L_1(M_{sa}, \tau)$ is M_{sa} then due to the duality theory the proved Theorem 4.1 holds true if one replaces L_1-space with M_{sa}.*

Acknowledgments

The first author acknowledges the MOE Grant FRGS13-071-0312.

References

[1] Bratteli O and Robinson D W 1979 *Operator algebras and quantum statistical mechanics. Part I* (New York-Heidelberg -Berlin: Springer)
[2] Derriennic Y 1976 *Ann. Inst. H. Poincaré Sec. B* **12** 111–129.
[3] Foguel S R 1971 *Israel J. Math.* **10** 275–280
[4] Foguel S R 1976 *Proc. Amer. Math. Soc.* **61** 262–264
[5] Foguel S R 1983 *Israel J. Math.* **45** (1983) 219–224
[6] Ganiev I and Mukhamedov F 2006 *Comment. Math. Univ. Carolinae* **47** 427–436
[7] Ganiev I, Mukhamedov F and Bekbaev D 2015 *Turk. J. Math.* **39** 583–594
[8] Lin M 1982 *Z. Wahrscheinlichkeitstheor. Verw. Geb.* **61** 513–525
[9] Mukhamedov F 2011 *Positivity* **15** 497-508
[10] Mukhamedov F 2015 *Bull. Korean Math. Soc.* **52** 1819–1826
[11] Mukhamedov F, Temir S and Akin H 2010 *Turkish Jour. Math.* **34** 85–93
[12] Nelson E 1974 *J. Funct. Anal.* **15** 103–117
[13] Neshveyev S and Tuset L 2004 *J. Reine und Angewandte Math* **538** 23–70
[14] Ornstein D and Sucheston L 1970 *Ann. Math. Statist.* **41** 1631–1639
[15] Takesaki M 1979 *Theory of operator algebras, I* (Berlin: Springer)
[16] Zaharopol R 1989 *Proc. Edin. Math. Soc.* **32** 363–370
[17] Wittmann R 1987 *Israel J. Math.* **59** 828

Extreme boundary of the space semi-additive functionals on the four-point set

Gayratbay Djabbarov

Department of Mathematics, Tashkent state pedagogical university, Tashkent, Uzbekistan

E-mail: gayrat_77@bk.ru

Abstract. The present paper devoted to study of the extreme boundary of the convex compact set of all semi-additive functionals on the four-point compactum. We shall find some classes of extreme points of the space semi-additive functionals $OS(\mathbf{4})$.

Keywords: semi-additive functional, extreme point, convex set.

AMS Subject Classification: 46A55; 46E10.

1. Introduction

The space $P(X)$ of all probability measures on a compactum X is well investigated. In [8], T. Radul introduced the space $O(X)$ of weakly additive order-preserving normalized functionals on a compactum X. Topological and geometric properties of weakly additive order-preserving normalized functionals were studied in [1,2,5,6]. In [4] the space of all of semi-additive positively-homogeneous weakly additive order-preserving normalized functionals was investigated and a general form of semi-additive functionals was given. Also categorical properties of the functor of semi-additive functionals OS have been investigated.

It is well-known that the extreme boundary of the space of probability measures on compactum coincides with the set of all Dirac measures on this compactum. This property plays a crucial rule in investigations of geometric properties of the space of probability measures on compactum. The structure of the extreme boundary of the space of semi-additive functionals on compactum is not yet described. In [7] it was described a general form of extreme points of space semi-additive functionals on three-point space.

The aim of the present paper is to study the extreme boundary of the convex compact set $OS(\mathbf{4})$.

2. Preliminaries

Let X be a compact set. Denote by $C(X)$ the algebra of all real-valued continuous functions $f : X \to \mathbb{R}$ with point-wise algebraic operations and sup-norm, i. e., with the norm $\|f\| = \max\{|f(x)| : x \in X\}$. For any $c \in \mathbb{R}$ by c_X we denote the constant function, defined by the formula $c_X(x) = c$, $x \in X$. Let φ, $\psi \in C(X)$. The inequality $\varphi \leq \psi$ means that $\varphi(x) \leq \psi(x)$ for all $x \in X$.

A functional $\nu : C(X) \to \mathbb{R}$ is said to be [8]:

(1) weakly additive if $\nu(\varphi + c_X) = \nu(\varphi) + c\nu(1_X)$ for all $\varphi \in C(X)$ and $c \in \mathbb{R}$;

(2) order-preserving, for any $\varphi,\ \psi \in C(X)$ with $\varphi \leq \psi$ we have $\nu(\varphi) \leq \nu(\psi)$;
(3) normalized if $\nu(1_X) = 1$;
(4) positively homogeneous if $\nu(t\varphi) = t\nu(\varphi)$ for all $\varphi, \in C(X), t \in \mathbb{R},\ t \geq 0$;
(5) semi-additive if $\nu(\varphi + \psi) \leq \nu(\varphi) + \nu(\psi)$ for all $\varphi, \psi \in C(X)$.

For every compactum X we denote

$$V(X) = \prod_{\varphi \in C(X)} [\min \varphi, \max \varphi].$$

For every map $f : X \to Y$ by $V(f)$ we denote the map from $V(X)$ to $V(Y)$ defined by

$$V(f)(\nu)(\varphi) = \nu(\varphi \circ f),\quad \nu \in V(X),\ \varphi \in C(X).$$

For a compactum X we denote by:
– $OS(X)$ the set of all weakly-additive, order-preserving, normalized, positive-homogeneous and semi-additive functionals on $C(X)$;
– $P(X)$ the set of all positive, normalized linear functionals on $C(X)$.

Let us consider these sets as subspaces of the space $V(X)$, equipped with the topology of point-wise convergence, in particularly, the base of neighborhoods for the functional $\nu \in \mathcal{F}(X)$, where $\mathcal{F} = OS, P$, is formed by the sets

$$\langle \nu;\ \varphi_1, \varphi_2, \cdots, \varphi_k, \varepsilon \rangle = \{\nu' \in \mathcal{F}(X) : |\nu'(\varphi_i) - \nu(\varphi_i)| < \varepsilon,\ i = \overline{1,k}\},$$

where $\varepsilon > 0$, $\varphi_i \in C(X)$, $i = \overline{1,k}, k \in \mathbb{N}$.

For every compactum X the spaces $OS(X)$ and $P(X)$ are convex compact.

Let F be a closed subspace of X. A functional $\nu \in OS(X)$ is said to be supported on F, if $\nu(f) = \nu(g)$ for all $f, g \in C(X)$ with $f|_F = g|_F$. The smallest closed set $F \subset X$ on which the functional μ is supported, is called the support of $\nu \in OS(X)$ and denoted by $\operatorname{supp} \nu$, i.e.,

$$\operatorname{supp} \nu = \bigcap \{F : \nu \text{ is supported on } F\}.$$

For every convex compactum K we denote by $cc(K)$ the space of all non empty convex compact subsets of K, equipped with the Vietoris topology. Recall [3] that a base of this topology is formed by the sets of the form

$$\langle U_1, \ldots, U_n \rangle = \{A \subseteq cc(K) : A \subseteq U_1 \cup \cdots \cup U_n \text{ and } A \cap U_i \neq \emptyset \text{ for every } i\},$$

where $U_1, \ldots, U_n$ run through the topology of K, $n \in \mathbb{N}$.

For $A \in cc(P(X))$ set

$$\nu_A(\varphi) = \sup\{\mu(\varphi) : \mu \in A\},\quad \varphi \in C(X). \tag{1}$$

Then $\nu_A \in OS(X)$. In [4, Proposition 4.4] it was shown that any functional from $OS(X)$ is represented in the form (1), moreover the mapping

$$A \in cc(P(X)) \mapsto \nu_A \in OS(X) \tag{2}$$

is an affine homeomorphism between spaces $cc(P(X))$ and $OS(X)$ (see [4, Theorem 1]).

Let X and Y be topological spaces and let $f : X \to Y$ be a continuous mapping. The mapping $\mathcal{F}(f) : \mathcal{F}(X) \to \mathcal{F}(Y)$, where $\mathcal{F} = OS, P$, is defined as the restriction of $V(f)$ on $\mathcal{F}(X)$.

In [4] it was proved that

$$OS(f)(\nu_A) = \nu_{P(f)(A)}. \tag{3}$$

Note that for the n-point compact $\mathbf{n} = \{0, 1, ..., n-1\}$, $n \in \mathbb{N}$, the space $C(\mathbf{n})$ is isomorphic to the space $\mathbb{R}^n$, moreover, isomorphism can be defined by

$$f \in C(\mathbf{n}) \to (f(0), f(1), ..., f(n-1)) \in \mathbb{R}^n.$$

In [4] it was shown that the space $OS(\mathbf{2})$ is affine isomorphic to the triangle

$$\triangle = \{(\alpha, \beta) : \alpha, \beta \in \mathbb{R}, 0 \leq \alpha \leq \beta \leq 1\},$$

moreover, this isomorphism can be established by a rule

$$(\alpha, \beta) \mapsto \lambda = \alpha\delta_0 + (1-\beta)\delta_1 + (\beta - \alpha)\delta_0 \vee \delta_1,$$

where δ_i is the Dirac functional on i, and a functional $\delta_0 \vee \delta_1 \in OS(\mathbf{2})$ is defined by

$$(\delta_0 \vee \delta_1)(f) = \max\{\delta_0(f), \delta_1(f)\},\ f \in C(\mathbf{2}).$$

Let K be a convex compact subset of a locally convex space E. Recall that the Minkovsky operation is defined by

$$\lambda_1 A_1 + \lambda_2 A_2 = \{\lambda_1 x_1 + \lambda_2 x_2 : x_1 \in A_1, x_2 \in A_2\},$$

where $\lambda_1, \lambda_2 \in \mathbb{R}, A_1, A_2 \in cc(K)$. According to [3] consider the equivalence relation $\sim$ on $cc(E) \times cc(E)$ defined by:

$$(A, B) \sim (C, D) \quad \text{if only if} \quad A + D = B + C.$$

Denote by L the space of the equivalence classes with respect to $\sim$ and let $[A, B]$ be the class containing (A, B). It is well-known that L is a linear space with respect to natural algebraic operations. For a convex neighborhood U of zero we put

$$U^* = \{[A, B] : A \subset B + U, B \subset A + U\}.$$

The sets of the form U^* are the base neighborhoods of zero in L. A mapping $\pi : cc(K) \to L$ defined by the rule

$$\pi(A) = [A, \{0\}]$$

is an embedding, moreover

$$\pi(\lambda_1 A_1 + \lambda_2 A_2) = \lambda_1 \pi(A_1) + \lambda_2 \pi(A_2)$$

for all $\lambda_1, \lambda_2 \in \mathbb{R}, A_1, A_2 \in cc(K)$.

Consider on $C(\mathbf{3})$ functionals of the following form:

$$\nu_1(f) = f(0), \tag{4}$$

$$\nu_2(f) = \max\{f(0), tf(1) + (1-t)f(2), \alpha f(0) + \beta f(1) + \gamma f(2)\}, \tag{5}$$

where $0 \leq t \leq 1$, $\alpha + \beta + \gamma = 1$, $\alpha, \beta, \gamma \geq 0$,

$$\nu_3(f) = \max\{\alpha f(0) + (1-\alpha)f(1), \beta f(1) + (1-\beta)f(2), \gamma f(2) + (1-\gamma)f(0)\}, \tag{6}$$

where $0 < \alpha, \beta, \gamma < 1$.

The functionals $\mu, \nu \in OS(X)$ are called *similar*, if there exists a homeomorphism $\Phi : X \to X$ such that $\nu = \mu \circ \Phi$.

The subsets $A, B \subseteq P(X)$ are called *similar*, if there exists a homeomorphism $\tau : X \to X$ such that $A = P(\tau)(B)$.

In [7] it was given a general form of extreme points of space semi-additive functionals on three-point space. Namely, a functional $\mu \in OS(\mathbf{3})$ is an extreme point in $OS(\mathbf{3})$ if and only if μ is similar to a functional of the form (4)–(6).

Journal of Physics: Conference Series **697** (2016) 012004 doi:10.1088/1742-6596/697/1/012004

3. Extreme points with dim ≤ 2

In this subsection we will find necessity and sufficiently conditions for functional of the form $\nu = \nu_A$, where $A \in cc(P(\mathbf{4}))$, $\dim A \leq 2$, to be an extreme point of $OS(\mathbf{4})$.

Consider on $P(\mathbf{4})$ convex subsets of the following forms:

$$A_1 = \{\delta_0\}, \tag{7}$$

$$A_2 = \mathrm{co}\{\alpha\delta_0 + (1-\alpha)\delta_1,\ \beta\delta_1 + (1-\beta)\delta_2,\ \gamma\delta_2 + (1-\gamma)\delta_0\}, \tag{8}$$

where $0 < \alpha, \beta, \gamma < 1$, (see Pic. 1),

$$A_3 = \mathrm{co}\{\delta_0, \alpha_1\delta_1 + \alpha_2\delta_2 + \alpha_3\delta_3, t_0\delta_0 + t_1\delta_1 + t_2\delta_2 + t_3\delta_3\}, \tag{9}$$

where $0 \leq \alpha_i, t_j \leq 1$, $\alpha_1 + \alpha_2 + \alpha_3 = 1$, $t_0 + t_1 + t_2 + t_3 = 1$, (see Pic. 2).

Let $A = A_2$. Then the restriction $\nu_A|_{C(\mathbf{3})}$ of ν_A is a functional of the form (6).

Let $A = A_3$ and H be a cross section of the simplex $P(\mathbf{4})$ containing A. Then H is a triangle affine homeomorphic to $P(\mathbf{3})$ and ν_A is a functional of the form (5).

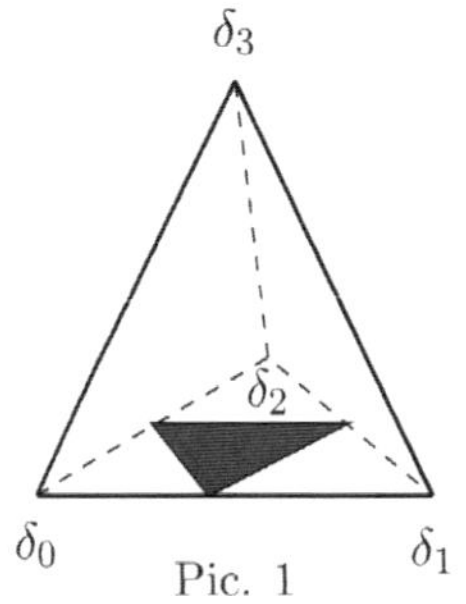

Pic. 1

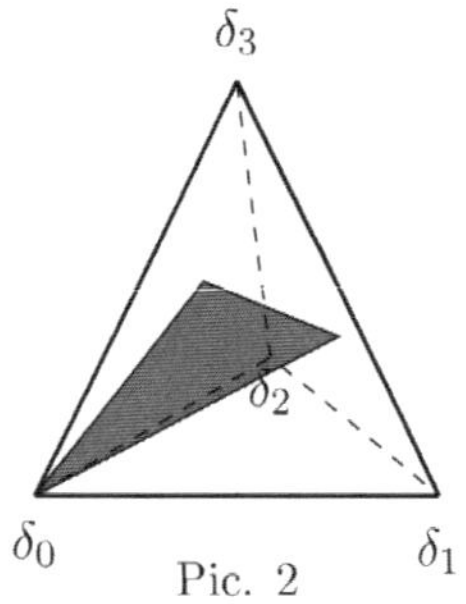

Pic. 2

The main result of this subsection is the following Theorem.

Theorem 3.1 *Let $A \in cc(P(\mathbf{4}))$ with $\dim A \leq 2$. Then ν_A is an extreme point in $OS(\mathbf{4})$ if and only if A is similar to a subset of the form (7)–(9).*

For the proof of the Theorem 3.1 we need several Lemmata.

Lemma 3.2 *Let A be a convex subset of $co\{\mu_1, \mu_2, \mu_3\}$, where $\mu_1 = t_1\delta_0 + (1-t_1)\delta_1$, $\mu_2 = t_2\delta_0 + (1-t_2)\delta_2$, $\mu_3 = t_3\delta_0 + (1-t_3)\delta_3$, $0 < t_1, t_2, t_3 < 1$. Then A is not an extreme point.*

Proof. Let us take $\varepsilon > 0$ such that $\varepsilon < \min\left\{1, \frac{t_i}{1-t_i} : i = 1, 2, 3\right\}$. Then

$$A_\varepsilon = \{(1+\varepsilon)\mu - \varepsilon\delta_0 : \mu \in A\},$$

$$A_{-\varepsilon} = \{(1-\varepsilon)\mu + \varepsilon\delta_0 : \mu \in A\}$$

both are convex subsets in $P(\mathbf{4})$.

Let $\mu \in A$ be an arbitrary functional. Since

$$\mu = \frac{(1+\varepsilon)\mu - \varepsilon\delta_0 + (1-\varepsilon)\mu + \varepsilon\delta_0}{2},$$

it follows that $A \subseteq (A_\varepsilon + A_{-\varepsilon})/2$.

On the other hand, for $\lambda_1, \lambda_2 \in A$, we have

$$\frac{(1+\varepsilon)\lambda_1 - \varepsilon\delta_0 + (1-\varepsilon)\lambda_2 + \varepsilon\delta_0}{2} = \frac{(1+\varepsilon)\lambda_1 + (1-\varepsilon)\lambda_2}{2} \in A.$$

This means that $A \supseteq (A_\varepsilon + A_{-\varepsilon})/2$. Thus $A = (A_\varepsilon + A_{-\varepsilon})/2$. The proof is complete. □

By the same argument, one can prove the following Lemma.

Lemma 3.3 *Let A be a convex subset in* $co\{\mu_1, \mu_2, \mu_3, \mu_4\}$, *where* $\mu_1 = t_1\delta_0 + (1 - t_1)\delta_1$, $\mu_2 = t_2\delta_0 + (1 - t_2)\delta_2$, $\mu_3 = t_3\delta_2 + (1 - t_3)\delta_3$, $\mu_4 = t_4\delta_1 + (1 - t_4)\delta_3$, $0 < t_1, t_2, t_3, t_4 < 1$. *Then A is not an extreme point.*

Lemma 3.4 *Let A be a convex subset in $P(\mathbf{4})$ with* $\dim A \leq 2$ *and let H be a cross section of the simplex $P(\mathbf{4})$ containing A. Suppose that $\delta_0 \in A$. Then A is an extreme point in $cc(P(\mathbf{4}))$ iff A is an extreme in H.*

Proof. Note that a cross section H is a triangle of the form $\mathrm{co}\{\delta_0, \mu_1, \mu_2\} \cong P(\mathbf{3})$, where $A \cong B$ means affine homeomorphism between convex sets A and B, $\mu_1 = t_1\delta_1 + (1 - t_1)\delta_3$, $\mu_2 = t_2\delta_2 + (1 - t_2)\delta_3$, $0 \leq t_1, t_2 \leq 1$.

It is clear that, if A is not an extreme point in $cc(\mathrm{co}\{\delta_0, \mu_1, \mu_2\})$, it follows that A is not extreme point in $cc(P(\mathbf{4}))$.

Suppose that A is an extreme point in $cc(\mathrm{co}\{\delta_0, \mu_1, \mu_2\})$.

Let $A = (B + C)/2$, where $B, C \in cc(P(\mathbf{4}))$. Since $\delta_0 \in A$, there exist $\lambda \in B$, $\mu \in C$ such that $\delta_0 = (\lambda + \mu)/2$. Then $\delta_0 = \lambda = \mu$, i.e. $\delta_0 \in B, C$.

Let $\nu \in B$. Since $\delta_0 \in C$ and $A = (B + C)/2$, it follows that $(\nu + \delta_0)/2 \in A$. Further,

$$\frac{3\nu + \delta_0}{4} = \frac{\nu + \frac{\nu + \delta_0}{2}}{2} \in A.$$

Continuing by this way we obtain

$$\frac{(2^n - 1)\nu + \delta_0}{2^n} \in A.$$

Thus $\nu \in \mathrm{co}\{\delta_0, \mu_1, \mu_2\}$. This means that $B \subseteq \mathrm{co}\{\delta_0, \mu_1, \mu_2\}$.

By a similar way we obtain that $C \subseteq \mathrm{co}\{\delta_0, \mu_1, \mu_2\}$. Thus

$$B \subseteq \mathrm{co}\{\delta_0, \mu_1, \mu_2\} \text{ and } C \subseteq \mathrm{co}\{\delta_0, \mu_1, \mu_2\}.$$

Since A is an extreme point in $cc(\mathrm{co}\{\delta_0, \mu_1, \mu_2\})$, it follows that $B = C = A$. The proof is complete. □

Proof of Theorem 3.1. Let $A \in cc(P(\mathbf{4}))$ with $\dim A \leq 2$. Let us show that ν_A is an extreme point in $OS(\mathbf{4})$ if and only if A is similar to a subset of the form (7)–(9).

Let H be a cross section of the simplex $P(\mathbf{4})$ containing A. Then H is a triangle or quadrangle.

Case 1. $H \cap \{\delta_0, \delta_1, \delta_2, \delta_3\} = \emptyset$. Then by Lemmata 3.2 and 3.3, it follows that A is not an extreme point.

Case 2. $H \cap \{\delta_0, \delta_1, \delta_2, \delta_3\} \neq \emptyset$. Without lost generality we can assume that $\delta_0 \in A$. By Lemma 3.4, A is an extreme point in $cc(P(\mathbf{4}))$ iff A is an extreme in $H \cong P(\mathbf{3})$,. Then [7, Theorem 3.1] implies that a functional ν_A is an extreme point in $cc(H) \cong OS(\mathbf{3})$ if and only if ν_A is similar to a functional of the form (4)–(6). This means that A is similar to a subset of the form (7)–(9). The proof is complete. □

4. Extreme points with dim = 3

In this subsection we will find sufficiently conditions for functional of the form $\nu = \nu_A$, where $A \in cc(P(\mathbf{4}))$, $\dim A = 3$, to be an extreme point of $OS(\mathbf{4})$.

Consider in $P(\mathbf{4})$ subsets of the following forms:

$$A = \mathrm{co}\{K_1, K_2\}, \tag{10}$$

where $K_1 \in cc(P(\mathbf{k}))$, $K_2 \in cc(P(\mathbf{4} \setminus \mathbf{k}))$, $k = 1, 2$;

$$A = \mathrm{co}\{K_1, \ldots, K_r\}, \tag{11}$$

where $K_i = A \cap F_i$ is an extreme point in $cc(F_i)$ and F_i is a 2-dimensional face of $P(\mathbf{4})$, $i = 1, \dots, r$, $2 \le r \le 4$;

$$A = \text{co}\{\mu_0, K\}, \tag{12}$$

where $K \in cc(P(\mathbf{3}))$ is an extreme point with $\dim K = 2$, $\mu_0 = (1 - \alpha_0)\nu_0 + \alpha_0 \delta_3$, $\nu_0 \in P(\mathbf{3})$, $0 < \alpha_0 < 1$.

Lemma 4.1 *Let A be a subset in $P(\mathbf{4})$ one of the forms (10)–(12). Then A is an extreme point in $cc(P(\mathbf{4}))$.*

Proof. Let A be a subset of the form (10) and let $A = (B + C)/2$, where $B, C \in cc(P(\mathbf{4}))$. Put

$$B_1 = B \cap P(\mathbf{k}),\ B_2 = B \cap P(\mathbf{4} \setminus \mathbf{k}),$$
$$C_1 = C \cap P(\mathbf{k}),\ C_2 = C \cap P(\mathbf{4} \setminus \mathbf{k}).$$

Since $P(\mathbf{k})$ and $P(\mathbf{4} \setminus \mathbf{k})$ both are faces in $P(\mathbf{4})$, it follows that $K_i = (B_i + C_i)/2$, $i = 1, 2$.

Let $\lambda \in B_1$, $\nu \in C_2$. Since $(\lambda + \nu)/2 \in A$, there exist $\mu_1 \in K_1, \mu_2 \in K_2$ such that $(\lambda + \nu)/2 = t_1\mu_1 + t_2\mu_2$, where $t_1, t_2 \ge 0, t_1 + t_2 = 1$. Take a characteristic function $\chi_{\mathbf{k}}$ of the set $\mathbf{k}$, i.e. $\chi_{\mathbf{k}}(i) = 1$ for $i \in \mathbf{k}$ and $\chi_{\mathbf{k}}(i) = 0$ for $i \notin \mathbf{k}$. From

$$(\lambda(\chi_{\mathbf{k}}) + \nu(\chi_{\mathbf{k}}))/2 = t_1\mu_1(\chi_{\mathbf{k}}) + t_2\mu_2(\chi_{\mathbf{k}}),$$

it follows that $1/2 = t_1$, and therefore $t_1 = t_2 = 1/2$. Thus $\lambda + \nu = \mu_1 + \mu_2$.

Now take an arbitrary function $f : \mathbf{4} \to \mathbb{R}$ such that $f|_{\mathbf{n} \setminus \mathbf{k}} \equiv 0$. Then $\lambda(f) = \mu_1(f)$, and therefore $\lambda = \mu_1$, $\nu = \lambda_2$. Thus $\lambda \in K_1$, $\nu \in K_2$. This means that $B_1 \subseteq K_1$, $C_2 \subseteq K_2$.

By a similar way we obtain that $B_2 \subseteq K_2$, $C_1 \subseteq K_1$. Thus

$$B \subseteq A \text{ and } C \subseteq A.$$

Let $\mu \in A$ be an arbitrary extreme point in A. Take $\lambda \in B, \nu \in C$, such that $\mu = (\lambda + \nu)/2$. Since μ is an extreme point, it follows that $\lambda = \nu = \mu$. So, $\mu \in B$ and $\mu \in C$. Since μ be an arbitrary extreme point, it follows that

$$A \subseteq B \text{ and } A \subseteq C.$$

Thus $B = C = A$. This means that A is an extreme point.

Let now A be a subset of the form (11) and $A = (B + C)/2$, where $B, C \in cc(P(\mathbf{4}))$. Put

$$B_i = B \cap F_i,\ C_i = C \cap F_i,\ i \in \overline{1, r}.$$

Since F_i is a face in $P(\mathbf{4})$, it follows that $K_i = (B_i + C_i)/2$, $i \in \overline{1, r}$. Taking into account that K_i is an extreme point in $cc(F_i)$ we have that $B_i = C_i = K_i$, $i \in \overline{1, r}$. Thus

$$A \subseteq B \text{ and } A \subseteq C.$$

Let $\lambda \in B$ be an arbitrary element. Take an arbitrary $\mu \in A$. Taking into account $A \subseteq C$ and $A = (B + C)/2$ we obtain that points

$$\lambda_1 = (\lambda + \mu)/2,\ \lambda_s = (\lambda + \lambda_{s-1})/2,\ s \ge 2$$

belong to A. Since $\lambda_s = \dfrac{(2^s - 1)\lambda + \mu}{2^s}$, it follows that $\lambda_s \to \lambda$ and therefore $\lambda \in A$. Thus $B \subseteq A$. So, $B = A$. By a similar way we obtain that $C = A$. Thus $B = C = A$.

Finally, let A be a subset of the form (12) and $A = (B + C)/2$, where $B, C \in cc(P(\mathbf{4}))$. Set

$$B_0 = B \cap P(\mathbf{3}),\ C_0 = C \cap P(\mathbf{3}).$$

Taking into account that $P(\mathbf{3})$ is a face in $P(\mathbf{4})$, we obtain that $K = (B_0 + C_0)/2$. Since K is an extreme point, we obtain that $K = B_0 = C_0$.

Take $\lambda_0 \in B, \nu_0 \in C$ such that $\mu_0 = (\lambda_0 + \nu_0)/2$. Let μ be an arbitrary extreme point of K. Since $(\lambda_0 + \mu)/2, (\nu_0 + \mu)/2 \in A$, there exist $\mu_1, \mu_2 \in K$ such that

Journal of Physics: Conference Series **697** (2016) 012004 doi:10.1088/1742-6596/697/1/012004

$$(\lambda_0+\mu)/2 = t_1\mu_1 + (1-t_1)\mu_0 \text{ and } (\nu_0+\mu)/2 = t_2\mu_2 + (1-t_2)\mu_0,$$

where $0 \le t_1 \le 1$, $0 \le t_2 \le 1$. Adding the last two equalities we get

$$\mu_0 + \mu = t_1\mu_1 + t_2\mu_2 + (2-t_1-t_2)\mu_0.$$

Take a function $f : \mathbf{4} \to \mathbb{R}$ defined by $f(3) = 1$ and $f|_{\mathbf{3}} \equiv 0$. From

$$\mu_0(f) + \mu(f) = t_1\mu_1(f) + t_2\mu_2(f) + (2-t_1-t_2)\mu_0(f)$$

it follows that $\alpha_0 = (2-t_1-t_2)\alpha_0$. Thus $t_1+t_2 = 1$. So, $\mu = t_1\mu_1 + t_2\mu_2$. Since μ is an extreme point, we obtain that $\mu = \mu_1 = \mu_2$. Thus $(\lambda_0+\mu)/2 = t_1\mu + (1-t_1)\mu_0$ and $(\nu_0+\mu)/2 = t_2\mu + (1-t_2)\mu_0$. This means that λ_0 and ν_0 both lie on the line passing through μ_0, μ. Since μ is an arbitrary and K is a non one-point, it follows that $\lambda_0 = \nu_0$. Thus $A = B = C$. The proof is complete. □

Consider the following system of elements in $P(\mathbf{4})$:

$$\mu_i = \sum_{j\neq i} \alpha_{ij}\delta_j, \tag{13}$$

where $\alpha_{ij} > 0$ for all $i \neq j$ and $\sum\limits_{j\neq i} \alpha_{ij} = 1$ for all $i \in \mathbf{4}$.

It is clear that $A = \mathrm{co}\{\mu_0, \mu_1, \mu_2, \mu_3\}$ is a 3-dimensional simplex with vertices $\mu_0, \mu_1, \mu_2, \mu_3$. In particular, if $\alpha_{ij} = \frac{1}{3}$ for all $i \neq j$, it follows that A is a simplex with vertices on the barycenters of 2-dimensional faces in $P(\mathbf{4})$. In this case A consists of points of the form: $\mu_i = \sum\limits_{i=0}^{3} t_i\delta_i$ with $0 \le t_i \le 1/3$ for all $i \in \mathbf{4}$.

Lemma 4.2 *Let $A = co\{\mu_0, \mu_1, \mu_2, \mu_3\}$. Then A is an extreme point.*

Proof. Let us consider the following two cases.

Case 1. $\alpha_{ij} = \frac{1}{3}$ for all $i \neq j$. Let $A = (B+C)/2$, where B, C are convex subsets in $P(\mathbf{4})$. Then there exist $\lambda_i \in B, \nu_i \in C$ such that $\mu_i = (\lambda_i + \nu_i)/2$ for all $i \in \mathbf{4}$. Let

$$\lambda_i = \sum_{j\neq i} \left(\frac{1}{3} + t_{ij}\right)\delta_j,$$

$$\nu_i = \sum_{j\neq i} \left(\frac{1}{3} - t_{ij}\right)\delta_j,$$

where $|t_{ij}| \le 1/3$ for all $i \neq j$ and $\sum\limits_{j\neq i} t_{ij} = 0$ for all $i \in \mathbf{4}$.

Since $(\lambda_k + \nu_p)/2 \in A$, $k, p \in \mathbf{n}$, it follows that

$$\frac{1}{3} + t_{kj} - t_{pj} \le \frac{1}{3},$$

i.e. $t_{kj} - t_{pj} \le 0$. Interchanging k and p we get $t_{pj} - t_{kj} \le 0$. Thus $t_{kj} = t_{pj}$ for all k, p, j with $k \neq j$, $p \neq j$. Denote $t_j = t_{ij}$. Then $\sum\limits_{j\neq i} t_j = 0$. Since

$$0 = \sum_{j\neq i} t_j = \sum_{j\neq k} t_j + t_i - t_k = t_i - t_k,$$

it follows that $t_1 = \ldots = t_n$. Thus $t_i = 0$ for all i. This means that $\lambda_i = \nu_i = \mu_i$ for all i. Thus $A = B = C$.

Case 2. Let $\mu_0, \mu_1, \mu_2, \mu_3$ be arbitrary points of the form (13).

Let $A = (B + C)/2$, where B, C are convex subsets in $P(\mathbf{4})$. Then there exist $\lambda_i \in B, \nu_i \in C$ such that $\mu_i = (\lambda_i + \nu_i)/2$ for all $i \in \mathbf{4}$. We can take the following representaions

$$\lambda_i = \sum_{j \neq i} (\alpha_{ij} + t_{ij})\,\delta_j,$$

$$\nu_i = \sum_{j \neq i} (\alpha_{ij} - t_{ij})\,\delta_j,$$

where $|t_{ij}| \leq 1/3$ for all $i \neq j$ and $\sum\limits_{j \neq i} t_{ij} = 0$ for all $i \in \mathbf{4}$.

Set

$$\lambda_i' = \sum_{j \neq i} (\alpha_{ij} + \varepsilon t_{ij})\,\delta_j,$$

$$\nu_i' = \sum_{j \neq i} (\alpha_{ij} - \varepsilon t_{ij})\,\delta_j,$$

where $0 < \varepsilon < 1$. Since $\mu_i = (\lambda_i' + \nu_i')/2$ for all $i \in \mathbf{4}$, it follows that

$$A \subseteq (\mathrm{co}\{\lambda_i'\} + \mathrm{co}\{\nu_i'\})/2.$$

Taking into account

$$\frac{\lambda_k' + \nu_p'}{2} = \frac{t_1\lambda_k + (1 - t_1)\lambda_p + t_2\nu_k + (1 - t_2)\nu_p}{2},$$

where $t_1 = (1 + \varepsilon)/2$, $t_2 = (1 - \varepsilon)/2$, we obtain that

$$A \supseteq (\mathrm{co}\{\lambda_i'\} + \mathrm{co}\{\nu_i'\})/2.$$

Thus

$$A = (\mathrm{co}\{\lambda_i'\} + \mathrm{co}\{\nu_i'\})/2. \tag{14}$$

So, if necessary, replacing t_{ij} with εt_{ij}, we can assume that $|t_{ij}| \leq \min\{\alpha_{ij}, 1/3\}$.

Set

$$\mu_i' = \sum_{j \neq i} \frac{1}{3}\delta_j,$$

$$\lambda_i' = \sum_{j \neq i} \left(\frac{1}{3} + t_{ij}\right)\delta_j,$$

$$\nu_i' = \sum_{j \neq i} \left(\frac{1}{3} - t_{ij}\right)\delta_j.$$

Since $\mathrm{co}\{\mu_i\}$ and $\mathrm{co}\{\mu_i\}$ both are simplex with the same dimensions, taking into account (14) we obtain that

$$\mathrm{co}\{\mu_i'\} = (\mathrm{co}\{\lambda_i'\} + \mathrm{co}\{\nu_i'\})/2.$$

By case 1 it follows that $t_{ij} = 0$. Thus $B = C = A$. The proof is complete. □

References

[1] Albeverio S, Ayupov Sh and Zaitov A 2008 *Topology and its Applications* **155** 1792–1799.
[2] Ayupov Sh and Zaitov A 2009 *Ukrainian J. Math.* **61** 1380–1386.
[3] Bazylevich L, Repovs D and Zarichniy M 2008 *Topology and its Applications* **155** 764–772.
[4] Davletov D and Djabbarov G 2008 *Methods of Funct. Anal. and Appl.* **14** 317–322.
[5] Djabbarov G 2005 *Uzb. Math. J.* **3** 17–25.
[6] Djabbarov G 2006 *Uzb. Math. J.* **2** 20–28.
[7] Djabbarov G *Contemporary Mathematics AMS* (is accepted).
[8] Radul T 1998 *Comment. Math. Univ. Carol.* **39** 609–615.

Algebra, Analysis and Quantum Probability IOP Publishing
Journal of Physics: Conference Series **697** (2016) 012005 doi:10.1088/1742-6596/697/1/012005

On a classification of finite dimensional algebras with respect to the orthogonal (unitary) changes of basis

Ural Bekbaev

Department of Science in Engineering, Faculty of Engineering, International Islamic University Malaysia, P.O. Box 10, 50728 Kuala Lumpur, Malaysia

E-mail: bekbaev@iium.edu.my

Abstract. In this paper, we consider a classification, with respect to the orthogonal (unitary) change of basis, of finite dimensional algebras. A finite system of invariants, which separates nonequivalent algebras, whose systems of structural constants are from an invariant, open, dense set, is given.

1. Introduction

The classification problem of finite dimensional algebras is one of the important problems in Algebra [1, 2]. One of the most used approaches to this problem is the structural (basis free, invariant) method. As an example of such method one can consider a proof of the famous Artin-Wedderburn theorem on classification (semi)simple associative algebras or Cartan's result on classification of (semi)simple Lie algebras. A structural approach to some more general, for example related to Genetics, algebras is not known. If even it can be done one should not expect to have an easy proof and simple formulation of the result. A disadvantage of the structural method is that the classification is understood only with respect to the general linear group.

Another approach to the classification problem is the coordinate (basis based, structural constants) method. In small dimensional cases, for such an approach we refer the reader to [3, 4, 5]. In case of any finite dimensional algebras, classifications with respect to the general linear groups and some other problems related to them are considered in [6]. In this paper, we also consider the classification problem for any finite dimensional algebras but only with respect to the orthogonal (unitary) change of basis. More exactly, we show how to construct an invariant, open, dense (in the Zariski topology) subset of the space of structural constants, and how to construct a finite system of invariants which separates any two nonequivalent algebras, with respect to the orthogonal (unitary) changes of basis, whose systems of structural constants are in that dense subset. To do it we consider the first the general classification problem of orbits, under one assumption on representation, which may be useful in many other classification problems as well. The main result concerning the classification of finite dimensional algebras (Theorem 2.5) will be derived from the general case though a proof of it may be presented straightforwardly also.

2. Classification and invariants under an Assumption

Let n, m be any natural numbers, $\tau : (G, V) \to V$ be a fixed linear algebraic representation of an algebraic subgroup G of $GL(m, F)$ over V, where F is any field and V is n-dimensional vector space over F. Further we consider this representation under the following assumption:

Assumption. There exists a nonempty open(in Zariski topology) G-invariant subset V_0 of V and an algebraic map $P : V_0 \to G$ such that $P(\tau(g, \mathbf{v})) = P(\mathbf{v})g^{-1}$ whenever $\mathbf{v} \in V_0$ and $g \in G$.

Theorem 2.1. *Elements $\mathbf{u}, \mathbf{v} \in V_0$ are G-equivalent, that is $\mathbf{u} = \tau(g, \mathbf{v})$ for some $g \in G$, if and only if*

$$\tau(P(\mathbf{u}), \mathbf{u}) = \tau(P(\mathbf{v}), \mathbf{v}).$$

Proof. If $\mathbf{u} = \tau(g, \mathbf{v})$ then

$$\begin{aligned}\tau(P(\mathbf{u}), \mathbf{u}) &= \tau(P(\tau(g, \mathbf{v})), \tau(g, \mathbf{v}))\\ &= \tau(P(\mathbf{v})g^{-1}, \tau(g, \mathbf{v})) = \tau(P(\mathbf{v}), \tau(g^{-1}, \tau(g, \mathbf{v})))\\ &= \tau(P(\mathbf{v}), \mathbf{v}).\end{aligned}$$

Visa versa, if $\tau(P(\mathbf{u}), \mathbf{u}) = \tau(P(\mathbf{v}), \mathbf{v})$ then

$$\tau(P(\mathbf{u})^{-1}P(\mathbf{v}), \mathbf{v}) = \tau((P(\mathbf{u}))^{-1}, \tau(P(\mathbf{v}), \mathbf{v})) = \tau((P(\mathbf{u}))^{-1}, \tau(P(\mathbf{u}), \mathbf{u})) = \mathbf{u}$$

that is $\mathbf{u} = \tau(g, \mathbf{v})$, where $g = P(\mathbf{u})^{-1}P(\mathbf{v})$. □

This theorem shows that the system of components of $\tau(P(\mathbf{x}), \mathbf{x})$ is a finite separating system of invariants for the G-orbits in V_0, where $\mathbf{x} = (x_1, x_2, ..., x_n)$ is an algebraic independent system of variables over F.

In what follows, it is assumed that F is an algebraically closed field.

Theorem 2.2. *The field of $G = GL(m, F)$-invariant rational functions $F(\mathbf{x})^{GL(m,F)}$ is generated over F by the system of components of $\tau(P(\mathbf{x}), \mathbf{x})$.*

Proof. It is evident that all components of $\tau(P(\mathbf{x}), \mathbf{x})$ are in $F(\mathbf{x})^{GL(m,F)}$. If $f(\mathbf{x}) = f(\tau(g, \mathbf{x}))$ for all $g \in G = GL(m, F)$ then

$$f(\mathbf{x}) = f(\tau(g, \mathbf{x})) = f(\tau(P(\tau(g, \mathbf{x}))^{-1}P(\mathbf{x}), \mathbf{x})) = f(\tau(P(\tau(g, \mathbf{x}))^{-1}, \tau(P(\mathbf{x}), \mathbf{x}))).$$

In particular it is true for $g = P(\mathbf{x})$. But the equality

$$P(\tau(P(\mathbf{x}), \mathbf{x})) = I_m$$

is valid as far as $P(\tau(g, \mathbf{v})) = P(\mathbf{v})g^{-1}$ whenever $\mathbf{v} \in V_0$ and $g \in GL(m, F)$. Therefore due to it one has $f(\mathbf{x}) = f(\tau(P(\mathbf{x}), \mathbf{x}))$. □

Corollary 2.3. *The field $F(\mathbf{x})$ is generated over $F(\mathbf{x})^{GL(m,F)}$ by the system of components of $P(\mathbf{x})$.*

Proof. Indeed $F(\mathbf{x})^{GL(m,F)}(P(\mathbf{x})) = F(\tau(P(\mathbf{x}), \mathbf{x}))(P(\mathbf{x})) = F(\tau(P(\mathbf{x}), \mathbf{x}), P(\mathbf{x}))$ and $\tau(P(\mathbf{x})^{-1}, \tau(P(\mathbf{x}), \mathbf{x})) = \mathbf{x}$ and therefore $F(\mathbf{x})^{GL(m,F)}(P(\mathbf{x})) = F(\mathbf{x})$. □

Theorem 2.4. *The transcendence degree of $F(\mathbf{x})^{GL(m,F)}$ over F equals to $n - m^2$ and the field extension $F(\mathbf{x})^{GL(m,F)} \subset F(\mathbf{x})$ is a pure transcendental extension.*

Proof. We show that the system of components of $P(\mathbf{x})$ is algebraic independent over $F(\mathbf{x})^{GL(m,F)}$. Let $f(y_{ij})_{i,j=1,2,...,m}$ be any polynomial over $F(\mathbf{x})^{GL(m,F)}$ and $f(P(\mathbf{x})) = 0$ which means that $f_{\mathbf{v}}(P(\mathbf{v})) = 0$ for all $\mathbf{v} \in V_1$, where V_1 is a $GL(m,F)$-invariant nonempty open subset of V_0, $f_{\mathbf{v}}(y_{ij})$ stands for the polynomial obtained from $f(y_{ij})_{i,j=1,2,...,m}$ by substitution $\mathbf{v}$ for $\mathbf{x}$. Therefore for any $\mathbf{v} \in V_1$ and $g \in GL(m,F)$ one has $0 = f_{\mathbf{v}}(P(\tau(g,\mathbf{v}))) = f_{\mathbf{v}}(P(\mathbf{v})g^{-1})$. In particular $0 = f_{\mathbf{v}}(g)$ for any $g \in GL(m,F)$ that is $0 = f_{\mathbf{v}}((y_{ij})_{i,j=1,2,...,m})$ whenever $\mathbf{v} \in V_1$. It implies that $f((y_{ij})_{i,j=1,2,...,m})$ is the zero polynomial. Now due to $F \subset F(\mathbf{x})^{GL(m,F)} \subset F(\mathbf{x})$, tr.deg.$F(\mathbf{x})/F = n$ and Corollary 2.3 implies the required result. □

Now consider any m dimensional algebra W, over the field F, the multiplication of which is given by a bilinear map $\cdot : W \times W \to W$ that is $(\mathbf{u},\mathbf{v}) \mapsto \mathbf{u}\cdot\mathbf{v}$, where $\mathbf{u},\mathbf{v} \in W$. If $e = (e^1, e^2, ..., e^m)$ is a basis for the vector space W then one can represent the product (bilinear map) by a matrix $A \in Mat(m \times m^2; F)$ such that $\mathbf{u}\cdot\mathbf{v} = eA(u \otimes v)$ for any $\mathbf{u} = eu, \mathbf{v} = ev$, where $u = (u_1, u_2, ..., u_m), v = (v_1, v_2, ..., v_m)$ are column vectors. So the entries of A are the structural constants of the algebra W with respect to the basis e.

If $e' = (e'^1, e'^2, ..., e'^m)$ is also a basis for W, $g \in GL(m,F)$, $e'g = e$ and $\mathbf{u}\cdot\mathbf{v} = e'B(u' \otimes v')$, where $\mathbf{u} = e'u', \mathbf{v} = e'v'$, then $\mathbf{u}\cdot\mathbf{v} = eA(u \otimes v) = e'B(u' \otimes v') = eg^{-1}B(gu \otimes gv) = eg^{-1}B(g \otimes g)(u \otimes v)$ as far as $\mathbf{u} = eu = e'u' = eg^{-1}u', \mathbf{v} = ev = e'v' = eg^{-1}v'$. Therefore the equality $B = gA(g^{-1})^{\otimes 2}$ is valid.

Let now τ stand for the representation of $GL(m,F)$ on $n = m^3$ dimensional vector space $V = Mat(m \times m^2; F)$ defined by $\tau : (g, A) \mapsto B = gA(g^{-1})^{\otimes 2}$.

We consider a subgroup $G = O(m,F)$ of $GL(m,F)$ defined by $G = \{g \in GL(m,F) : g^t g = I_m\}$, where g^t stands for the transpose of g.

We will construct an algebraic map $P : V_0 \to GL(m,F)$, for which the equality of the Assumption holds true for any $\mathbf{v} \in V_0$ and $g \in G$, in the following way. It is known that

$$\mathbf{Tr_1}(B) = \mathbf{Tr_1}(A)g^{-1},\ \mathbf{Tr_2}(B) = \mathbf{Tr_2}(A)g^{-1},$$

where $\mathbf{Tr_1}(A)$ stands for the row vector with entries $\sum_{j=1}^{n} A^j_{j,i}$ -the contraction on the first upper and lower indices, $\mathbf{Tr_2}(A)$ stands for the row vector with entries $\sum_{j=1}^{n} A^j_{i,j}$ -the contraction on the first upper and second lower indices.

By induction it is easy to show that $(BB^t)^k B = g((AA^t)^k A)(g^{-1})^{\otimes 2}$ for any nonnegative integer k. Therefore one has the following row equalities

$$\mathbf{Tr_i}((BB^t)^k B) = \mathbf{Tr_i}((AA^t)^k A)g^{-1},\ i = 1, 2, k \in N.$$

It implies that if $P(A)$ is a matrix, for example, consisting of the first m these rows then $P(\tau(g,A)) = P(A)g^{-1}$ for any $g \in O(m,F)$. So we can state the following result.

Theorem 2.5. *Two algebras with structural constants $A, B \in Mat(m \times m^2, F)$ for which $\det P(A) \neq 0, \det P(B) \neq 0$, are $O(m,F)$- equivalent if and only if*

$$P(B)B(P(B)^{-1})^{\otimes 2} = P(A)A(P(A)^{-1})^{\otimes 2} \text{ and } P(B)P(B)^t = P(A)P(A)^t.$$

Proof. Due to Theorem 2.1 the first equality is equivalent to $B = \tau(g, A)$, where $g = P(B)^{-1}P(A)$. The second equality is equivalent to $g = P(B)^{-1}P(A) \in O(m,F)$. □

Corollary 2.6. *Two algebras with structural constants $A, B \in Mat(m \times m^2, F)$ for which $\det P(A) \neq 0, \det P(B) \neq 0$, are $G = SO(m,F)$- equivalent if and only if*

$$P(B)B(P(B)^{-1})^{\otimes 2} = P(A)A(P(A)^{-1})^{\otimes 2},\ P(B)P(B)^t = P(A)P(A)^t$$

and $\det(P(B)) = \det(P(A))$.

Remark 2.7. *It is evident that when F is the field of complex numbers Theorem 2.5 and Corollary 2.6 hold true if one changes in them t to $*$-the transpose with conjugation.*

Theorem 2.5 shows that any $O(m, F)$-invariant function defined on V_0 is a function of the system of components $P(\mathbf{x})\mathbf{x}(P(\mathbf{x})^{-1})^{\otimes 2}$ and $P(\mathbf{x})P(\mathbf{x})^t$. But all these components are rational functions of variables $\mathbf{x} = (x^i_{jk})_{i,j,k=1,2,...,m}$ therefore one can ask the following question.

Problem. Let $f(\mathbf{x}), f_1(\mathbf{x}), f_2(\mathbf{x}), ..., f_k(\mathbf{x})$ be rational functions in $x_1, x_2, ..., x_k$ over F. If it is known that the function $f(\mathbf{x})$ is a function, in the broad sense of function, of the functions $f_1(\mathbf{x}), f_2(\mathbf{x}), ..., f_k(\mathbf{x})$ over a dense subset of F^k, that is there exists function h for which $f(\mathbf{u}) = h(f_1(\mathbf{u}), f_2(\mathbf{u}), ..., f_k(\mathbf{u}))$ for any $\mathbf{u}$ from that subset, then does it imply that $f(\mathbf{x})$ is algebraic over the field $F(f_1(\mathbf{x}), f_2(\mathbf{x}), ..., f_k(\mathbf{x}))$?

Remark 2.8. *The same classification problem can be considered for all finite dimensional commutative (anti-commutative) algebras as well. Similar to Theorem 2.5 result for them can be obtained by appropriate construction of the map P.*

Example 2.9. *As an application let us consider two dimensional algebra case for $G = O(2, F)$. Let*

$$A = \begin{pmatrix} A^1_{1,1} & A^1_{1,2} & A^1_{2,1} & A^1_{2,2} \\ A^{21}_{1,1} & A^2_{1,2} & A^2_{2,1} & A^2_{2,2} \end{pmatrix}$$

be a matrix of structural constants. In this case

$$\mathbf{Tr_1}(A) = (A^1_{1,1} + A^2_{2,1}, A^1_{1,2} + A^2_{2,2}),\ \mathbf{Tr_2}(A) = (A^1_{1,1} + A^2_{1,2}, A^1_{2,1} + A^2_{2,2})$$

and for $P(A)$ one can take the matrix $P(A) = \begin{pmatrix} A^1_{1,1} + A^2_{2,1} & A^1_{1,2} + A^2_{2,2} \\ A^1_{1,1} + A^2_{1,2} & A^1_{2,1} + A^2_{2,2} \end{pmatrix}$ and for V_0 all A for which $\det P(A) = (A^1_{1,1} + A^2_{2,1})(A^1_{2,1} + A^2_{2,2}) - (A^1_{1,2} + A^2_{2,2})(A^1_{1,1} + A^2_{1,2}) \neq 0$. In this particular case the equality $P(\tau(g, A)) = P(A)g^{-1}$ is true whenever $g \in GL(2, F)$. To see the corresponding separating system of invariants in orthogonal group case one should evaluate $P(\mathbf{x})P(\mathbf{x})^t$, $P(\mathbf{x})\mathbf{x}(P(\mathbf{x})^{-1})^{\otimes 2}$ and for $G = GL(2, F)$ case only $P(\mathbf{x})\mathbf{x}(P(\mathbf{x})^{-1})^{\otimes 2}$, where $\mathbf{x} = \begin{pmatrix} x^1_{1,1} & x^1_{1,2} & x^1_{2,1} & x^1_{2,2} \\ x^2_{1,1} & x^2_{1,2} & x^2_{2,1} & x^2_{2,2} \end{pmatrix}$.

Acknowledgments

This research is supported by IIUM grant END B13-033-0918.

References

[1] Albert A A 2003 *Structure of algebras, Colloquium publications 24*, (AMS), 1–113.
[2] Erdmann K and Wildon M 2006 *Introduction to Lie Algebras*, (1st ed.) (Springer-Berlin)
[3] Michel G and Elisabeth R 2011 *African J. Math. Phys.* **10** 81–91.
[4] Durán Díaz R et al 2003 *Lin. Alg. Appl.* **364** 1–12.
[5] Hernáandez Encinas L et al 2004 *Lin. Alg. Appl.*, **387** 69–82.
[6] Popov V 2011 *arXiv: 1411.6570*

Leibniz algebras associated with some finite-dimensional representation of Diamond Lie algebra

Luisa M. Camacho[1], **Manuel Ladra**[2], **Iqboljon A. Karimjanov**[2], **Bakhrom A. Omirov**[3]

[1] Departamento de Matemática Aplicada I, Universidad de Sevilla, Avda. Reina Mercedes, s/n, 41012 Sevilla, Spain
[2] Department of Algebra, University of Santiago de Compostela, 15782, Santiago de Compostela, Spain
[3] Institute of Mathematics, National University of Uzbekistan, Dormon Yoli str. 29, 100125, Tashkent, Uzbekistan

E-mail: lcamacho@us.es

Abstract. In this paper we classify Leibniz algebras whose associated Lie algebra is four-dimensional Diamond Lie algebra $\mathfrak{D}$ and the ideal generated by squares of elements is represented by one of the finite-dimensional indecomposable $\mathfrak{D}$-modules U_n^1, U_n^2, W_n^1 or W_n^2.

1. Introduction

Leibniz algebras were discovered by A. Bloh in 1965 who called them D-algebras [1]. They attracted interest after Jean-Louis Loday [2] noticed that the classical Chevalley-Eilenberg boundary map in the exterior module of a Lie algebra can be lifted to the tensor module which yields a new chain complex. In fact this complex is well defined for any Leibniz algebra. This motivated him to introduce the notion of right (equivalently, left) Leibniz algebra, which is a nonassociative algebra with the right (equivalently, left) multiplication operator being a derivation. Thus a Leibniz algebra satisfies all defining properties of a Lie algebra except the antisymmetry of its product.

During the last 20 years the theory of Leibniz algebras has been actively investigated and numerous papers devoted to the study of these algebras. In fact, many results of theory of Lie algebras have been extended to Leibniz algebras case. For instance, the classical results on Cartan subalgebras [3, 4], Engel's theorem [5], Levi's decomposition [6], properties of solvable algebras with given nilradical [7] and others from the theory of Lie algebras are also true for Leibniz algebras.

Namely, the analogue of Levi's decomposition for Leibniz algebras asserts that any Leibniz algebra is decomposed into a semidirect sum of its solvable radical and a semisimple Lie algebra. Therefore, the main problem of the description of finite-dimensional Leibniz algebras consists of the study of solvable Leibniz algebras.

In fact, each non-Lie Leibniz algebra L contains a non-trivial ideal (further denoted by I), which is the subspace spanned by the squares of elements of the algebra L. Moreover, it is readily to see that this ideal belongs to the right annihilator of L, that is $[L, I] = 0$. Note also that the ideal I is the minimal ideal with the property that the quotient algebra L/I is a Lie algebra (the quotient algebra is said to be associated Lie algebra to a Leibniz algebra L).

One of the approaches to the investigation of Leibniz algebras is a description of such algebras whose quotient algebra with respect to the ideal I is a given Lie algebra [8, 9, 10, 11].

The map $I \times L/I \to I$ defined as $(i, \overline{x}) \mapsto [i, x]$ endows I with a structure of L/I-module. Considering the direct sum of vector spaces $Q(L) = L/I \oplus I$, then the operation $(-, -)$ defines the Leibniz algebra structure on $Q(L)$ with multiplication

$$[\overline{x}, \overline{y}] = \overline{[x, y]}, \quad [\overline{x}, i] = [x, i], \quad [i, \overline{x}] = 0, \quad [i, j] = 0, \qquad x, y \in L, \ i, j \in I.$$

Therefore, for given a Lie algebra G and a G-module M, we can construct a Leibniz algebra $L = G \oplus M$ by the above construction.

In the paper [12] the notion of irreducible Leibniz representation is introduced and it is shown that there are only two kinds irreducible representations. Namely, one of them coincides with Lie representation (it said to be Lie representation) and another one has trivial action on the left side, while the action on the right side is irreducible (it said to be Leibniz representation). Following these concepts we shall say Leibniz representation (Leibniz module) for that representation (respectively, module) which has trivial action on the left side.

In this paper we study Leibniz algebras whose associated Lie algebra is the four-dimensional Diamond Lie algebra $\mathfrak{D}$ and the ideal I is one of its finite-dimensional indecomposable Leibniz representations of $\mathfrak{D}$ which are described in [13].

Actually, for a Leibniz algebra L with associated Diamond Lie algebra $\overline{\mathfrak{D}} = L/I$ we could decompose it into direct sum of vector spaces $L = \mathfrak{D} \oplus I$, where $\mathfrak{D}$ is the preimage of $\overline{\mathfrak{D}}$ under the natural homomorphism $\varphi \colon L \to \overline{\mathfrak{D}}$. Clearly, the ideal I can be considered as Leibniz $\overline{\mathfrak{D}}$-module. Taking into account that the ideal I is contained in right annihilator of the algebra L, the multiplications in L are determined from the products $[\mathfrak{D}, \mathfrak{D}]$ and $[I, \overline{\mathfrak{D}}]$. Since I is a Leibniz module over the Lie algebra $\overline{\mathfrak{D}}$, then the product $[I, \overline{\mathfrak{D}}]$ corresponds to a chosen right Lie $\overline{\mathfrak{D}}$-module. Thus, the main problem of the description of Leibniz algebras with associated Lie algebra $\overline{\mathfrak{D}}$ and with the ideal I chosen by specific right $\overline{\mathfrak{D}}$-module consists to identifying the product $[\mathfrak{D}, \mathfrak{D}]$.

Throughout the paper all vector spaces and algebras are finite-dimensional over the field of the complex numbers.

2. Preliminaries

In this section we give necessary definitions and preliminary results.

Definition 2.1. *An algebra $(L, [-, -])$ over a field $\mathbb{F}$ is called a Leibniz algebra if for any $x, y, z \in L$, the so-called Leibniz identity*

$$[x, [y, z]] = [[x, y], z] - [[x, z], y]$$

holds.

The real Diamond Lie algebra $\mathfrak{D}$ is a four-dimensional Lie algebra with basis $\{J, P_1, P_2, T\}$ and non-zero relations

$$[J, P_1] = P_2, \qquad [J, P_2] = -P_1, \qquad [P_1, P_2] = T.$$

The complexification of the Diamond Lie algebra: $\mathfrak{D} \otimes_{\mathbb{R}} \mathbb{C}$ displays the following complex basis:

$$\{P_+ = P_1 - iP_2, \quad P_- = P_1 + iP_2, \quad T, \quad J\},$$

where i is the imaginary unit, whose nonzero commutators are

$$[J, P_+] = iP_+, \qquad [J, P_-] = -iP_-, \qquad [P_+, P_-] = 2iT.$$

Since throughout the paper we shall consider only complex Diamond algebra, the same notation $\mathfrak{D}$ does not lead to confusion.

In the paper [13] the authors construct, for any $n \in \mathbb{N}$, a $(3n+3)$-dimensional Lie module V_n over the algebra $\mathfrak{D}$, which is endowed with a basis $\{v_k^j\}_{k=0,\cdots,n}^{j=0,1,2}$. In the paper they use the action $\mathfrak{D} \cdot V_n$ (the action $V_n \cdot \mathfrak{D}$ evidently is defined by the antisymmetric law).

In the Leibniz algebra $L = \mathfrak{D} \oplus V_n$ we shall identify the action $V_n \cdot \mathfrak{D}$ with the product $[V_n, D]$. In order to have compatibility instead of the action $\mathfrak{D} \cdot V_n$ we will use $V_n \cdot \mathfrak{D}$ as follows:

$$\begin{array}{llll}
v_k^j \cdot J = -\frac{i}{2}(n-2k)v_k^j, & & k = 0, \dots, n, & j = 0,1,2, \\
v_k^j \cdot P_+ = -(n-k+1)v_{k-1}^{j+1}, & v_0^j \cdot P_+ = 0, \quad v_k^2 \cdot P_+ = 0, & k = 1, \dots, n, & j = 0,1, \\
v_k^j \cdot P_- = -(k+1)v_{k+1}^{j+1}, & v_n^j \cdot P_- = 0, \quad v_k^2 \cdot P_- = 0, & k = 0, \dots, n-1, & j = 0,1, \\
v_k^0 \cdot T = \frac{i}{2}(n-2k)v_k^2, & v_k^j \cdot T = 0, & k = 0, \dots, n, & j = 1,2.
\end{array} \tag{1}$$

Remark 2.2. *Here we changed* $T \cdot v_k^0 = 2i(n-2k)v_k^2$*, misprint of [13], to the correct expression* $T \cdot v_k^0 = -\frac{i}{2}(n-2k)v_k^2$.

In order to have the compatibility with the above representations, in the law of Diamond algebra we use the following table of multiplication (we make a change of basis $P'_+ = P_-$, $P'_- = P_+$):

$$[J, P_+] = -iP_+, \quad [J, P_-] = iP_-, \quad [P_+, P_-] = -2iT.$$

In the next proposition two different decompositions of module V_n are presented.

Proposition 2.3 ([13]). *Let* V_n *be the* $\mathfrak{D}$*-module constructed above and let us denote by* $\lfloor x \rfloor$ *the integer part of* x*, then*

- *if* $n = 0$*, then* V_n *decomposes in the direct sum of three trivial one-dimensional modules.*
- *If* $n = 2j$, $j \in \mathbb{N}, j \geq 1$*, then* V_n *decomposes into the direct sum* $V_n = U_n^1 \oplus U_n^2$ *of two modules, respectively, of dimension* $3(n/2)+2$ *and* $3(n/2)+1$*, given by* $U_n^1 = \mathrm{span}\{v_0^0, v_{2k}^0, v_{2k-1}^1, v_0^2, v_{2k}^2\}_{k=1,\dots,\lfloor n/2 \rfloor}$ *and* $U_n^2 = \mathrm{span}\{v_{2k-1}^0, v_0^1, v_{2k}^1, v_{2k-1}^2\}_{k=1,\dots,\lfloor n/2 \rfloor}$.
- *If* $n = 2j+1$, $j \in \mathbb{N}, j \geq 0$*, then* V_n *decomposes into the direct sum* $V_n = W_n^1 \oplus W_n^2$ *of two modules of equal dimension* $3n/2$*, respectively, given by* $W_n^1 = \mathrm{span}\{v_{2k}^0, v_{2k+1}^1, v_{2k}^2\}_{k=0,\dots,\lfloor n/2 \rfloor}$ *and* $W_n^2 = \mathrm{span}\{v_{2k+1}^0, v_{2k}^1, v_{2k+1}^2\}_{k=0,\dots,\lfloor n/2 \rfloor}$.

In the paper [13] it is proved that the terms of the above decompositions are indecomposable $\mathfrak{D}$-modules.

Theorem 2.4. *The modules U_n^1, U_n^2, W_n^1 and W_n^2 are indecomposable $\mathfrak{D}$-modules.*

Further, we shall need the result that specifies the simplest case, that is $[\mathfrak{D},\mathfrak{D}] \subseteq \mathfrak{D}$, with some conditions on the ideal I.

Lemma 2.5 ([11]). *Let L be a Leibniz algebra such that $L/I \cong \overline{\mathfrak{D}}$ and I a Leibniz $\overline{\mathfrak{D}}$-module. If there exists a basis $\{X_1, X_2, \dots, X_n\}$ of I such that $[X_i, J] = \alpha_i X_i$, $\alpha_i \notin \{-2; 0; 2\}$, where $1 \le i \le n$, then $[\mathfrak{D},\mathfrak{D}] \subseteq \mathfrak{D}$.*

3. Main result

This section is devoted to the study of Leibniz algebras with associated Diamond Lie algebra and with the condition that the ideal I is one of indecomposable $\mathfrak{D}$-modules U_n^1, U_n^2, W_n^1 or W_n^2.

3.1. Leibniz algebras whose ideal I is the $\mathfrak{D}$-module U_n^1.

Let $\{v_0^0, v_{2k}^0, v_{2k-1}^1, v_0^2, v_{2k}^2\}_{k=1,\dots,n/2}$ be the basis of module U_n^1 chosen in Proposition 2.3 with even n. Then from Proposition 2.3 and action 1 we have the products in the Leibniz algebra L:

$$\begin{cases} [v_{2k}^0, J] = \frac{i}{2}(n-4k)v_{2k}^0, & k = 0,\dots,\frac{n}{2},\\ [v_{2k-1}^1, J] = \frac{i}{2}(n-4k+2)v_{2k-1}^1, & k = 1,\dots,\frac{n}{2},\\ [v_{2k}^2, J] = \frac{i}{2}(n-4k)v_{2k}^2, & k = 0,\dots,\frac{n}{2},\\ [v_{2k}^0, P_+] = (n-2k+1)v_{2k-1}^1, & k = 1,\dots,\frac{n}{2},\\ [v_{2k-1}^1, P_+] = (n-2k+2)v_{2k-2}^2, & k = 1,\dots,\frac{n}{2},\\ [v_{2k}^0, P_-] = (2k+1)v_{2k+1}^1, & k = 0,\dots,\frac{n}{2}-1,\\ [v_{2k-1}^1, P_-] = 2kv_{2k}^2, & k = 1,\dots,\frac{n}{2},\\ [v_{2k}^0, T] = -\frac{i}{2}(n-4k)v_{2k}^2, & k = 0,\dots,\frac{n}{2}. \end{cases} \tag{2}$$

Theorem 3.1. *An arbitrary Leibniz algebra with corresponding Lie algebra $\overline{\mathfrak{D}}$ and I associated with $\overline{\mathfrak{D}}$-module defined by (2) admits a basis $\{J, P_+, P_-, T, v_0^0, v_{2k}^0, v_{2k-1}^1, v_0^2, v_{2k}^2\}_{k=1,\dots,n/2}$, where n is even and the table of multiplication $[\mathfrak{D},\mathfrak{D}]$ has the following form:*

- $n = 4s$

$$\begin{cases} [J,P_+] = -iP_+, & [J,P_-] = iP_-, & [P_+,P_-] = -2iT,\\ [P_+,J] = iP_+, & [P_-,J] = -iP_-, & [P_-,P_+] = 2iT + 2\alpha_1 v_{2s}^2,\\ [J,T] = \alpha_1 v_{2s}^2 & [J,J] = \alpha_2 v_{2s}^2, & [P_+,P_+] = \alpha_3 v_{2s-2}^2,\\ [P_-,P_-] = \alpha_4 v_{2s+2}^2. \end{cases}$$

- $n = 4s-2$

$$\begin{cases} [J,P_+] = -iP_+, & [P_+,J] = iP_+ + 2is\beta_1 v_{2s-2}^2,\\ [J,P_-] = iP_-, & [P_-,J] = -iP_- - 2is\beta_1 v_{2s}^2,\\ [P_+,P_-] = -2iT, & [P_-,P_+] = 2iT + 2\beta_2 v_{2s-1}^1,\\ [J,J] = \beta_1 v_{2s-1}^1, & [J,T] = \beta_2 v_{2s-1}^1,\\ [P_+,P_+] = \beta_3 v_{2s-3}^1, & [P_-,P_-] = \beta_4 v_{2s+1}^1,\\ [P_+,T] = 2is\beta_2 v_{2s-2}^2, & [T,P_+] = -i(2s\beta_2 - (s-1)\beta_3)v_{2s-2}^2,\\ [P_-,T] = -2is\beta_2 v_{2s}^2, & [T,P_-] = i(4s\beta_2 - (s-1)\beta_4)v_{2s}^2. \end{cases}$$

where $\alpha_i,\ \beta_i \in \mathbb{C},\ 1 \le i \le 4.$

Proof. We will consider two cases $n = 4s$ and $n = 4s - 2$.

Let us introduce notation

$$[J,J] = a_0^0 v_0^0 + \sum_{k=1}^{n/2} a_{2k}^0 v_{2k}^0 + \sum_{k=1}^{n/2} a_{2k-1}^1 v_{2k-1}^1 + a_0^2 v_0^2 + \sum_{k=1}^{n/2} a_{2k}^2 v_{2k}^2.$$

Case 1. Let $n = 4s$. Taking the following change of basis:

$$J' = J + \frac{ia_0^0}{2s} v_0^0 + \sum_{k=1}^{s-1} \frac{ia_{2k}^0}{2s-2k} v_{2k}^0 + \sum_{k=s+1}^{2s} \frac{ia_{2k}^0}{2s-2k} v_{2k}^0 + \sum_{k=1}^{2s} \frac{ia_{2k-1}^1}{2s-2k+1} v_{2k-1}^1$$
$$+ \frac{ia_0^2}{2s} v_0^2 + \sum_{k=1}^{s-1} \frac{ia_{2k}^2}{2s-2k} v_{2k}^2 + \sum_{k=s+1}^{2s} \frac{ia_{2k}^2}{2s-2k} v_{2k}^2,$$

we can assume that $[J,J] = a_{2s}^0 v_{2s}^0 + a_{2s}^2 v_{2s}^2$.

Lifting from the quotient Lie algebra $\overline{\mathfrak{D}}$ to the Leibniz algebra L we have

$$[J,P_+] = -iP_+ + b_0^0 v_0^0 + \sum_{k=1}^{2s} b_{2k}^0 v_{2k}^0 + \sum_{k=1}^{2s} b_{2k-1}^1 v_{2k-1}^1 + b_0^2 v_0^2 + \sum_{k=1}^{2s} b_{2k}^2 v_{2k}^2,$$
$$[J,P_-] = iP_- + c_0^0 v_0^0 + \sum_{k=1}^{2s} c_{2k}^0 v_{2k}^0 + \sum_{k=1}^{2s} c_{2k-1}^1 v_{2k-1}^1 + c_0^2 v_0^2 + \sum_{k=1}^{2s} c_{2k}^2 v_{2k}^2,$$
$$[P_+,P_-] = -2iT + d_0^0 v_0^0 + \sum_{k=1}^{2s} d_{2k}^0 v_{2k}^0 + \sum_{k=1}^{2s} d_{2k-1}^1 v_{2k-1}^1 + d_0^2 v_0^2 + \sum_{k=1}^{2s} d_{2k}^2 v_{2k}^2.$$

Making the change of basis elements as follows:

$$P'_+ = P_+ + ib_0^0 v_0^0 + \sum_{k=1}^{2s} ib_{2k}^0 v_{2k}^0 + \sum_{k=1}^{2s} ib_{2k-1}^1 v_{2k-1}^1 + ib_0^2 v_0^2 + \sum_{k=1}^{2s} ib_{2k}^2 v_{2k}^2,$$
$$P'_- = P_- - ic_0^0 v_0^0 - \sum_{k=1}^{2s} ic_{2k}^0 v_{2k}^0 - \sum_{k=1}^{2s} ic_{2k-1}^1 v_{2k-1}^1 - ic_0^2 v_0^2 - \sum_{k=1}^{2s} ic_{2k}^2 v_{2k}^2,$$
$$T' = T + \tfrac{i}{2}(d_0^0 v_0^0 + \sum_{k=1}^{2s} d_{2k}^0 v_{2k}^0 + (d_1^1 + ib_0^0)v_1^1 + \sum_{k=2}^{2s}(d_{2k-1}^1 + i(2k-1)b_{2k-2}^0)v_{2k-1}^1 + d_0^2 v_0^2$$
$$+ \sum_{k=1}^{2s}(d_{2k}^2 + 2ikb_{2k-1}^1)v_{2k}^2),$$

we derive the products

$$[J,P_+] = -iP_+, \quad [J,P_-] = iP_-, \quad [P_+,P_-] = -2iT.$$

Considering the chain of equalities

$$P_+ = i[J,P_+] = [J,[P_+,J]] = [[J,P_+],J] - [[J,J],P_+],$$
$$P_- = -i[J,P_-] = [J,[P_-,J]] = [[J,P_-],J] - [[J,J],P_-],$$

we conclude $[P_+, J] = iP_+ + ia^0_{2s}(2s+1)v^1_{2s-1}$ and $[P_-, J] = -iP_- - ia^0_{2s}(2s+1)v^1_{2s+1}$, respectively.

We set

$$[P_+, P_+] = q^0_0v^0_0 + \sum_{k=1}^{2s} q^0_{2k}v^0_{2k} + \sum_{k=1}^{2s} q^1_{2k-1}v^1_{2k-1} + q^2_0v^2_0 + \sum_{k=1}^{2s} q^2_{2k}v^2_{2k},$$

$$[P_-, P_-] = l^0_0v^0_0 + \sum_{k=1}^{2s} l^0_{2k}v^0_{2k} + \sum_{k=1}^{2s} l^1_{2k-1}v^1_{2k-1} + l^2_0v^2_0 + \sum_{k=1}^{2s} l^2_{2k}v^2_{2k},$$

$$[J, T] = r^0_0v^0_0 + \sum_{k=1}^{2s} r^0_{2k}v^0_{2k} + \sum_{k=1}^{2s} r^1_{2k-1}v^1_{2k-1} + r^2_0v^2_0 + \sum_{k=1}^{2s} r^2_{2k}v^2_{2k}.$$

Applying the Leibniz identity to the following triples we get further constraints on the products.

Leibniz identity	Constraint
$\{P_+, J, P_+\}$	$\Rightarrow\ a^0_{2s} = q^0_0 = q^2_0 = 0,\ q^1_{2k-1} = 0,\ 1 \le k \le 2s,\ q^0_{2k} = q^2_{2k} = 0,\ k \ne s-1,\ s > 1$
$\{P_-, J, P_-\}$	$\Rightarrow\ l^0_0 = l^2_0 = 0,\ l^1_{2k-1} = 0,\ 1 \le k \le 2s,\ l^0_{2k} = l^2_{2k} = 0 = 0,\ k \ne s+1.$

Thus, we obtain

$$[J, J] = a^2_{2s}v^2_{2s}, \qquad [P_+, J] = iP_+, \qquad [P_-, J] = -iP_-,$$
$$[P_+, P_+] = q^0_{2s-2}v^0_{2s-2} + q^2_{2s-2}v^2_{2s-2}, \quad [P_-, P_-] = l^0_{2s+2}v^0_{2s+2} + l^2_{2s+2}v^2_{2s+2}.$$

Note that for $s = 1$, we have $[P_+, P_+] = q^0_0v^0_0 + q^2_0v^2_0$, which agrees with the case $s > 1$.

Considering the Leibniz identity to the following triples implies the following constraints:

Leibniz identity	Constraint
$\{J, T, J\}$	$\Longrightarrow\ [J, T] = r^0_{2s}v^0_{2s} + r^2_{2s}v^2_{2s},$
$\{P_+, J, P_-\}$	$\Longrightarrow\ [T, J] = 0,$
$\{J, P_+, T\}$	$\Longrightarrow\ [P_+, T] = i(2s+1)r^0_{2s}v^1_{2s-1},$
$\{J, P_-, T\}$	$\Longrightarrow\ [P_-, T] = -i(2s+1)r^0_{2s}v^1_{2s+1}.$

Taking into account that $[P_+, P_-] = -2iT$ in the following chain of equalities

$$-2i[J, T] = [J, [P_+, P_-]] = [[J, P_+], P_-] - [[J, P_-], P_+] = -i[P_+, P_-] - i[P_-, P_+],$$

we deduce $[P_-, P_+] = 2iT + 2r^0_{2s}v^0_{2s} + 2r^2_{2s}v^2_{2s}$.

In order to identify the products $[T, P_+]$ and $[T, P_-]$, we introduce the notations:

$$[T, P_+] = m^0_0v^0_0 + \sum_{k=1}^{2s} m^0_{2k}v^0_{2k} + \sum_{k=1}^{2s} m^1_{2k-1}v^1_{2k-1} + m^2_0v^2_0 + \sum_{k=1}^{2s} m^2_{2k}v^2_{2k},$$

$$[T, P_-] = t^0_0v^0_0 + \sum_{k=1}^{2s} t^0_{2k}v^0_{2k} + \sum_{k=1}^{2s} t^1_{2k-1}v^1_{2k-1} + t^2_0v^2_0 + \sum_{k=1}^{2s} t^2_{2k}v^2_{2k}.$$

In a similar way as above we obtain

Leibniz identity		Constraint
$\{T, P_+, J\}$	$\Longrightarrow$	$[T, P_+] = m^1_{2s-1} v^1_{2s-1}$,
$\{T, P_-, J\}$	$\Longrightarrow$	$[T, P_-] = t^1_{2s+1} v^1_{2s+1}$,
$\{P_+, P_-, T\}$	$\Longrightarrow$	$[T, T] = -s(2s+1) r^0_{2s} v^2_{2s}$,
$\{P_+, P_+, T\}$	$\Longrightarrow$	$q^0_{2s-2} = -(s+1)(2s+1) r^0_{2s}$,
$\{P_+, P_+, P_-\}$	$\Longrightarrow$	$m^1_{2s-1} = -i/2(2s+1)(2s^2+s+1) r^0_{2s}$,
$\{P_-, P_-, T\}$	$\Longrightarrow$	$l^0_{2s+2} = -(2s+1)(s+1) r^0_{2s}$,
$\{T, P_+, P_-\}$	$\Longrightarrow$	$t^1_{2s+1} = -i/2(2s+1)(2s^2+s+3) r^0_{2s}$.

From the restrictions, we derive

$$
\begin{aligned}
[P_+, P_+] &= -(s+1)(2s+1) r^0_{2s} v^0_{2s-2} + q^2_{2s-2} v^2_{2s-2}, \\
[T, P_+] &= -i/2(2s+1)(2s^2+s+1) r^0_{2s} v^1_{2s-1}, \\
[P_-, P_-] &= -(2s+1)(s+1) r^0_{2s} v^0_{2s+2} + l^2_{2s+2} v^2_{2s+2}, \\
[T, P_-] &= -i/2(2s+1)(2s^2+s+3) r^0_{2s} v^1_{2s+1}.
\end{aligned}
$$

Finally, if we apply the Leibniz identity to the triple of elements $\{P_-, P_+, P_-\}$, we obtain $r^0_{2s} = 0$. Thus, by assuming $(r^2_{2s}, a^2_{2s}, q^2_{2s-2}, l^2_{2s+2}) = (\alpha_1, \alpha_2, \alpha_3, \alpha_4)$, we get the first family of the theorem.

Case 2. Let $n = 4s - 2$. Taking the change of basis in the following form:

$$
\begin{aligned}
J' = J &+ \frac{i a^0_0}{2s-1} v^0_0 + \sum_{k-1}^{2s-1} \frac{i a^0_{2k}}{2s-2k-1} v^0_{2k} + \sum_{k=1}^{s-1} \frac{i a^1_{2k-1}}{2s-2k} v^1_{2k-1} \\
&+ \sum_{k=s+1}^{2s-1} \frac{i a^1_{2k-1}}{2s-2k} v^1_{2k-1} + \frac{i a^2_0}{2s-1} v^2_0 + \sum_{k=1}^{2s-1} \frac{i a^2_{2k}}{2s-2k-1} v^2_{2k},
\end{aligned}
$$

we can assume that $[J, J] = a^1_{2s-1} v^1_{2s-1}$.

Analogously to the previous case, we can get the products

$$
[J, P_+] = -iP_+, \qquad [J, P_-] = iP_-, \qquad [P_+, P_-] = -2iT.
$$

Verifying the Leibniz identity on triples we have the following restrictions:

Leibniz identity		Constraint
$\{J, J, P_+\}$	$\Longrightarrow$	$[P_+, J] = iP_+ + 2isa^1_{2s-1} v^2_{2s-2}$,
$\{J, J, P_-\}$	$\Longrightarrow$	$[P_-, J] = -iP_- - 2isa^1_{2s-1} v^2_{2s}$.

We put

$$[P_+,P_+]=q_0^0v_0^0+\sum_{k=1}^{2s-1}q_{2k}^0v_{2k}^0+\sum_{k=1}^{2s-1}q_{2k-1}^1v_{2k-1}^1+q_0^2v_0^2+\sum_{k=1}^{2s-1}q_{2k}^2v_{2k}^2,$$

$$[P_-,P_-]=l_0^0v_0^0+\sum_{k=1}^{2s-1}l_{2k}^0v_{2k}^0+\sum_{k=1}^{2s-1}l_{2k-1}^1v_{2k-1}^1+l_0^2v_0^2+\sum_{k=1}^{2s-1}l_{2k}^2v_{2k}^2,$$

$$[J,T]=r_0^0v_0^0+\sum_{k=1}^{2s-1}r_{2k}^0v_{2k}^0+\sum_{k=1}^{2s-1}r_{2k-1}^1v_{2k-1}^1+r_0^2v_0^2+\sum_{k=1}^{2s-1}r_{2k}^2v_{2k}^2.$$

From the Leibniz identity, we have

Leibniz identity		Constraint
$\{P_+,J,P_+\}$	$\Longrightarrow$	$[P_+,P_+]=q_{2s-3}^1v_{2s-3}^1,$
$\{P_-,J,P_-\}$	$\Longrightarrow$	$[P_-,P_-]=l_{2s+1}^1v_{2s+1}^1,$
$\{J,J,T\}$	$\Longrightarrow$	$[J,T]=r_{2s-1}^1v_{2s-1}^1,$
$\{P_+,J,P_-\}$	$\Longrightarrow$	$[T,J]=0,$
$\{J,P_+,T\}$	$\Longrightarrow$	$[P_+,T]=2isr_{2s-1}^1v_{2s-2}^2,$
$\{J,P_-,T\}$	$\Longrightarrow$	$[P_-,T]=-2isr_{2s-1}^1v_{2s}^2,$
$\{J,P_+,P_-\}$	$\Longrightarrow$	$[P_-,P_+]=2iT+2r_{2s-1}^1v_{2s-1}^1.$

Setting

$$[T,P_+]=m_0^0v_0^0+\sum_{k=1}^{2s-1}m_{2k}^0v_{2k}^0+\sum_{k=1}^{2s-1}m_{2k-1}^1v_{2k-1}^1+m_0^2v_0^2+\sum_{k=1}^{2s-1}m_{2k}^2v_{2k}^2,$$

$$[T,P_-]=t_0^0v_0^0+\sum_{k=1}^{2s-1}t_{2k}^0v_{2k}^0+\sum_{k=1}^{2s-1}t_{2k-1}^1v_{2k-1}^1+t_0^2v_0^2+\sum_{k=1}^{2s-1}t_{2k}^2v_{2k}^2,$$

and applying the Leibniz identity to the following triples of elements:

$$\{T,P_+,J\},\ \{T,P_-,J\},\ \{P_+,P_-,T\},\ \{P_+,P_+,P_-\},\ \{T,P_+,P_-\},\ \{P_-,P_+,P_-\},$$

we derive

$$[T,P_+]=(i(s-1)q_{2s-3}^1-2isr_{2s-1}^1)v_{2s-2}^2,$$
$$[T,P_-]=(4isr_{2s-1}^1-i(s-1)l_{2s-2}^1)v_{2s}^2,$$
$$[T,T]=0.$$

Finally, by denoting $(a_{2s-1}^1,r_{2s-1}^1,q_{2s-3}^1,l_{2s+1}^1)=(\beta_1,\beta_2,\beta_3,\beta_4)$, we have the second family. □

Journal of Physics: Conference Series **697** (2016) 012006 doi:10.1088/1742-6596/697/1/012006

3.2. Leibniz algebras whose ideal I is the $\mathfrak{D}_C$-module U_n^2.

Suppose that the ideal I is defined as a Leibniz $\mathfrak{D}$-module by the irreducible representation U_n^2 and $\{v_{2k-1}^0, v_0^1, v_{2k}^1, v_{2k-1}^2\}_{k=1,\dots,n/2}$ for even n is the basis of I chosen as in Proposition 2.3. Then the products $[I, \mathfrak{D}]$ have the form:

$$\begin{cases} [v_{2k-1}^0, J] = \frac{i}{2}(n-4k+2)v_{2k-1}^0, & k=1,\dots,\frac{n}{2},\\ [v_{2k}^1, J] = \frac{i}{2}(n-4k)v_{2k}^1, & k=0,\dots,\frac{n}{2},\\ [v_{2k-1}^2, J] = \frac{i}{2}(n-4k+2)v_{2k-1}^2, & k=1,\dots,\frac{n}{2},\\ [v_{2k-1}^0, P_+] = (n-2k+2)v_{2k-2}^1, & k=1,\dots,\frac{n}{2},\\ [v_{2k}^1, P_+] = (n-2k+1)v_{2k-1}^2, & k=1,\dots,\frac{n}{2},\\ [v_{2k-1}^0, P_-] = 2kv_{2k}^1, & k=1,\dots,\frac{n}{2},\\ [v_{2k}^1, P_-] = (2k+1)v_{2k+1}^2, & k=0,\dots,\frac{n}{2}-1,\\ [v_{2k-1}^0, T] = -i/2(n-4k+2)v_{2k-1}^2, & k=1,\dots,\frac{n}{2}. \end{cases}$$

Theorem 3.2. *An arbitrary Leibniz algebra with corresponding Lie algebra $\overline{\mathfrak{D}}$ and with the ideal I defined as a Leibniz $\overline{\mathfrak{D}}$-module U_n^2 admits a basis $\{J, P_+, P_-, T, v_{2k-1}^0, v_0^1, v_{2k}^1, v_{2k-1}^2\}_{k=1,\dots,n/2}$, where n is even such that the table of multiplication $[\mathfrak{D}, \mathfrak{D}]$ has the following form:*

- $n = 4s$

$$\begin{cases} [J,P_+] = -iP_+, & [P_+,J] = iP_+ + i(2s+1)\gamma_1 v_{2s-1}^2,\\ [J,P_-] = iP_-, & [P_-,J] = -iP_- - i(2s+1)\gamma_1 v_{2s+1}^2,\\ [P_+,P_-] = -2iT, & [P_-,P_+] = 2iT + 2\gamma_2 v_{2s}^1,\\ [J,J] = \gamma_1 v_{2s}^1, & [J,T] = \gamma_2 v_{2s}^1,\\ [P_+,P_+] = \gamma_3 v_{2s-2}^1, & [P_-,P_-] = \gamma_4 v_{2s+2}^1,\\ [P_+,T] = i(2s+1)\gamma_2 v_{2s-1}^2, & [T,P_+] = -i((2s+1)\gamma_2 - \frac{(2s-1)\gamma_3}{2})v_{2s-1}^2,\\ [P_-,T] = -i(2s+1)\gamma_2 v_{2s+1}^2, & [T,P_-] = i(2(2s+1)\gamma_2 - \frac{(2s-1)\gamma_4}{2})v_{2s+1}^2, \end{cases}$$

- $n = 4s-2$

$$\begin{cases} [J,P_+] = -iP_+, & [J,P_-] = iP_-, & [P_+,P_-] = -2iT,\\ [P_+,J] = iP_+, & [P_-,J] = -iP_-, & [P_-,P_+] = 2iT + 2\delta_1 v_{2s-1}^2,\\ [J,T] = \delta_1 v_{2s-1}^2 & [J,J] = \delta_2 v_{2s-1}^2, & [P_+,P_+] = \delta_3 v_{2s-3}^2,\\ [P_-,P_-] = \delta_4 v_{2s+1}^2, \end{cases}$$

where $\gamma_i,\ \delta_i \in \mathbb{C},\ 1 \le i \le 4$.

Proof. Let us denote

$$[J,J] = \sum_{k=1}^{n/2} a_{2k-1}^0 v_{2k-1}^0 + a_0^1 v_0^1 + \sum_{k=1}^{n/2} a_{2k}^1 v_{2k}^1 + \sum_{k=1}^{n/2} a_{2k-1}^2 v_{2k-1}^2.$$

In a similar way to the proof of Theorem 3.1 we will consider the cases $n = 4s$ and $n = 4s-2$.

Case 1. Let $n = 4s$. Taking the change of basis elements as follows:

$$J' = J + \sum_{k=1}^{2s} \frac{ia^0_{2k-1}}{2s-2k+1} v^0_{2k-1} + \frac{ia^1_0}{2s} v^1_0 + \sum_{k=1}^{s-1} \frac{ia^1_{2k}}{2s-2k} v^1_{2k} + \sum_{k=s+1}^{2s} \frac{ia^1_{2k}}{2s-2k} v^1_{2k} + \sum_{k=1}^{2s} \frac{ia^2_{2k-1}}{2s-2k+1} v^2_{2k-1},$$

we can assume that $[J, J] = a^1_{2s} v^1_{2s}$.

Applying similar arguments as in the proof of Theorem 3.1, we derive

$$[J, P_+] = -iP_+, \quad [J, P_-] = iP_-, \quad [P_+, P_-] = -2iT.$$

We set

$$[P_+, P_+] = \sum_{k=1}^{2s} q^0_{2k-1} v^0_{2k-1} + q^1_0 v^1_0 + \sum_{k=1}^{2s} q^1_{2k} v^1_{2k} + \sum_{k=1}^{2s} q^2_{2k-1} v^2_{2k-1},$$

$$[P_-, P_-] = \sum_{k=1}^{2s} l^0_{2k-1} v^0_{2k-1} + l^1_0 v^1_0 + \sum_{k=1}^{2s} l^1_{2k} v^1_{2k} + \sum_{k=1}^{2s} l^2_{2k-1} v^2_{2k-1},$$

$$[J, T] = \sum_{k=1}^{2s} r^0_{2k-1} v^0_{2k-1} + r^1_0 v^1_0 + \sum_{k=1}^{2s} r^1_{2k} v^1_{2k} + \sum_{k=1}^{2s} r^2_{2k-1} v^2_{2k-1}.$$

Applying the Leibniz identity to the following triples of elements:

$$\{J, P_+, J\}, \ \{J, P_-, J\}, \ \{P_+, J, P_+\}, \ \{P_-, J, P_-\}, \ \{J, J, T\}, \ \{P_+, J, P_-\},$$

we deduce restrictions which imply the following expressions for the products

$$[P_+, J] = iP_+ + ia^1_{2s}(2s+1)v^2_{2s-1}, \quad [P_-, J] = -iP_- - ia^1_{2s}(2s+1)v^2_{2s+1},$$
$$[P_+, P_+] = q^1_{2s-2} v^1_{2s-2}, \quad [P_-, P_-] = l^1_{2s+2} v^1_{2s+2}, \quad [J, T] = r^1_{2s} v^1_{2s}, \quad [T, J] = 0.$$

Moreover, we have

Leibniz identity	Constraint
$\{J, P_+, T\}$	$\Longrightarrow$ $[P_+, T] = ir^1_{2s}(2s+1)v^2_{2s-1},$
$\{J, P_-, T\}$	$\Longrightarrow$ $[P_-, T] = -ir^1_{2s}(2s+1)v^2_{2s+1},$
$\{J, P_+, P_-\}$	$\Longrightarrow$ $[P_-, P_+] = 2iT + 2r^1_{2s} v^1_{2s}.$

We also denote

$$[T, P_+] = \sum_{k=1}^{2s} m^0_{2k-1} v^0_{2k-1} + m^1_0 v^1_0 + \sum_{k=1}^{2s} m^1_{2k} v^1_{2k} + \sum_{k=1}^{2s} m^2_{2k-1} v^2_{2k-1},$$

$$[T, P_-] = \sum_{k=1}^{2s} t^0_{2k-1} v^0_{2k-1} + t^1_0 v^1_0 + \sum_{k=1}^{2s} t^1_{2k} v^1_{2k} + \sum_{k=1}^{2s} t^2_{2k-1} v^2_{2k-1}.$$

Applying the Leibniz identity to the elements $\{T, P_+, J\}$, $\{T, P_-, J\}$, $\{P_+, P_-, P_-\}$, we get

$$[T, P_+] = m^0_{2s-1}v^0_{2s-1} + m^2_{2s-1}v^2_{2s-1},$$
$$[T, P_-] = t^0_{2s+1}v^0_{2s+1} + t^2_{2s+1}v^2_{2s+1},$$
$$[T, T] = 0.$$

Finally, we have

Leibniz identity		Constraint
$\{P_+, P_+, P_-\}$	$\Longrightarrow$	$m^0_{2s-1} = 0, \quad m^2_{2s-1} = 1/2i(2s-1)q^1_{2s-2} - i(2s+1)r^1_{2s},$
$\{T, P_+, P_-\}$	$\Longrightarrow$	$t^0_{2s+1} = 0,$
$\{P_-, P_+, P_-\}$	$\Longrightarrow$	$t^2_{2s+1} = 2i(2s+1)r^1_{2s} - 1/2i(2s-1)l^1_{2s+2}.$

Denoting the parameters $(a^1_{2s}, r^1_{2s}, q^1_{2s-2}, l^1_{2s+2})$ by $(\gamma_1, \gamma_2, \gamma_3, \gamma_4)$, we obtain the first family of the theorem.

Case 2. Let $n = 4s - 2$. The second family of the theorem is obtained by applying similar arguments as in the previous case. □

3.3. Leibniz algebras whose ideal I is the Leibniz $\mathfrak{D}$-module either W^1_n or W^2_n.

Let L be a Leibniz algebra such that the ideal I is defined as a Leibniz $\mathfrak{D}$-module by indecomposable Lie representation W^1_n of the algebra $\mathfrak{D}$ [13]. Then one can assume that $I = \mathrm{span}\{v^0_{2k}, v^1_{2k+1}, v^2_{2k}\}_{k=0,\dots,\lfloor n/2\rfloor}$ where n is odd and

$$\begin{cases} [v^0_{2k}, J] = \frac{i}{2}(n-4k)v^0_{2k}, & k = 0, \dots, \lfloor n/2\rfloor, \\ [v^1_{2k+1}, J] = \frac{i}{2}(n-4k-2)v^1_{2k-1}, & k = 0, \dots, \lfloor n/2\rfloor, \\ [v^2_{2k}, J] = \frac{i}{2}(n-4k)v^2_{2k}, & k = 0, \dots, \lfloor n/2\rfloor, \\ [v^0_{2k}, P_+] = (n-2k+1)v^1_{2k-1}, & k = 1, \dots, \lfloor n/2\rfloor, \\ [v^1_{2k+1}, P_+] = (n-2k)v^2_{2k}, & k = 1, \dots, \lfloor n/2\rfloor, \\ [v^0_{2k}, P_-] = (2k+1)v^1_{2k+1}, & k = 0, \dots, \lfloor (n-1)/2\rfloor, \\ [v^1_{2k+1}, P_-] = (2k+2)v^2_{2k+2}, & k = 0, \dots, \lfloor n/2\rfloor, \\ [v^0_{2k}, T] = -i/2(n-4k)v^2_{2k}, & k = 0, \dots, \lfloor n/2\rfloor \end{cases} \tag{3}$$

Let now the ideal I defined as a Leibniz $\mathfrak{D}$-module by the indecomposable Lie representation W^2_n of the algebra $\mathfrak{D}$ [13]. Then one can assume $I = \mathrm{span}\{v^0_{2k+1}, v^1_{2k}, v^2_{2k+1}\}_{k=0,\dots,\lfloor n/2\rfloor}$, where n is odd and

$$\begin{cases} [v^0_{2k+1}, J] = \frac{i}{2}(n-4k-2)v^0_{2k+1}, & k = 0, \dots, \lfloor n/2\rfloor, \\ [v^1_{2k}, J] = \frac{i}{2}(n-4k)v^1_{2k}, & k = 0, \dots, \lfloor n/2\rfloor, \\ [v^2_{2k+1}, J] = \frac{i}{2}(n-4k-2)v^2_{2k+1}, & k = 0, \dots, \lfloor n/2\rfloor, \\ [v^0_{2k+1}, P_+] = (n-2k)v^1_{2k}, & k = 1, \dots, \lfloor n/2\rfloor, \\ [v^1_{2k}, P_+] = (n-2k+1)v^2_{2k-1}, & k = 1, \dots, \lfloor n/2\rfloor, \\ [v^0_{2k+1}, P_-] = (2k+2)v^1_{2k+2}, & k = 0, \dots, \lfloor (n-1)/2\rfloor, \\ [v^1_{2k}, P_-] = (2k+1)v^2_{2k+1}, & k = 0, \dots, \lfloor n/2\rfloor, \\ [v^0_{2k+1}, T] = -i/2(n-4k-2)v^2_{2k+1}, & k = 0, \dots, \lfloor n/2\rfloor. \end{cases} \tag{4}$$

Theorem 3.3. *Let L be a Leibniz algebra with associated Diamond Lie algebra $\overline{\mathfrak{D}}$ and the ideal I is defined as a Leibniz $\mathfrak{D}$-module by indecomposable Lie representation either W_n^1 or W_n^2. Then*

$$[\mathfrak{D}, \mathfrak{D}] \subseteq \mathfrak{D}.$$

Proof. The proof of the theorem follows from the products (3)–(4) and Lemma 2.5. □

Acknowledgments

The authors were supported by Ministerio de Economía y Competitividad (Spain), grant MTM2013-43687-P (European FEDER support included). The second and third authors were supported by Xunta de Galicia, grant GRC2013-045 (European FEDER support included).

References

[1] Bloh A 1965 *Dokl. Akad. Nauk SSSR* **165** 471–473
[2] Loday J-L 1993 *Enseign. Math. (2)* **39** 269–293
[3] Albeverio S A, Ayupov Sh A and Omirov B A 2006 *Rev. Mat. Complut.* **19** 183–195
[4] Omirov B A 2006 *J. Algebra* **302** 887–896
[5] Ayupov Sh A and Omirov B A 1998 *Algebra and operator theory (Tashkent, 1997)* (Dordrecht: Kluwer Acad. Publ.) pp 1–12
[6] Barnes D W 2012 *Bull. Austral. Math. Soc.* **86** 184–185
[7] Casas J M, Ladra M, Omirov B A and Karimjanov I A 2013 *Linear Multilinear Algebra* **61** 758–774
[8] Ayupov Sh A, Camacho L M, Khudoyberdiyev A K and Omirov B A 2015 *J. Geom. Phys.* **98** 181–195
[9] Calderón A J, Camacho L and Omirov B A 2014 Leibniz algebras of Heisenberg type arXiv:1411.3861
[10] Omirov B A, Rakhimov I S and Turdibaev R M 2013 *Algebr. Represent. Theory* **16** 1507–1519
[11] Uguz S, Karimjanov A and Omirov B A 2015 Leibniz algebras associated with representations of the Diamond Lie algebra arXiv:1507.01349
[12] Fialowski A and Mihálka É Z 2015 *Algebr. Represent. Theory* **18** 477–490
[13] Casati P, Minniti S and Salari V 2010 *J. Math. Phys.* **51** 033515, 20

Universal enveloping crossed module of Leibniz crossed modules and representations

Rafael F. Casado[1], Xabier García-Martínez[1], Manuel Ladra[1]

[1] Department of Algebra, Universidade de Santiago de Compostela, 15782, Santiago de Compostela, Spain

E-mail: xabier.garcia@usc.es

Abstract. The universal enveloping algebra functor $\mathtt{UL}\colon \mathsf{Lb} \to \mathsf{Alg}$, defined by Loday and Pirashvili [1], is extended to crossed modules. Then we construct an isomorphism between the category of representations of a Leibniz crossed module and the category of left modules over its universal enveloping crossed module of algebras. Note that the procedure followed in the proof for the Lie case cannot be adapted, since the actor in the category of Leibniz crossed modules does not always exist.

1. Introduction

Leibniz algebras, which are a non-antisymmetric generalization of Lie algebras, were introduced in 1965 by Bloh [2], who called them D-algebras and referred to the well-known *Leibniz identity* as *differential identity*. In 1993 Loday [3] made them popular and studied their (co)homology. From that moment, many authors have studied this structure, obtaining very relevant algebraic results [1, 4] and applications to Geometry [5, 6] and Physics [7].

Crossed modules of groups were described for the first time by Whitehead in the late 1940s [8] as an algebraic model for path-connected CW-spaces whose homotopy groups are trivial in dimensions greater than 2. From that moment, crossed modules of different algebraic objects, not only groups, have been considered, either as tools or as algebraic structures in their own right. For instance, in [1] crossed modules of Leibniz algebras were defined in order to study cohomology.

Observe that in Ellis's PhD thesis [9] it is proved that, given a category of Ω-groups $\mathcal{C}$ such as the categories of associative and Leibniz algebras, crossed modules, cat^1-objects, internal categories and simplicial objects in $\mathcal{C}$ whose normal complexes are of length 1 are equivalent structures. Internal categories can be described in terms of what Baez calls strict 2-dimensional objects (see [10] for groups and [11] for Lie algebras). By analogy to Baez's terminology, crossed modules of associative algebras (respectively Leibniz algebras) can be viewed as strict associative 2-algebras [12] (respectively strict Leibniz 2-algebras [13, 14]).

In the case of Lie algebras, the universal enveloping algebra plays two important roles: the category of representations of a Lie algebra is isomorphic to the category of left modules over its universal enveloping algebra and it is right adjoint to the Liezation functor. For Leibniz algebras, these roles are played by two different functors: Loday and Pirashvili [1] proved that, given a Leibniz algebra $\mathfrak{p}$, the category of left $\mathtt{UL}(\mathfrak{p})$-modules is isomorphic to the category of $\mathfrak{p}$-representations, where $\mathtt{UL}(\mathfrak{p})$ is the universal enveloping algebra of $\mathfrak{p}$. On the other hand, if associative algebras are replaced by dialgebras, there exist a universal enveloping dialgebra functor [15], which is right adjoint to the functor that assigns to every dialgebra its corresponding Leibniz algebra.

As Norrie states in [16], it is surprising the ease of the generalization to crossed modules of many properties satisfied by the objects in the base category. In [17], the universal enveloping dialgebra functor is extended to crossed modules. The aim of this article is to extend to crossed modules the functor UL and the aforementioned isomorphism between representations of a Leibniz algebra and left modules over its universal enveloping algebra. Observe that the analogous isomorphism in the case of Lie algebras can be easily proved via the actor, but this method cannot be applied in our case, since the actor of a Leibniz crossed module does not necessarily exist [13] (see [18] for the the 1-dimensional case). This makes our approach especially interesting.

In Section 2 we recall some basic definitions and properties, such as the concept of crossed module of associative and Leibniz algebras, along with the notions of the corresponding cat^1-objects. In Section 3 we give proper definitions of left modules over a crossed module of associative algebras and representations of a Leibniz crossed module. In Section 4 we describe the generalization to crossed modules of the functor $\mathtt{UL}\colon \mathbf{Lb} \to \mathbf{Alg}$, that is $\mathtt{XUL}\colon \mathbf{XLb} \to \mathbf{XAlg}$, which assigns to every Leibniz crossed module its corresponding universal enveloping crossed module of algebras. Finally, we construct an isomorphism between the categories of representations of a Leibniz crossed module and the left modules over its universal enveloping crossed module of algebras.

Notations and conventions
Throughout the paper, we fix a commutative ring K with unit. All algebras are considered over K. The categories of Leibniz and (non-unital) associative algebras will be denoted by **Lb** and **Alg**, respectively.

2. Preliminaries

Definition 2.1 ([3]). *A Leibniz algebra $\mathfrak{p}$ over K is a K-module together with a bilinear operation $[\ ,\ ]\colon \mathfrak{p}\times\mathfrak{p}\to\mathfrak{p}$, called the Leibniz bracket, which satisfies the Leibniz identity:*

$$[[p_1,p_2],p_3] = [p_1,[p_2,p_3]] + [[p_1,p_3],p_2],$$

for all $p_1,p_2,p_3 \in \mathfrak{p}$. A morphism of Leibniz algebras is a K-linear map that preserves the bracket.

We will denote by **Lb** the category of Leibniz algebras and morphisms of Leibniz algebras. These are in fact right Leibniz algebras. For the opposite structure, that is $[p_1,p_2]' = [p_2,p_1]$, the left Leibniz identity is

$$[p_1,[p_2,p_3]']' = [[p_1,p_2]',p_3]' + [p_2,[p_1,p_3]']',$$

for all $p_1,p_2,p_3 \in \mathfrak{p}$.

If the bracket of a Leibniz algebra $\mathfrak{p}$ happens to be antisymmetric, then $\mathfrak{p}$ is a Lie algebra. Furthermore, every Lie algebra is a Leibniz algebra. For more examples, see [3].

Recall that a Leibniz algebra $\mathfrak{p}$ acts on another Leibniz algebra $\mathfrak{q}$ if there are two bilinear maps $\mathfrak{p} \times \mathfrak{q} \to \mathfrak{q}$, $(p, q) \mapsto [p, q]$ and $\mathfrak{q} \times \mathfrak{p} \to \mathfrak{q}$, $(q, p) \mapsto [q, p]$, satisfying six relations, which are obtained from the Leibniz identity by taking two elements in $\mathfrak{p}$ and one in $\mathfrak{q}$ (three identities) and one element in $\mathfrak{p}$ and two elements in $\mathfrak{q}$ (three identities). Given an action of a Leibniz algebra $\mathfrak{p}$ on another Leibniz algebra $\mathfrak{q}$, it is possible to consider the semidirect product $\mathfrak{q} \rtimes \mathfrak{p}$, whose Leibniz structure is given by:

$$[(q_1, p_1), (q_2, p_2)] = ([q_1, q_2] + [p_1, q_2] + [q_1, p_2], [p_1, p_2]),$$

for all (q_1, p_1), $(q_2, p_2) \in \mathfrak{q} \oplus \mathfrak{p}$.

Definition 2.2 ([1]). *A* representation of a Leibniz algebra $\mathfrak{p}$ *is a K-module M equipped with two actions $\mathfrak{p} \times M \to M$, $(p, m) \mapsto [p, m]$ and $M \times \mathfrak{p} \to M$, $(m, p) \mapsto [m, p]$, satisfying the following three axioms:*

$$\begin{aligned}
[m, [p_1, p_2]] &= [[m, p_1], p_2] - [[m, p_2], p_1], \\
[p_1, [m, p_2]] &= [[p_1, m], p_2] - [[p_1, p_2], m], \\
[p_1, [p_2, m]] &= [[p_1, p_2], m] - [[p_1, m], p_2],
\end{aligned}$$

for all $m \in M$ and $p_1, p_2 \in \mathfrak{p}$.

A morphism $f\colon M \to N$ of $\mathfrak{p}$-representations is a K-linear map which is compatible with the left and right actions of $\mathfrak{p}$.

Remark 2.3. *Given a $\mathfrak{p}$-representation M, we can endow the direct sum of K-modules $M \oplus \mathfrak{p}$ with a Leibniz structure such that M is an abelian ideal and $\mathfrak{p}$ is a subalgebra. The converse statement is also true. It is evident that the Leibniz structure of $M \oplus \mathfrak{p}$ is the one of $M \rtimes \mathfrak{p}$, as described previously.*

Definition 2.4 ([1]). *Let $\mathfrak{p}^l$ and $\mathfrak{p}^r$ be two copies of a Leibniz algebra $\mathfrak{p}$. We will denote by l_x and r_x the elements of $\mathfrak{p}^l$ and $\mathfrak{p}^r$ corresponding to $x \in \mathfrak{p}$. Consider the tensor K-algebra $\mathrm{T}(\mathfrak{p}^l \oplus \mathfrak{p}^r)$, which is associative and unital. Let I be the two-sided ideal corresponding to the relations:*

$$\begin{aligned}
&[x, y]_r = y_r x_r - x_r y_r, \\
&[x, y]_l = y_r x_l - x_l y_r, \\
&x_l(y_r + y_l) = 0.
\end{aligned}$$

for all $x, y \in \mathfrak{p}$. The universal enveloping algebra of the Leibniz algebra $\mathfrak{p}$ *is the associative and unital algebra*

$$\mathtt{UL}(\mathfrak{p}) := \mathrm{T}(\mathfrak{p}^l \oplus \mathfrak{p}^r)/I.$$

This construction defines a functor $\mathtt{UL}\colon$ **Lb** $\to$ **Alg**.

Theorem 2.5 ([1]). *The category of representations of the Leibniz algebra $\mathfrak{p}$ is isomorphic to the category of left modules over $\mathtt{UL}(\mathfrak{p})$.*

Proof. Let M be a representation of $\mathfrak{p}$. It is possible to define a left action of $\mathtt{UL}(\mathfrak{p})$ on the K-module M as follows. Given $x_l \in \mathfrak{p}^l$, $x_r \in \mathfrak{p}^r$ and $m \in M$,

$$x_l \cdot m = [x, m], \qquad x_r \cdot m = [m, x].$$

These actions can be extended to an action of $\mathrm{T}(\mathfrak{p}^l \oplus \mathfrak{p}^r)$ by composition and linearity. It is not complicated to check that this way M is equipped with a structure of left $\mathtt{UL}(\mathfrak{p})$-module.

Regarding the converse statement, it is immediate that, starting with a left $\mathtt{UL}(\mathfrak{p})$-module, the restrictions of the actions to $\mathfrak{p}^l$ and $\mathfrak{p}^r$ give two actions of $\mathfrak{p}$ which make M into a representation. □

Recall that a left module over an associative algebra A can be described as a morphism $\alpha\colon A \to \mathrm{End}(M)$, where M is a K-module.

Both **Lb** and **Alg** are categories of interest, notion introduced by Orzech in [19]. See [20] for a proper definition and more examples. Categories of interest are a particular case of categories of groups with operations, for which Porter [21] described the notion of crossed module. The following definitions agree with the one given by Porter.

Definition 2.6. *A* crossed module of Leibniz algebras *(or* Leibniz crossed module*)* $(\mathfrak{q}, \mathfrak{p}, \eta)$ *is a morphism of Leibniz algebras* $\eta\colon \mathfrak{q} \to \mathfrak{p}$ *together with an action of* $\mathfrak{p}$ *on* $\mathfrak{q}$ *such that*

$$\eta([p,q]) = [p, \eta(q)] \quad \textit{and} \quad \eta([q,p]) = [\eta(q), p],$$
$$[\eta(q_1), q_2] = [q_1, q_2] = [q_1, \eta(q_2)],$$

for all $q, q_1, q_2 \in \mathfrak{q}$, $p \in \mathfrak{p}$.

A *morphism of Leibniz crossed modules* (φ, ψ) from $(\mathfrak{q}, \mathfrak{p}, \eta)$ to $(\mathfrak{q}', \mathfrak{p}', \eta')$ is a pair of Leibniz homomorphisms, $\varphi\colon \mathfrak{q} \to \mathfrak{q}'$ and $\psi\colon \mathfrak{p} \to \mathfrak{p}'$, such that

$$\psi\eta = \eta'\varphi,$$
$$\varphi([p,q]) = [\psi(p), \varphi(q)] \quad \text{and} \quad \varphi([q,p]) = [\varphi(q), \psi(p)],$$

for all $q \in \mathfrak{q}$, $p \in \mathfrak{p}$.

Definition 2.7. *A* crossed module of algebras (B, A, ρ) *is an algebra homomorphism* $\rho\colon B \to A$ *together with an action of* A *on* B *such that*

$$\rho(ab) = a\rho(b) \quad \textit{and} \quad \rho(ba) = \rho(b)a,$$
$$\rho(b_1)b_2 = b_1b_2 = b_1\rho(b_2),$$

for all $a \in A$, $b_1, b_2 \in B$.

A *morphism of crossed modules of algebras* $(\varphi, \psi)\colon (B, A, \rho) \to (B', A', \rho')$ is a pair of algebra homomorphisms, $\varphi\colon B \to B'$ and $\psi\colon A \to A'$, such that

$$\psi\rho = \rho'\varphi,$$
$$\varphi(ba) = \varphi(b)\psi(a) \quad \text{and} \quad \varphi(ab) = \psi(a)\varphi(b),$$

for all $b \in B$, $a \in A$.

We will denote by **XLb** and **XAlg** the categories of Leibniz crossed modules and crossed modules of associative algebras, respectively. Crossed modules can be alternatively describe as cat^1-objects, namely:

Definition 2.8. *A* cat^1-Leibniz algebra $(\mathfrak{p}_1, \mathfrak{p}_0, s, t)$ *consists of a Leibniz algebra* $\mathfrak{p}_1$ *together with a Leibniz subalgebra* $\mathfrak{p}_0$ *and the structural morphisms* $s, t\colon \mathfrak{p}_1 \to \mathfrak{p}_0$ *such that*

$$s|_{\mathfrak{p}_0} = t|_{\mathfrak{p}_0} = \mathrm{id}_{\mathfrak{p}_0}, \tag{CLb1}$$
$$[\operatorname{Ker} s, \operatorname{Ker} t] = 0 = [\operatorname{Ker} t, \operatorname{Ker} s], \tag{CLb2}$$

Definition 2.9. *A* cat^1-algebra (A_1, A_0, σ, τ) *consists of an algebra* A_1 *together with a subalgebra* A_0 *and the structural morphisms* $\sigma, \tau\colon A_1 \to A_0$ *such that*

$$\sigma|_{A_0} = \tau|_{A_0} = \mathrm{id}_{A_0}, \tag{CAs1}$$
$$\operatorname{Ker}\sigma \operatorname{Ker}\tau = 0 = \operatorname{Ker}\tau \operatorname{Ker}\sigma. \tag{CAs2}$$

It is a well-known fact (see for instance [9]) that the category of crossed modules of Leibniz algebras (resp. associative algebras) is equivalent to the category of cat^1-Leibniz algebras (resp. associative algebras).

Given a crossed module of Leibniz algebras $(\mathfrak{q}, \mathfrak{p}, \eta)$, the corresponding cat^1-Leibniz algebra is $(\mathfrak{q} \rtimes \mathfrak{p}, \mathfrak{p}, s, t)$, where $s(q,p) = p$ and $t(q,p) = \eta(q)+p$ for all $(q,p) \in \mathfrak{q} \rtimes \mathfrak{p}$. Conversely, given a cat^1-Leibniz algebra $(\mathfrak{p}_1, \mathfrak{p}_0, s, t)$, the corresponding Leibniz crossed module is $t|_{\operatorname{Ker} s}\colon \operatorname{Ker} s \to \mathfrak{p}_0$, with the action of $\mathfrak{p}_0$ on $\operatorname{Ker} s$ induced by the bracket in $\mathfrak{p}_1$. The equivalence for associative algebras is analogous.

3. Representations of crossed modules

Since our intention is to extend Theorem 2.5 to crossed modules, it is necessary to give a proper definition of representations over Leibniz crossed modules and left modules over crossed modules of algebras.

In the case of crossed modules of associative algebras, by analogy to the 1-dimensional situation, left modules can be described via the endomorphism crossed module:

Definition 3.1. *Let* (B, A, ρ) *be a crossed module of algebras. A* left (B, A, ρ)-module *is an abelian crossed module of algebras* (V, W, δ), *that is* δ *is simply a morphism of* K*-modules and the action of* W *on* V *is trivial, together with a morphism of crossed modules of algebras* $(\varphi, \psi)\colon (B, A, \rho) \to (\operatorname{Hom}_K(W, V), \operatorname{End}(V, W, \delta), \Gamma)$.

Note that $\operatorname{End}(V, W, \delta)$ is the algebra of all pairs (α, β), with $\alpha \in \operatorname{End}_K(V)$ and $\beta \in \operatorname{End}_K(W)$, such that $\beta\delta = \delta\alpha$. Furthermore, $\Gamma(d) = (d\delta, \delta d)$ for all $d \in \operatorname{Hom}_K(W, V)$. The action of $\operatorname{End}(V, W, \delta)$ on $\operatorname{Hom}_K(W, V)$ is given by

$$(\alpha, \beta) \cdot d = \alpha d \qquad \text{and} \qquad d \cdot (\alpha, \beta) = d\beta,$$

for all $d \in \operatorname{Hom}_K(W, V)$, $(\alpha, \beta) \in \operatorname{End}(V, W, \delta)$. See [22, 23] for further details.

For the categories of crossed modules of groups and Lie algebras, representations can be defined via an object called the actor (see [24, 16]). However this is not the case for Leibniz crossed modules (see [13]). Nevertheless, it is possible to give a definition by equations:

Definition 3.2. *A* representation of a Leibniz crossed module $(\mathfrak{q}, \mathfrak{p}, \eta)$ *is an abelian Leibniz crossed module* (N, M, μ) *endowed with:*

(i) *Actions of the Leibniz algebra* $\mathfrak{p}$ *(and so* $\mathfrak{q}$ *via* η*) on* N *and* M*, such that the homomorphism* μ *is* $\mathfrak{p}$*-equivariant, that is*

$$\mu([p,n]) = [p,\mu(n)], \tag{LbEQ1}$$
$$\mu([n,p]) = [\mu(n),p], \tag{LbEQ2}$$

for all $n \in N$ *and* $p \in \mathfrak{p}$.

(ii) *Two* K*-bilinear maps* $\xi_1 \colon \mathfrak{q} \times M \to N$ *and* $\xi_2 \colon M \times \mathfrak{q} \to N$ *such that*

$$\mu\xi_2(m,q) = [m,q], \tag{LbM1a}$$
$$\mu\xi_1(q,m) = [q,m], \tag{LbM1b}$$
$$\xi_2(\mu(n),q) = [n,q], \tag{LbM2a}$$
$$\xi_1(q,\mu(n)) = [q,n], \tag{LbM2b}$$
$$\xi_2(m,[p,q]) = \xi_2([m,p],q) - [\xi_2(m,q),p], \tag{LbM3a}$$
$$\xi_1([p,q],m) = \xi_2([p,m],q) - [p,\xi_2(m,q)], \tag{LbM3b}$$
$$\xi_2(m,[q,p]) = [\xi_2(m,q),p] - \xi_2([m,p],q), \tag{LbM3c}$$
$$\xi_1([q,p],m) = [\xi_1(q,m),p] - \xi_1(q,[m,p]), \tag{LbM3d}$$
$$\xi_2(m,[q,q']) = [\xi_2(m,q),q'] - [\xi_2(m,q'),q], \tag{LbM4a}$$
$$\xi_1([q,q'],m) = [\xi_1(q,m),q'] - [q,\xi_2(m,q')], \tag{LbM4b}$$
$$\xi_1(q,[p,m]) = -\xi_1(q,[m,p]), \tag{LbM5a}$$
$$[p,\xi_1(q,m)] = -[p,\xi_2(m,q)], \tag{LbM5b}$$

for all $q,q' \in \mathfrak{q}$, $p \in \mathfrak{p}$, $n \in N$ $m,m' \in M$.

Remark 3.3. *As in Remark 2.3, given a* $(\mathfrak{q},\mathfrak{p},\eta)$*-representation* (N,M,μ)*, we can obtain a Leibniz crossed module structure on* $(N \oplus \mathfrak{q}, M \oplus \mathfrak{p}, \mu \oplus \eta)$ *where* N *and* M *are abelian ideals and* $\mathfrak{q}$ *and* $\mathfrak{p}$ *are subalgebras of* $N \oplus \mathfrak{q}$, $M \oplus \mathfrak{p}$ *respectively. Conversely it is also true. Moreover, a representation can be seen as an action of* $(\mathfrak{q},\mathfrak{p},\eta)$ *over an abelian Leibniz crossed module* (N,M,μ) *in the sense of [13].*

4. Isomorphism between the categories of representations

In this section, we extend to crossed modules the universal enveloping algebra functor and then we give the construction of an isomorphism between the categories of representations of a Leibniz crossed module and left modules over its corresponding universal enveloping crossed module of algebras. Recall that the method used in the proof of the equivalent result in the case of Lie algebras cannot be applied in our case due to the lack of actor in the category of Leibniz crossed modules.

Let $(\mathfrak{q},\mathfrak{p},\eta)$ be a Leibniz crossed module and consider its corresponding cat^1-Leibniz algebra

$$\mathfrak{q} \rtimes \mathfrak{p} \overset{s}{\underset{t}{\rightrightarrows}} \mathfrak{p}\ ,$$

with $s(q,p) = p$ and $t(q,p) = \eta(q) + p$ for all $(q,p) \in \mathfrak{q} \rtimes \mathfrak{p}$. Now, if we apply $\mathtt{UL}$ to the previous diagram, we get

$$\mathtt{UL}(\mathfrak{q} \rtimes \mathfrak{p}) \overset{\mathtt{UL}(s)}{\underset{\mathtt{UL}(t)}{\rightrightarrows}} \mathtt{UL}(\mathfrak{p})\ .$$

Although it is true that $\mathtt{UL}(s)|_{\mathtt{UL}(\mathfrak{p})} = \mathtt{UL}(t)|_{\mathtt{UL}(\mathfrak{p})} = \mathrm{id}_{\mathtt{UL}(\mathfrak{p})}$, in general, the second condition for cat^1-algebras (CAs2) is not satisfied. Nevertheless, we can consider the quotient $\overline{\mathtt{UL}}(\mathfrak{q} \rtimes \mathfrak{p}) = \mathtt{UL}(\mathfrak{q} \rtimes \mathfrak{p})/X$, where $X = \operatorname{Ker}\mathtt{UL}(s)\operatorname{Ker}\mathtt{UL}(t) + \operatorname{Ker}\mathtt{UL}(t)\operatorname{Ker}\mathtt{UL}(s)$, and the induced morphisms $\overline{\mathtt{UL}}(s)$ and $\overline{\mathtt{UL}}(t)$. In this way, the diagram

$$\overline{\mathtt{UL}}(\mathfrak{q} \rtimes \mathfrak{p}) \underset{\overline{\mathtt{UL}}(t)}{\overset{\overline{\mathtt{UL}}(s)}{\rightrightarrows}} \mathtt{UL}(\mathfrak{p})$$

is clearly a cat^1-algebra. Note that $\mathtt{UL}(\mathfrak{p})$ can be regarded as a subalgebra of $\overline{\mathtt{UL}}(\mathfrak{q} \rtimes \mathfrak{p})$.

We can now define $\mathtt{XUL}(\mathfrak{q}, \mathfrak{p}, \eta)$ as the crossed module of associative algebras given by $(\operatorname{Ker}\overline{\mathtt{UL}}(s), \mathtt{UL}(\mathfrak{p}), \overline{\mathtt{UL}}(t)|_{\operatorname{Ker}\overline{\mathtt{UL}}(s)})$. This construction defines a functor $\mathtt{XUL}\colon \mathbf{XLb} \to \mathbf{XAlg}$.

Theorem 4.1. *The category of representations of a Leibniz crossed module* $(\mathfrak{q}, \mathfrak{p}, \eta)$ *is isomorphic to the category of left modules over its universal enveloping crossed module of algebras* $\mathtt{XUL}(\mathfrak{q}, \mathfrak{p}, \eta)$.

Proof. Let (N, M, μ) be a left $(\operatorname{Ker}\overline{\mathtt{UL}}(s), \mathtt{UL}(\mathfrak{p}), \overline{\mathtt{UL}}(t)|_{\operatorname{Ker}\overline{\mathtt{UL}}(s)})$-module. Then we have a homomorphism

$$(\varphi, \psi)\colon (\operatorname{Ker}\overline{\mathtt{UL}}(s), \mathtt{UL}(\mathfrak{p}), \overline{\mathtt{UL}}(t)|_{\operatorname{Ker}\overline{\mathtt{UL}}(s)}) \to (\operatorname{Hom}_K(M, N), \operatorname{End}(N, M, \mu), \Gamma),$$

such that $\psi \circ \overline{\mathtt{UL}}(t)|_{\operatorname{Ker}\overline{\mathtt{UL}}(s)} = \Gamma \circ \varphi$, $\varphi(ba) = \varphi(b)\psi(a)$ and $\varphi(ab) = \psi(a)\varphi(b)$. We need to define actions of $\mathfrak{p}$ on N and M satisfying (LbEQ1) and (LbEQ2) and we need to define $\xi_1\colon \mathfrak{q} \times M \to N$ and $\xi_2\colon M \times \mathfrak{q} \to N$ satisfying identities (LbM1a)–(LbM5b).

We define the actions of $\mathfrak{p}$ on N and M as those induced by $\psi\colon \mathtt{UL}(\mathfrak{p}) \to \operatorname{End}(N, M, \mu)$ as in Theorem 2.5. The identities (LbEQ1) and (LbEQ2) are followed by the properties of $\operatorname{End}(N, M, \mu)$. We define the morphisms ξ_1 and ξ_2 by $\xi_1(q, m) = \varphi\big((q, 0)_l\big)(m)$ and $\xi_2(m, q) = \varphi\big((q, 0)_r\big)(m)$. The identities (LbM1a), (LbM2a) and (LbM1b), (LbM2b) are followed by the commutative square $\psi \circ \overline{\mathtt{UL}}(t)|_{\operatorname{Ker}\overline{\mathtt{UL}}(s)} = \Gamma \circ \varphi$ applied to the elements $(q, 0)_r$ and $(q, 0)_l$ respectively. Given the element $([p, q], 0)_r \in \operatorname{Ker}\overline{\mathtt{UL}}(s)$, we have that

$$([p, q], 0)_r = [(0, p)_r, (q, 0)_r] = (q, 0)_r(0, p)_r - (0, p)_r(q, 0)_r.$$

Applying φ to this relation and using that $\varphi(ba) = \varphi(b)\psi(a)$ and $\varphi(ab) = \psi(a)\varphi(b)$ we obtain $\varphi([p, q], 0)_r = \varphi(q, 0)_r\psi_2(p)_r - \psi_1(p)_r\varphi(q, 0)_r$ which implies (LbM3a). Proceeding in the same way for the elements $([p, q], 0)_l$, $([q, p], 0)_r$ and $([q, p], 0)_l$ we check that identities (LbM3b), (LbM3c) and (LbM3d) are satisfied. Doing a similar argument on the elements $([q, q'], 0)_r$ and $([q, q'], 0)_l$ we obtain identities (LbM4a) and (LbM4b). Applying φ to the relations

$$(q, 0)_l(0, p)_l = -(q, 0)_l(0, p)_r \qquad \text{and} \qquad (0, p)_l(q, 0)_l = -(0, p)_l(q, 0)_r$$

we have identities (LbM5a) and (LbM5b) respectively.

Conversely, let (N, M, μ) be a $(\mathfrak{q}, \mathfrak{p}, \eta)$-representation. We need to construct a morphism of crossed modules of algebras (φ, ψ) from $\mathtt{XUL}(\mathfrak{q}, \mathfrak{p}, \eta)$ to $(\operatorname{Hom}_K(M, N), \operatorname{End}(N, M, \mu), \Gamma)$. The homomorphism $\psi = (\psi_1, \psi_2)\colon \mathtt{UL}(\mathfrak{p}) \to \operatorname{End}(N, M, \mu)$ is the homomorphism induced by the actions of $\mathfrak{p}$ on N and M as in Theorem 2.5. It is well defined by identities (LbEQ1) and

(LbEQ2). Let be the homomorphism $\Phi\colon (\mathfrak{q} \rtimes \mathfrak{p})^l \oplus (\mathfrak{q} \rtimes \mathfrak{p})^r \to \operatorname{Hom}_K(N \oplus M, N \oplus M)$ defined by

$$\Phi(q,p)_l(n,m) = ([q,n] + [p,n] + \xi_1(q,m), [p,m]),$$
$$\Phi(q,p)_r(n,m) = ([n,q] + [n,p] + \xi_2(m,q), [m,p]).$$

Note that they can also be rewritten as

$$\Phi(q,p)_l(n,m) = \big(t(q,p)_l(n) + \xi_1(q,m), s(q,p)_l(m)\big),$$

$$\Phi(q,p)_r(n,m) = \big(t(q,p)_r(n) + \xi_2(m,q), s(q,p)_r(m)\big).$$

By the universal property of the tensor algebra, there is a unique homomorphism

$$\mathrm{T}(\Phi)\colon \mathrm{T}\left((\mathfrak{q} \rtimes \mathfrak{p})^l \oplus (\mathfrak{q} \rtimes \mathfrak{p})^r\right) \to \operatorname{Hom}_K(N \oplus M, N \oplus M),$$

commuting with the inclusion.

We consider the projection $\pi\colon \operatorname{Hom}_K(N \oplus M, N \oplus M) \to \operatorname{Hom}_K(M, N \oplus M)$, where $\pi(f)(m) = f(0,m)$, and denote $\varphi' = \pi \circ \mathrm{T}(\Phi)$. Given an element of the form $(q,p)_r(q',p')_r - (q',p')_r(q,p)_r + [(q,p)_r, (q',p')_r]$ we obtain that

$$\begin{aligned}
\varphi'\big((q,p)_r(q',p')_r - (q',p')_r(q,p)_r + [(q,p)_r,(q',p')_r]\big)(m)&\\
= [\xi_2(m,q'),q] + \big[[m,p'],q\big] + [\xi_2(m,q'),p]&\\
- [\xi_2(m,q),q'] - \big[[m,p],q'\big] - [\xi_2(m,q),p']&\\
+ \xi_2(m,[q,q']) + \xi_2(m,[q,p']) + \xi_2(m,[q',p]) = 0,&
\end{aligned}$$

by the properties of ξ_2.

Analogously, it is possible to prove that φ' vanishes on the other two relations of the universal enveloping algebra. Then φ' factors through $\mathtt{UL}(\mathfrak{q} \rtimes \mathfrak{p})$. In order to ease notation we will refer to it as φ' as well.

By definition it is clear that $\varphi'|_{\operatorname{Ker}\mathtt{UL}(s)}(M) \subseteq N$ and $\mathrm{T}(\Phi)\big((\mathfrak{q} \rtimes \mathfrak{p})^l \oplus (\mathfrak{q} \rtimes \mathfrak{p})^r\big)(N) = \mathrm{T}(t)\big((\mathfrak{q} \rtimes \mathfrak{p})^l \oplus (\mathfrak{q} \rtimes \mathfrak{p})^r\big)(N)$. Then φ' factors through $\operatorname{Ker}\mathtt{UL}(t)\operatorname{Ker}\mathtt{UL}(s)$. Moreover, we have that

$$\begin{aligned}
\varphi'(q',p')_r(q,p)_r(m) &= \big([\xi_2(m,q),q'] + [\xi_2(m,q),p'] + \xi_2([m,p],q)\big)\\
&= \big(\xi_2(m,[t(p,q)_r,q']) + [\xi_2(m,q'),t(p,q)_r] + [\xi_2(m,q),s(q',p')_r], s(q',p')_r s(q,p)_r m\big),
\end{aligned}$$

extending this argument we see that φ' also factors through $\operatorname{Ker}\mathtt{UL}(s)\operatorname{Ker}\mathtt{UL}(t)$.

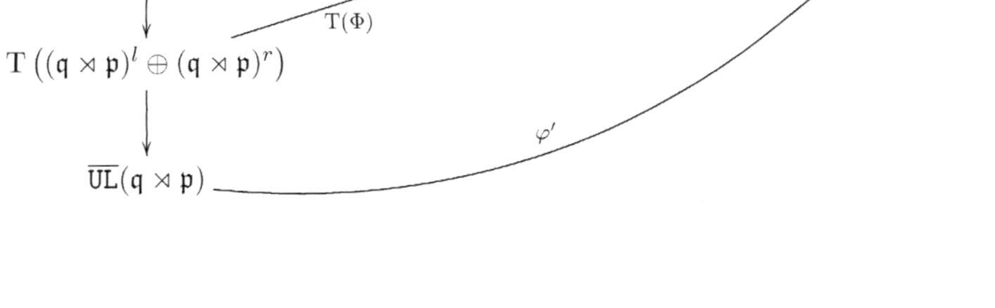

Therefore, φ will be the restriction of φ' to $\operatorname{Ker}\overline{\mathtt{UL}}(\mathfrak{q} \rtimes \mathfrak{p})$ and it will take values in $\operatorname{Hom}_K(M, N)$, that is

$$\varphi\colon \operatorname{Ker}\overline{\mathtt{UL}}(\mathfrak{q} \rtimes \mathfrak{p}) \to \operatorname{Hom}_K(M, N).$$

With these definitions of φ and ψ, to check that

$$(\varphi, \psi)\colon (\operatorname{Ker}\overline{\mathtt{UL}}(s), \mathtt{UL}(\mathfrak{p}), \overline{\mathtt{UL}}(t)|_{\operatorname{Ker}\overline{\mathtt{UL}}(s)}) \to (\operatorname{Hom}_K(M, N), \operatorname{End}(N, M, \mu), \Gamma)$$

is a morphism of crossed modules of algebras is now a matter of straightforward computations. □

Acknowledgments
The authors were supported by Ministerio de Economía y Competitividad (Spain), grant MTM2013-43687-P (European FEDER support included). The second and third authors were supported by Xunta de Galicia, grant GRC2013-045 (European FEDER support included). The second author is also supported by an FPU scholarship, Ministerio de Educación, Cultura y Deporte (Spain).

References

[1] Loday J L and Pirashvili T 1993 *Math. Ann.* **296** 139–158
[2] Bloh A 1965 *Dokl. Akad. Nauk SSSR* **165** 471–473
[3] Loday J L 1993 *Enseign. Math. (2)* **39** 269–293
[4] Loday J L and Pirashvili T 1998 *Georgian Math. J.* **5** 263–276
[5] Kinyon M and Weinstein A 2001 *Amer. J. Math.* **123** 525–550
[6] Lodder J M 1998 *Ann. Inst. Fourier (Grenoble)* **48** 73–95
[7] Felipe R, López-Reyes N and Ongay F 2003 *Lett. Math. Phys.* **63** 157–164
[8] Whitehead J H C 1949 *Bull. Amer. Math. Soc.* **55** 453–496
[9] Ellis G 1984 *Crossed modules and their higher dimensional analogues* Ph.D. thesis University of Wales
[10] Baez J C and Lauda A D 2004 *Theory Appl. Categ.* **12** 423–491
[11] Baez J C and Crans A S 2004 *Theory Appl. Categ.* **12** 492–538
[12] Khmaladze E 2012 *Proc. A. Razmadze Math. Inst.* **159** 57–64
[13] Casado R F 2015 *Relations between crossed modules of different algebras* Ph.D. thesis Universidade de Santiago de Compostela
[14] Sheng Y and Liu Z 2013 *Comm. Algebra* **41** 1929–1953
[15] Loday J L 2001 *Dialgebras and related operads* (*Lecture Notes in Math.* vol 1763) (Springer, Berlin) pp 7–66
[16] Norrie K 1990 *Bull. Soc. Math. France* **118** 129–146
[17] Casas J M, Casado R F, Khmaladze E and Ladra M 2015 More on crossed modules of Lie, Leibniz, associative and diassociative algebras arXiv:1311.7079
[18] Casas J M, Datuashvili T and Ladra M 2010 *Appl. Categ. Structures* **18** 85–114
[19] Orzech G 1972 *J. Pure Appl. Algebra* **2** 287–314; ibid. 2 (1972), 315–340
[20] Montoli A 2010 *Theory Appl. Categ.* **23** 7–21
[21] Porter T 1987 *Proc. Edinburgh Math. Soc. (2)* **30** 373–381
[22] Casas J M, Casado R F, Khmaladze E and Ladra M 2014 *Homology Homotopy Appl.* **16** 143–158
[23] Casas J M, Inassaridze N, Khmaladze E and Ladra M 2014 *J. Homotopy Relat. Struct.* **9** 223–237
[24] Casas J M and Ladra M 1998 *Comm. Algebra* **26** 2065–2089

Journal of Physics: Conference Series **697** (2016) 012008 doi:10.1088/1742-6596/697/1/012008

On $HL^2(L, L)$ for semisimple Leibniz algebras

L M Camacho[1], M Ladra[2] and R M Turdibaev[3]

[1] Dpto. Matemática Aplicada I. University of Sevilla, 41012, Spain,
[2,3]Department of Algebra, University of Santiago de Compostela, 15782, Spain.

E-mail: `lcamacho@us.es`, `manuel.ladra@usc.es`, `rustamtm@yahoo.com`

Abstract. In this paper we present a decomposition of $HL^n(L, L)$ into a direct sum of some subspaces for a finite dimensional complex semisimple Leibniz algebra L. Furthermore, we provide a more specific decomposition in case $n = 2$ into two subspaces. We verify that one of those subspaces annihilates for specific Leibniz algebras with liezation $\mathfrak{sl}_2$ and some others.

1. Introduction

Leibniz algebras were rediscovered by A. Bloh [3] and later were given another impulse of investigation due to works of J.–L. Loday [7] who discovered that the classical Chevalley–Eilenberg boundary map in the exterior module of a Lie algebra can be lifted to the tensor module. To prove that it becomes a chain complex the only identity needed is a so called Leibniz identity which yields to definition of Leibniz algebra and Leibniz (co)homology.

In this work we study Leibnz algebra cohomology and using an idea from [1] we decompose the cochain space $CL^n(L, L)$ into a direct sub of special subspaces that we call level spaces. Further we concentrate on so called semisimple Leibniz algebras. They are Leibniz algebras with corresponding Lie algebras being semisimple. Semisimple Leibniz algebra L can be naturally consider as a $\mathbb{Z}_2$-graded algebra with L_0 being its liezation and $L_1 = I$. We notice that due to structural properties of semisimple Leibniz algebras, coboundary operator preserves the level spaces. Thus we are able to establish in Theorem 4.3 a decomposition of $HL^n(L, L)$ into direct sum of n subspaces, that we label from $-n + 1$ to 0.

Our goal is to check proposed conjecture in [2] that $HL^2(L, L) = 0$ for any finite-dimensional complex semisimple Leibniz algebra L. Authors in [2] validate the claim for simple Leibniz algebra with liezation $L/I \cong \mathfrak{sl}_2$ over $\mathbb{C}$.

In order to conjecture to hold, we elaborate on each of the subspaces in the decomposition of $HL^2(L, L)$ to be zero and present equivalent conditions to the conjecture to be true in Proposition 5.6 and Proposition 5.7 in terms of Lie algebra and their representations.

In the last section, we establish that space of level -1 is zero for all finite-dimensional Leibniz algebras with liezation $\mathfrak{sl}_2$ in Theorem 6.1 and we check analogous statement for another Leibniz algebra with liezation $\mathfrak{sl}_2 \oplus \mathfrak{sl}_2$ in Theorem 6.2.

2. Preliminaries

Given a Lie algebra and its non-trivial module a different algebraic structure arises on the direct sum.

Example 2.1. *([5]) Let $\mathfrak{g}$ be a Lie algebra and M be a $\mathfrak{g}$-module. Consider $L = \mathfrak{g} \oplus M$ with a bracket $[(g_1, m_1), (g_2, m_2)] := ([g_1, g_2], -g_2.m_1)$. Then*

$$[[x, y], z] = [[x, z], y] + [x, [y, z]] \tag{1}$$

holds for any $x, y, z \in L$ and this algebra is not a Lie algebra if the action of $\mathfrak{g}$ on M is not trivial.

Looking to the above identity (1) one notices that the right multiplication operator $[-, z]$ by any element z is a derivation (see [6]) and satisfies a so called Leibniz rule. Although defined earlier by A. Bloh [3] these objects attracted more attention after series of J.-L. Loday and his collaborators' works.

Definition 2.2. *An algebra $(L, [-, -])$ over a field $\mathbb{K}$ is called a Leibniz algebra if identity (1) holds.*

Leibniz algebra of Example 2.1 is denoted as $\mathfrak{g} \dot{+} M$ and is called hemisemidirect product Leibniz algebra in [5].

Any Lie algebra is a Leibniz algebra but not the converse. Given a Leibniz algebra L its two sided ideal generated by elements $[x, x]$ for all $x \in L$ is very important. In this work we denote this ideal by I. The quotient Leibniz algebra L/I is easily seen to be a Lie algebra, called the liezation of Leibniz algebra L.

Definition 2.3. *The set $\mathrm{Ann}_R(L) = \{x \in L \,|\, [L, x] = 0\}$ of a Leibniz algebra L is called the right annihilator of L.*

One can show that $\mathrm{Ann}_R(L)$ is an ideal of L. Note that due to Leibniz identity it follows that $[L, I] = 0$. Thus I is a subset of a right annihilator $\mathrm{Ann}_R(L) = \{x \in L \mid [L, x] = 0\}$. The center is defined as $Z(L) = \{x \in L \mid [x, L] = [L, x] = 0\}$. If $I = L$ then Leibniz algebra is a trivial algebra with all products being zero. If $I = 0$ then L is a Lie algebra. In this work we assume I to be non trivial, eliminating Lie and trivial Leibniz algebras from the study. One has a short exact sequence $0 \to I \to L \xrightarrow{f} \mathfrak{g} \to 0$. Note that projection f is universal in the sense that a Leibniz map from L to any Lie algebra factors through f.

A *representation* M of a Leibniz algebra L is introduced in [6].

Definition 2.4. *A vector space M is called a representation or bimodule over a Leibniz algebra L if there are two bilinear maps:*

$$[-,-]\colon L \times M \to M \qquad \text{and} \qquad [-,-]\colon M \times L \to M$$

satisfying the following three axioms

$$\begin{aligned}
[m,[x,y]] &= [[m,x],y] - [[m,y],x], \\
[x,[m,y]] &= [[x,m],y] - [[x,y],m], \\
[x,[y,m]] &= [[x,y],m] - [[x,m],y],
\end{aligned}$$

for any $m \in M$, $x, y \in L$.

A Leibniz bimodule is called *anti-symmetric* when $[x,m] = 0$ for $x \in L$, $m \in M$. Provided that, the last two required identities vanish. Since ideal I is in the right-annihilator, if $x \in I$ or $y \in I$ the first identity vanishes as well. Therefore, the action of L on anti-symmetric L-bimodule M is determined by the action of complement of I that is required to satisfy only the first axiom.

Now for $g \in \mathfrak{g}$ define $g \circ m := -[m, \phi(g)]$, where ϕ is an inverse map of f. These are well-defined, since if $g = g'$ then $\phi(g - g') \in I$ and I is in the right annihilator. Moreover, the first identity turns into

$$\begin{aligned}
[g_1, g_2] \circ m = -[m, \phi([g_1, g_2])] = -[m, [\phi(g_1), \phi(g_2)]] = \\
-[[m, \phi(g_1)], \phi(g_2)] + [[m, \phi(g_2)], \phi(g_1)] = g_2 \circ [m, \phi(g_1)] - g_1 \circ [m, \phi(g_2)] \\
= g_1 \circ (g_2 \circ m) - g_2 \circ (g_1 \circ m),
\end{aligned}$$

i.e. M becomes a left $\mathfrak{g}$-module (or a right $\mathfrak{g}$-module with action $m \circ g = [m, \phi(g)]$).

Conversely, if M is a left $\mathfrak{g}$-module with an action $\circ : \mathfrak{g} \times M \to M$, then by defining $[m, l] := -f(l) \circ m$ for $l \in L$ implies by the arguments above that M is an anti-symmetric L-bimodule.

Thus, anti-symmetric Leibniz L-bimodule is equivalent to a Lie algebra $\mathfrak{g}$-module of it liezation $\mathfrak{g}$. For instance, an ideal generated by squares I can be considered as a Lie algebra module over liezation $\mathfrak{g}$.

Call Leibniz algebra L *semisimple* if $\mathfrak{g}$ is semisimple. L is called *simple* if the only non-trivial ideal of L is $I \neq [L, L]$. These agrees with suggested definition in [1]. T. Pirashvili proved the following statement.

Proposition 2.5. *([10]) Let $f : L \to \mathfrak{g}$ be an epimorphism from an arbitrary finite dimensional Leibniz algebra L to semisimple Lie algebra $\mathfrak{g}$. Then f admits a section.*

Considering a semisimple Leibniz algebra L its liezation by definition is a semisimple Lie algebra. The fundamental projection onto its liezation admits a section due to the above proposition. This leads to a key fact that is the base of the study on decomposition of cohomology of semisimple Leibniz algebras in the following sections.

Journal of Physics: Conference Series **697** (2016) 012008 doi:10.1088/1742-6596/697/1/012008

Corollary 2.6. *Let L be a finite dimensional semisimple Leibniz algebra with liezation $\mathfrak{g}$. Then $L \cong \mathfrak{g} \dotplus I$.*

A *cohomology* of a Leibniz algebra L with coefficients in the representation M are defined in [8] as follows.

Define the space of *n-cochains* $CL^n(L, M) = \mathrm{Hom}_{\mathbb{K}}(L^{\otimes n}, M)$ for $n \geq 0$ and a $\mathbb{K}$-homomorphism $\partial^n : CL^n(L, M) \to CL^{n+1}(L, M)$ by

$$\begin{aligned}(\partial^n f)(x_1, \dots, x_{n+1}) := &[x_1, f(x_2, \dots, x_{n+1})]+\\ &+\sum_{i=2}^{n+1}(-1)^i[f(x_1, \dots, \widehat{x}_i, \dots, x_{n+1}), x_i]+\\ &+\sum_{1\leq i<j\leq n+1}(-1)^{j+1}f(x_1, \dots, x_{i-1}, [x_i, x_j], x_{i+1}, \dots, \widehat{x}_j, \dots, x_{n+1}).\end{aligned}$$

This $(CL^*(L, M), \partial)$ is a cochain complex. Its n-th cohomology group is well defined by $HL^n(L, M) := ZL^n(L, M)/BL^n(L, M)$, where the elements $ZL^n(L, M) := \ker \partial^n$ and $BL^n(L, M) := \mathrm{im}\, \partial^{n-1}$ are called *n-cocycles* and *n-coboundaries*, respectively.

Recall that it is noted in [8] that

$$M^L := \{m \in M \mid [l, m] = 0, \forall l \in L\} = HL^0(L, M)$$

which is called *bisubmodule of left invariants* of M.

In Leibniz algebras a derivation is defined as usual.

Definition 2.7. *A linear map $d\colon L \to M$ is called a derivation of L in M if*

$$d([x, y]) = [d(x), y] + [x, d(y)] \quad \textit{for any } x, y \in L.$$

Space of all derivations from L to M is denoted by $\mathrm{Der}(L, M)$.

Moreover, for a given $m \in M$ a map $R_m : L \to M$ defined by $R_m(l) = [l, m]$ is a derivation. It is called an inner derivation and running through all of the elements of the bimodule we obtain space of inner derivation $\mathcal{I}\mathrm{nnder}(L, M)$.

It is known that $HL^1(L, M) = \mathrm{Der}(L, M)/\mathcal{I}\mathrm{nnder}(L, M)$.

If M is an anti-symmetric L-bimodule then inner derivations are zero and

$$HL^1(L, M) = \mathrm{Der}(L, M) = \{f : L \to M \mid f([l_1, l_2]) = [f(l_1), l_2]\}.$$

Further, in [8] considering L^a as the antisymmetric representation whose underlying $\mathbb{K}$ module is L and action $L \times L^a \to L$ is Leibniz bracket of L, it is claimed that $HL^1(L, L^a) \neq 0$. Indeed, taking an identity map $id|_L$ we see that $id \in HL^1(L, L^a)$. With similar arguments once can obtain that $HL^1(L, I) \neq 0$ where I is an L-module by the algebra bracket action.

Note that when M is symmetric L-bimodule [8] establishes

$$HL^1(L, M) = HL^1(\mathfrak{g}, M) = H^1(\mathfrak{g}, M)$$

for liezation $\mathfrak{g}$ of Leibniz algebra L.

However, similar statement for anti-symmetric M does not hold.

One of the main tools in this work relies on T. Pirashvili's result [10] on Leibniz cohomology of semisimple Lie algebras.

Theorem 2.8. *[10] Let $\mathfrak{g}$ be a finite dimensional semisimple Lie algebra and M be a Leibniz representation of $\mathfrak{g}$. Then $HL^n(\mathfrak{g}, M) = 0$ for all $n \geq 2$.*

It was proved by constructing some spectral sequences. The same result was proved by P. Ntolo [9] using Casimir element and constructing the explicit homotopy.

3. Decomposition of $CL^n(L, L)$ for a Leibniz algebra

Let V be a vector space over a field $\mathbb{K}$ and $V = V_0 \oplus V_1$ be a non-trivial decomposition. The goal of this section is to present the decomposition of $\mathrm{Hom}_{\mathbb{K}}(V^{\otimes n}, V)$ that will be heavily used in the next sections.

For a natural number n and an integer k define $\mathcal{I}(n, k)$ to be a set of bijections from $\{1, 2, \ldots, n\}$ to a set of k ones and $n-k$ zeros for $0 \leq k \leq n$ and $\mathcal{I}(n, k) = \emptyset$ otherwise.

For a bijection $\pi \in \mathcal{I}(n, k)$ and a number j equal to 0 or 1 let us denote for short $H^j_{\pi(1)\ldots\pi(n)} := \mathrm{Hom}_{\mathbb{K}}(V_{\pi(1)} \otimes \cdots \otimes V_{\pi(n)}, V_j)$.

Denote by $CL^n(V, V) = \mathrm{Hom}_{\mathbb{K}}(V^{\otimes n}, V)$ and define for $-n \leq k \leq 1$ a subspace $CL^n(V, V)_{(-k)} = \left(\underset{\pi\in\mathcal{I}(n,k)}{\oplus} H^0_{\pi(1)\ldots\pi(n)}\right) \oplus \left(\underset{\pi\in\mathcal{I}(n,k+1)}{\oplus} H^1_{\pi(1)\ldots\pi(n)}\right)$. Note that there are $\binom{n}{k} + \binom{n}{k+1} = \binom{n+1}{k+1}$ summands in the above decomposition. Moreover, for any $H^j_{\pi(1)\ldots\pi(n)}$ in this decomposition one has $k = j - (\pi(1) + \cdots + \pi(n))$.

Let us call subspace $CL^n(V, V)_{(k)}$ of *level* k. The following statement presents a decomposition isomorphic to $CL^n(V, V)$ into subspaces of all possible levels $-n, \ldots, 0, 1$.

Proposition 3.1. $CL^n(V, V) \cong \bigoplus_{k=-n}^{1} CL^n(V, V)_{(-k)}$.

Proof. We have

$$\begin{aligned}
CL^n(V, V) = {} & \mathrm{Hom}(V^{\otimes n}, V) = \mathrm{Hom}((V_0 \oplus V_1)^{\otimes n}, V_0 \oplus V_1) \\
\cong {} & \mathrm{Hom}((V_0 \oplus V_1)^{\otimes n}, V_0) \oplus \mathrm{Hom}((V_0 \oplus V_1)^{\otimes n}, V_1) \\
\cong {} & \mathrm{Hom}(V_1 \otimes V_1 \otimes \cdots \otimes V_1, V_0) \bigoplus \\
& \bigoplus \left(\underset{i\in\mathcal{I}(n,n-1)}{\oplus} H^0_{i(1)\ldots i(n)} \oplus \underset{i\in\mathcal{I}(n,n)}{\oplus} H^1_{i(1)\ldots i(n)}\right) \bigoplus \\
& \vdots \\
& \bigoplus \left(\underset{i\in\mathcal{I}(n,0)}{\oplus} H^0_{i(1)\ldots i(n)} \oplus \underset{i\in\mathcal{I}(n,1)}{\oplus} H^1_{i(1)\ldots i(n)}\right) \bigoplus \\
& \bigoplus \mathrm{Hom}(V_0 \otimes V_0 \otimes \cdots \otimes V_0, V_1),
\end{aligned}$$

or in more compact form

$$CL^n(V, V) \cong \bigoplus_{k=-1}^{n} \left(\underset{i\in\mathcal{I}(n,k)}{\oplus} H^0_{i(1)\ldots i(n)}\right) \oplus \left(\underset{i\in\mathcal{I}(n,k+1)}{\oplus} H^1_{i(1)\ldots i(n)}\right).$$

Hence, the claim is proved. □

4. Decomposition of $HL^n(L,L)$ for semisimple Leibniz algebra

Now consider a semi-simple Leibniz algebra L as an *adjoint* representation of L, i.e. L acting on itself by Leibniz algebra bracket. Due to Corollary 2.6 we have $L \cong I + \mathfrak{g}$. Setting $L_0 = \mathfrak{g}$ and $L_1 = I$ cochain spaces $CL^n(L,L)$ admits the decomposition of Proposition 3.1. However, due to $[L,I] = 0$ one can obtain more properties of the coboundary operator.

Proposition 4.1. *Coboundary operator ∂ preserves the level spaces.*

Proof. For a non-zero $\varphi \in CL^n(L,L)_{(k)}$ consider $\partial\varphi(x_1,\dots,x_n,x_{n+1})$, where exactly m of $x_1,\dots,x_{n+1}$ belong to I. It is evident that if $m < k$ or $m > k+1$ then $\partial\varphi(x_1,\dots,x_n,x_{n+1}) = 0$. We will consider the other possible cases.

Let $\varphi \in H^0_{i(1)\dots i(n)}$ for some $i \in \mathcal{I}(n,k)$. If $m = k$ then due to $\operatorname{im}\varphi \in \mathfrak{g}$ it follows that $\partial\varphi(x_1,\dots,x_n,x_{n+1}) \in \mathfrak{g}$. If $m = k+1$ then $\partial\varphi(x_1,\dots,x_n,x_{n+1})$ belongs to I if $x_1 \in I$ and zero otherwise. Hence,

$$\partial(H^0_{i(1)\dots i(n)}) \subseteq \left(\underset{j\in\mathcal{I}(n+1,k)}{\oplus} H^0_{j(1)\dots j(n+1)}\right) \oplus H^1_{1i(1)\dots i(n)}.$$

Now consider $\varphi \in H^1_{i(1)\dots i(n)}$ for some $i \in \mathcal{I}(n,k+1)$. It is obvious that $m = k$ implies $\partial\varphi(x_1,\dots,x_n,x_{n+1}) = 0$. One verifies the last possibility $m = k+1$ to yield $\partial(H^1_{i(1)\dots i(n)}) \subseteq \underset{j\in\mathcal{I}(n+1,k+1)}{\oplus} H^1_{j(1)\dots j(n+1)}$.

Thus $\partial(CL^n(L,L)_{(-k)}) \subseteq CL^{n+1}(L,L)_{(-k)}$ for all $-1 \le k \le n$. □

Let us denote by $ZL^n(L,L)_{(k)}$ the kernel of restriction of ∂ on $CL^n(L,L)_{(n)}$ and by $BL^n(L,L)_{(k)}$ the image of restriction of ∂ on $CL^{n-1}(L,L)_{(k)}$ for all $-n \le k \le 1$.

Proposition 4.2. *Let L be a finite-dimensional semisimple Leibniz algebra. Then $ZL^n(L,L)_{(-n)} = 0$ for all positive integers n.*

Proof. Let $\varphi \in ZL^n(L,L)_{(-n)}$. Then $\varphi : \otimes^n I \to \mathfrak{g}$ and $\partial\varphi(g,i_1,\dots,i_n) = [g,\varphi(i_1,\dots,i_n)] = 0$ for all $g \in \mathfrak{g}, i_1,\dots,i_n \in I$. This yields $\operatorname{im}\varphi \in Z(\mathfrak{g})$ which is zero since $\mathfrak{g}$ is semi-simple. Therefore, $\varphi = 0$ and we are done. □

Provided with Proposition 4.1 one has $BL^n(L,L)_{(k)} \subseteq ZL^n(L,L)_{(k)}$ and consider subspace $ZL^n(L,L)_{(k)}/BL^n(L,L)_{(k)}$. Let us denote it by $HL^n(L,L)_{(k)}$, for all $-n \le k \le 1$. As a consequence on Propositions 3.1 and 4.1 we have the following

Theorem 4.3. *Let L be a finite-dimensional semisimple Leibniz algebra. Then*

$$HL^n(L,L) \cong HL^n(L,L)_{(-n+1)} \oplus \cdots \oplus HL^n(L,L)_{(-1)} \oplus HL^n(L,L)_{(0)}$$

for $n \ge 2$ and $HL^1(L,L) \cong HL^1(L,L)_{(0)} \oplus HL^1(\mathfrak{g},I)$.

Journal of Physics: Conference Series **697** (2016) 012008 doi:10.1088/1742-6596/697/1/012008

Proof. Clearly, we have

$$BL^n(L,L) \cong \bigoplus_{k=-n}^{1} BL^n(L,L)_{(-k)},$$
$$ZL^n(L,L) \cong \bigoplus_{k=-n}^{1} ZL^n(L,L)_{(-k)}.$$

Note that by definition $BL^n(L,L)_{(-n)} = 0$ and together with Proposition 4.2 we obtain $HL^n(L,L)_{(-n)} = 0$.

Observe that $HL^n(L,L)_{(1)} = HL^n(\mathfrak{g}, I)$. If $n > 1$ by Theorem 2.8 it is zero. Hence, the result follows. □

5. $HL^2(L,L)$ for semisimple Leibniz algebra

Let us describe $BL^2(L,L) = BL^2(L,L)_{(-1)} \oplus BL^2(L,L)_{(0)} \oplus BL^2(L,L)_{(1)}$ first.

Proposition 5.1. *A coboundary operator ∂ acts on $CL^1(L,L)$ as follows:*

$$H_1^0 \hookrightarrow H_{10}^0 \oplus H_{01}^0 \oplus H_{11}^1, \; H_1^1 \oplus H_0^0 \to H_{10}^0 \oplus H_{00}^0, \; H_0^1 \to H_{00}^1.$$

Proof. Consider ∂d for $d \in CL^1(L,L)$. We list the only non-zero actions of ∂d on an element $(x,y) \in (L_{i(1)}, L_{i(2)})$ depending on d.

1. Let $d \in CL^1(L,L)_{(-1)} = H_1^0$. Then $\partial(H_1^0) \subseteq H_{11}^1 \oplus H_{01}^0 \oplus H_{10}^0$. Indeed,
$\partial d(x_0, y_1) = [x_0, d(y_1)] \in L_0$,
$\partial d(x_1, y_0) = [d(x_1), y_0] - d([x_1, y_0]) \in L_0$,
$\partial d(x_1, y_1) = [x_1, d(y_1)] \in L_1$.
Moreover, if $\partial d = 0$ then from $[x_0, d(y_1)] = 0$ we obtain $d = 0$, i.e. ∂ sends H_1^0 to $H_{11}^1 \oplus H_{01}^0 \oplus H_{10}^0$ injectively. In particular, this implies $\dim BL^2_{(-1)}(L,L) = \dim H_1^0 = \dim \mathrm{Hom}(L_1, L_0) = \dim L_1 \cdot \dim L_0$.

2. Let $d \in CL^1(L,L)_{(0)} = H_1^1 \oplus H_0^0$. Then we analyse separate cases.
2.1 Let $d_1 \in H_1^1$. Then $\partial(H_1^1) \subseteq H_{10}^1$. Indeed,
$\partial d_1(x_1, y_0) = [d_1(x_1), y_0] - d_1([x_1, y_0]) \in L_1$.
2.2 Let $d_0 \in H_0^0$. Then $\partial(H_0^0) \subseteq H_{00}^0 \oplus H_{10}^1$. Indeed,
$\partial d_0(x_0, y_0) = [d_0(x_0), y_0] + [x_0, d_0(y_0)] - d_0([x_0, y_0]) \in L_0$,
$\partial d_0(x_1, y_0) = [x_1, d_0(y_0)] \in L_1$.

3. Let $d \in CL^1(L,L)_{(1)} = H_0^1$. Then $\partial(H_0^1) \subseteq H_{00}^1$. Indeed, $\partial d(x_0, y_0) = [d(x_0), y_0] - d([x_0, y_0]) \subseteq L_1$. □

Now we will concentrate on $ZL^2(L,L)$. For a $\varphi \in CL^2(L,L)$ we have

$$\partial\varphi(x,y,z) = [x, \varphi(y,z)] - [\varphi(x,y), z] + [\varphi(x,z), y] + \varphi(x, [y,z]) - \varphi([x,y], z) + \varphi([x,z], y) \quad (2)$$

By Proposition 4.2 there is a decomposition $ZL^2(L,L) = ZL^2(L,L)_{(-1)} \oplus ZL^2(L,L)_{(0)} \oplus ZL^2(L,L)_{(1)}$. We have $ZL^2(L,L)_{(-1)} = \ker \partial|_{H_{10}^0 \oplus H_{01}^0 \oplus H_{11}^1}$ and proposition below shows that coboundary operator takes every component of $CL^2(L,L)_{(-1)}$ injectively.

Proposition 5.2. *A coboundary operator ∂ acts on $CL^2(L,L)_{(-1)}$ as follows*

$$\begin{array}{rcl} H^0_{01} & \hookrightarrow & H^1_{101} \oplus H^0_{010} \oplus H^0_{001}, \\ H^0_{10} & \hookrightarrow & H^1_{110} \oplus H^0_{100} \oplus H^0_{010}, \\ H^1_{11} & \hookrightarrow & H^1_{101} \oplus H^1_{110}. \end{array}$$

Proof. We have $CL^2(L,L)_{(-1)} = H^0_{10} \oplus H^0_{01} \oplus H^1_{11}$. Now let us examine how ∂ acts on each of the subspaces.

Case 1. Let $\varphi \in H^0_{01}$. Then in the system (2) the only non-zero equalities are

$$\begin{array}{rl} \partial\varphi(x_1,y_0,z_1) = & [x_1,\varphi(y_0,z_1)] \in L_1, \\ \partial\varphi(x_0,y_0,z_1) = & [x_0,\varphi(y_0,z_1)] + [\varphi(x_0,z_1),y_0] - \varphi([x_0,y_0],z_1) \in L_0, \\ \partial\varphi(x_0,y_1,z_0) = & -[\varphi(x_0,y_1),z_0] + \varphi(x_0,[y_1,z_0]) + \varphi([x_0,z_0],y_1) \in L_0. \end{array}$$

Hence, $H^0_{01} \to H^1_{101} \oplus H^0_{010} \oplus H^0_{001}$. Now if $\varphi \in \ker \partial|_{H^0_{01}}$ then $[x_1,\varphi(y_0,z_1)] = 0$ implies $\varphi(y_0,z_1) \subseteq I = L_1$ while $\operatorname{im}\varphi \subseteq L_0$.

Thus $\varphi = 0$ and $H^0_{01} \hookrightarrow H^1_{101} \oplus H^0_{010} \oplus H^0_{001}$.

Case 2. Let $\varphi \in H^0_{10}$. Similarly, we have the following equalities from system (2):

$$\begin{array}{rl} \partial\varphi(x_0,y_1,z_0) = & [x_0,\varphi(y_1,z_0)] \in L_0 \\ \partial\varphi(x_1,y_0,z_0) = & -[\varphi(x_1,y_0),z_0] + [\varphi(x_1,z_0),y_0] \\ & +\varphi(x_1,[y_0,z_0]) - \varphi([x_1,y_0],z_0) + \varphi([x_1,z_0],y_0) \in L_0 \\ \partial\varphi(x_1,y_1,z_0) = & [x_1,\varphi(y_1,z_0)] \in L_1 \end{array}$$

Hence, $H^0_{10} \to H^1_{110} \oplus H^0_{100} \oplus H^0_{010}$. Analogously as above, if $\varphi \in \ker \partial|_{H^0_{10}}$ then equality $[x_1,\varphi(y_1,z_0)] = 0$ implies $\varphi = 0$. Therefore, $H^0_{10} \hookrightarrow H^1_{110} \oplus H^0_{100} \oplus H^0_{010}$.

Case 3. Let $\varphi \in H^1_{11}$, then system (2) implies

$$\begin{array}{rl} \partial\varphi(x_1,y_0,z_1) = & [\varphi(x_1,z_1),y_0] - \varphi([x_1,y_0],z_1) \in L_1, \\ \partial\varphi(x_1,y_1,z_0) = & -[\varphi(x_1,y_1),z_0] + \varphi(x_1,[y_1,z_0]) + \varphi([x_1,z_0],y_1) \in L_1. \end{array}$$

Assuming $\varphi \in \ker \partial|_{H^1_{11}}$ from the equalities $\partial\varphi(x_1,z_0,y_1) = \partial\varphi(x_1,y_1,z_0) = 0$ we obtain $\varphi(x_1,[y_1,z_0]) = 0$. Due to $[I,\mathfrak{g}] = I$ it implies $\varphi = 0$. Hence, $H^1_{11} \hookrightarrow H^1_{101} \oplus H^1_{110}$.

□

We have $\partial(H^0_{00} \oplus H^1_{01} \oplus H^1_{10}) \subseteq H^0_{000} \oplus H^1_{100} \oplus H^1_{010} \oplus H^1_{001}$. Now let us examine how ∂ acts on each of the subspaces.

Proposition 5.3. *Coboundary operator acts on $CL^2(L,L)_{(0)}$ as follows*

$$\partial : H^0_{00} \hookrightarrow H^1_{100} \oplus H^0_{000}$$

$$\partial : H^1_{10} \to H^1_{100}$$

$$\partial : H^1_{01} \hookrightarrow H^1_{010} \oplus H^1_{001}$$

and $ZL^2(L,L)_{(0)} = \ker \partial|_{H^0_{00} \oplus H^1_{10}}$.

Proof. *Case 1.* Assume $\varphi \in H^0_{00}$, by system (2) we have

$$\begin{aligned}\partial\varphi(x_1,y_0,z_0) = & [x_1,\varphi(y_0,z_0)] \in L_1,\\ \partial\varphi(x_0,y_0,z_0) = & [x_0,\varphi(y_0,z_0)] - [\varphi(x_0,y_0),z_0] + [\varphi(x_0,z_0),y_0]\\ & +\varphi(x_0,[y_0,z_0]) - \varphi([x_0,y_0],z_0) + \varphi([x_0,z_0],y_0) \in L_0.\end{aligned}$$

Given that $\varphi \in \ker \partial|_{H^0_{00}}$ then $[x_1,\varphi(y_0,z_0)] = 0$ which implies $\varphi = 0$. Thus, $\partial : H^0_{00} \hookrightarrow H^1_{100} \oplus H^0_{000}$.

Case 2. Assuming $\varphi \in H^1_{10}$ non-zero equalities of system (2) yields

$$\begin{aligned}\partial\varphi(x_1,y_0,z_0) = & -[\varphi(x_1,y_0),z_0] + [\varphi(x_1,z_0),y_0]\\ & + \varphi(x_1,[y_0,z_0]) - \varphi([x_1,y_0],z_0) + \varphi([x_1,z_0],y_0) \subseteq L_1.\end{aligned}$$

Hence, $\partial : H^1_{10} \to H^1_{100}$.

Case 3. Let $\varphi \in H^1_{01}$. System (2) provides

$$\begin{aligned}\partial\varphi(x_0,y_1,z_0) = & -[\varphi(x_0,y_1),z_0] + \varphi(x_0,[y_1,z_0]) + \varphi([x_0,z_0],y_1) \subseteq L_1;\\ \partial\varphi(x_0,y_0,z_1) = & [\varphi(x_0,z_1),y_0] - \varphi([x_0,y_0],z_1) \in L_1.\end{aligned}$$

Then $\partial : H^1_{01} \hookrightarrow H^1_{010} \oplus H^1_{001}$ since assuming $\varphi \in \ker \partial|_{H^1_{01}}$ the last two equalities yield $[x_1,\varphi(y_0,z_0)] = 0$ which implies $\varphi = 0$.

From above it follows that

$$ZL^2(L,L)_{(0)} = \ker \partial|_{H^0_{00}\oplus H^1_{01}\oplus H^1_{10}} = \ker \partial|_{H^0_{00}\oplus H^1_{10}} \oplus \ker \partial|_{H^1_{01}} = \ker \partial|_{H^0_{00}\oplus H^1_{10}},$$

which finishes the proof. □

As a consequence of the last two propositions we obtain

Theorem 5.4. *For a finite dimensional semisimple Leibniz algebra L over $\mathbb{C}$ we have an isomorphism*

$$HL^2(L,L) \cong HL^2(L,L)_{(-1)} \oplus HL^2(L,L)_{(0)},$$

where

$$\begin{aligned}HL^2(L,L)_{(-1)} &= \ker \partial|_{H^0_{10}\oplus H^0_{01}\oplus H^1_{11}}/\partial(H^0_1) \text{ and}\\ HL^2(L,L)_{(0)} &= \ker \partial|_{H^0_{00}\oplus H^1_{10}}/\partial(H^1_1 \oplus H^0_0).\end{aligned}$$

From this point we will concentrate our study on each of the subspaces of level -1 and 0 separately.

Proposition 5.5. *Any 2-cocycle $\varphi = \varphi^0_{10} + \varphi^0_{01} + \varphi^1_{11} \in \ker \partial|_{H^0_{10}\oplus H^0_{01}\oplus H^1_{11}}$ is uniquely determined by a map $\phi \in \mathrm{Hom}(I \otimes \mathfrak{g}) \to \mathfrak{g}$ that satisfies*

$$\phi(x_1,[y_0,z_0]) = \phi([x_1,y_0],z_0) - \phi([x_1,z_0],y_0).$$

Moreover, φ is determined by the following equalities

$$\begin{cases}\varphi^0_{01}(x_0,[z_1,y_0]) = & -[x_0,\phi(z_1,y_0)]\\ \varphi^1_{11}(x_1,[z_1,y_0]) = & -[x_1,\phi(z_1,y_0)]\\ \varphi^0_{10}([x_1,y_0],z_0) = & \phi([x_1,y_0],z_0) - [\phi(x_1,y_0),z_0]\end{cases} \tag{3}$$

Proof. Let $\varphi = \varphi_{11}^1 + \varphi_{01}^0 + \varphi_{10}^0 \in \ker \partial|_{H_{11}^1 \oplus H_{01}^1 \oplus H_{10}^1} = ZL_{-1}^2$. Then $\partial\varphi = 0$ implies the following:

$$0 = \partial\varphi(x_1, y_0, z_1) = [\varphi_{11}^1(x_1, z_1), y_0] - \varphi_{11}^1([x_1, y_0], z_1) + [x_1, \varphi_{01}^0(y_0, z_1)],$$

$$0 = \partial\varphi(x_0, y_0, z_1) = [x_0, \varphi_{01}^0(y_0, z_1)] + [\varphi_{01}^0(x_0, z_1), y_0] - \varphi_{01}^0([x_0, y_0], z_1),$$

$$\begin{aligned}0 = \partial\varphi(x_0, z_1, y_0) = & -[\varphi_{01}^0(x_0, z_1), y_0] + \varphi_{01}^0(x_0, [z_1, y_0]) + \varphi_{01}^0([x_0, y_0], z_1) \\ & + [x_0, \varphi_{10}^0(z_1, y_0)],\end{aligned}$$

$$\begin{aligned}0 = \partial\varphi(x_1, y_0, z_0) = & -[\varphi_{10}^0(x_1, y_0), z_0] + [\varphi_{10}^0(x_1, z_0), y_0] \\ & + \varphi_{10}^0(x_1, [y_0, z_0]) - \varphi_{10}^0([x_1, y_0], z_0) + \varphi_{10}^0([x_1, z_0], y_0),\end{aligned}$$

$$\begin{aligned}0 = \partial\varphi(x_1, z_1, y_0) = & -[\varphi_{11}^1(x_1, z_1), y_0] + \varphi_{11}^1(x_1, [z_1, y_0]) + \varphi_{11}^1([x_1, y_0], z_1) \\ & + [x_1, \varphi_{10}^0(z_1, y_0)].\end{aligned}$$

Adding the first and the last equalities we obtain

$$\varphi_{11}^1(x_1, [z_1, y_0]) = -[x_1, \varphi_{01}^0(y_0, z_1) + \varphi_{10}^0(z_1, y_0)].$$

Adding the second and the third equalities we obtain

$$\varphi_{01}^0(x_0, [z_1, y_0]) = -[x_0, \varphi_{01}^0(y_0, z_1) + \varphi_{10}^0(z_1, y_0)].$$

Hence, these are defining relations for $\varphi = \varphi_{11}^1 + \varphi_{01}^0 + \varphi_{10}^0 \in ZL_{-1}^2$:

$$\begin{cases} 0 = & [\varphi_{11}^1(x_1, z_1), y_0] - \varphi_{11}^1([x_1, y_0], z_1) + [x_1, \varphi_{01}^0(y_0, z_1)] \\ 0 = & [x_0, \varphi_{01}^0(y_0, z_1)] + [\varphi_{01}^0(x_0, z_1), y_0] - \varphi_{01}^0([x_0, y_0], z_1) \\ \varphi_{01}^0(x_0, [z_1, y_0]) = & -[x_0, \varphi_{01}^0(y_0, z_1) + \varphi_{10}^0(z_1, y_0)] \\ 0 = & -[\varphi_{10}^0(x_1, y_0), z_0] + [\varphi_{10}^0(x_1, z_0), y_0] \\ & + \varphi_{10}^0(x_1, [y_0, z_0]) - \varphi_{10}^0([x_1, y_0], z_0) + \varphi_{10}^0([x_1, z_0], y_0) \\ \varphi_{11}^1(x_1, [z_1, y_0]) = & -[x_1, \varphi_{01}^0(y_0, z_1) + \varphi_{10}^0(z_1, y_0)]. \end{cases}$$

Taking third and last equations one can deduce the first equation using the fact that any element $z_1 \in L_1$ has a decomposition $z_1 = \sum\limits_{l \leq k \leq m} [z_k^1, z_k^0]$ for some $z_k^i \in L_i$ for $i = 0, 1$ and $1 \leq k \leq m$, $m \in \mathbb{N}$. Indeed,

$$\begin{aligned}[\varphi_{11}^1(x_1, z_1), y_0] & - \varphi_{11}^1([x_1, y_0], z_1) + [x_1, \varphi_{01}^0(y_0, z_1)] = \\ = & - \Big[[x_1, \textstyle\sum_{l \leq k \leq m} \varphi_{01}^0(z_k^0, z_k^1) + \varphi_{10}^0(z_k^1, z_k^0)], y_0\Big] \\ & + \left[[x_1, y_0], \sum_{l \leq k \leq m} \varphi_{01}^0(z_k^0, z_k^1) + \varphi_{10}^0(z_k^1, z_k^0) \right] \\ & - \left[x_1, [y_0, \sum_{l \leq k \leq m} \varphi_{01}^0(z_k^0, z_k^1) + \varphi_{10}^0(z_k^1, z_k^0)] \right] = 0.\end{aligned}$$

Similarly, we can check that third equality implies the second:

$$[x_0,\varphi^0_{01}(y_0,z_1)]+[\varphi^0_{01}(x_0,z_1),y_0]-\varphi^0_{01}([x_0,y_0],z_1)$$
$$=-[x_0,[y_0,\sum_{l\le k\le m}\varphi^0_{01}(z^0_k,z^1_k)+\varphi^0_{10}(z^1_k,z^0_k)]]$$
$$-\left[\left[x_0,\sum_{l\le k\le m}\varphi^0_{01}(z^0_k,z^1_k)+\varphi^0_{10}(z^1_k,z^0_k)\right],y_0\right]$$
$$+\left[[x_0,y_0],\sum_{l\le k\le m}\varphi^0_{01}(z^0_k,z^1_k)+\varphi^0_{10}(z^1_k,z^0_k)\right]=0.$$

Define $\phi\in\mathrm{Hom}(L_1\otimes L_0,L_0)$ by $\phi(i,g)=\varphi^0_{01}(g,i)+\varphi^0_{10}(i,g)$. Then substituting $\varphi^0_{10}(i,g)=\phi(i,g)-\varphi^0_{01}(g,i)$ into the fourth equation we obtain the following:

$$0=-\left[\phi(x_1,y_0)-\varphi^0_{01}(y_0,x_1),z_0\right]+\left[\phi(x_1,z_0)-\varphi^0_{01}(z_0,x_1),y_0\right]+$$
$$+\left(\phi(x_1,[y_0,z_0])-\varphi^0_{01}([y_0,z_0],x_1)\right)-\left(\phi([x_1,y_0],z_0)-\varphi^0_{01}(z_0,[x_1,y_0])\right)$$
$$+\left(\phi([x_1,z_0],y_0)-\varphi^0_{01}(y_0,[x_1,z_0])\right)=$$

$$=-\left[\phi(x_1,y_0),z_0\right]+\left[\phi(x_1,z_0),y_0\right]+\phi(x_1,[y_0,z_0])-\phi([x_1,y_0],z_0)$$
$$+\phi([x_1,z_0],y_0)+[\varphi^0_{01}(y_0,x_1),z_0]-\left[\varphi^0_{01}(z_0,x_1),y_0\right]-\varphi^0_{01}([y_0,z_0],x_1)$$
$$+\varphi^0_{01}(z_0,[x_1,y_0])-\varphi^0_{01}(y_0,[x_1,z_0])=$$
$$=\left(-[\phi(x_1,y_0),z_0]+\varphi^0_{01}(z_0,[x_1,y_0])\right)+\left([\phi(x_1,z_0),y_0]-\varphi^0_{01}(y_0,[x_1,z_0])\right)$$
$$+\left([\varphi^0_{01}(y_0,x_1),z_0]-\left[\varphi^0_{01}(z_0,x_1),y_0\right]-\varphi^0_{01}([y_0,z_0],x_1)\right)$$
$$+\left(\phi(x_1,[y_0,z_0])-\phi([x_1,y_0],z_0)+\phi([x_1,z_0],y_0)\right)=$$

(since the bilinear forms involved here map to the Lie algebra, we can use antisymmetry)

$$=\left([z_0,\phi(x_1,y_0)]+\varphi^0_{01}(z_0,[x_1,y_0])\right)-\left([y_0,\phi(x_1,z_0)]+\varphi^0_{01}(y_0,[x_1,z_0])\right)$$
$$+\left(\left[y_0,\varphi^0_{01}(z_0,x_1)\right]+[\varphi^0_{01}(y_0,x_1),z_0]-\varphi^0_{01}([y_0,z_0],x_1)\right)$$
$$+\left(\phi(x_1,[y_0,z_0])-\phi([x_1,y_0],z_0)+\phi([x_1,z_0],y_0)\right).$$

Now the first two expressions in parentheses are zero due to third equality. The third expression in parentheses is also zero due to second equality. Hence, we obtain

$$\phi(x_1,[y_0,z_0])-\phi([x_1,y_0],z_0)+\phi([x_1,z_0],y_0)=0.$$

Since any element $z_1\in L_1$ admits a decomposition $z_1=\sum_{l\le k\le m}[z^1_k,z^0_k]$ for some $z^i_k\in L_i$ for $i=0,1$ and $1\le k\le m$, $m\in\mathbb{N}$, we have

$$\varphi^0_{10}(z_1,x_0)=\phi(z_1,x_0)-\varphi^0_{01}(x_0,z_1)=\phi(z_1,x_0)-\sum_{l\le k\le m}\varphi^0_{01}(x_0,[z^1_k,z^0_k])$$

$$= \phi(z_1, x_0) + [x_0, \sum_{l \leq k \leq m} \phi(z_k^1, z_k^0)].$$

In particular,

$$\varphi_{10}^0([x_1, y_0], z_0) = \phi([x_1, y_0], z_0) + [z_0, \phi(x_1, y_0)] = \phi([x_1, y_0], z_0) - [\phi(x_1, y_0), z_0].$$

Summarizing, we can determine all components of $\varphi = \varphi_{11}^1 + \varphi_{01}^0 + \varphi_{10}^0 \in ZL_{-1}^2$ in terms of $\phi \in \mathrm{Hom}(L_1 \otimes L_0) \to L_0$

$$\begin{cases} \varphi_{01}^0(x_0, [z_1, y_0]) = & -[x_0, \phi(z_1, y_0)] \\ \varphi_{11}^1(x_1, [z_1, y_0]) = & -[x_1, \phi(z_1, y_0)] \\ \varphi_{10}^0([x_1, y_0], z_0) = & \phi([x_1, y_0], z_0) - [\phi(x_1, y_0), z_0] \end{cases}$$

where ϕ satisfies

$$\phi(x_1, [y_0, z_0]) = \phi([x_1, y_0], z_0) - \phi([x_1, z_0], y_0).$$

For the sake of convenience, let us re-denote ϕ by $\phi(i, g) = -\varphi_{01}^0(g, i) - \varphi_{10}^0(i, g)$. Then

$$\begin{cases} \varphi_{01}^0(x_0, [z_1, y_0]) = & [x_0, \phi(z_1, y_0)] \\ \varphi_{11}^1(x_1, [z_1, y_0]) = & [x_1, \phi(z_1, y_0)] \\ \varphi_{10}^0([x_1, y_0], z_0) = & [\phi(x_1, y_0), z_0] - \phi([x_1, y_0], z_0) \end{cases}$$

which finishes the proof of the proposition. □

It is conjectured in [2] that $HL^2(L, L) = 0$ for any semisimple Leibniz algebra L. Authors in [2] validate the claim for simple Leibniz algebra with $L/I \cong \mathfrak{sl}_2$. Armed with the proposition above in order to check if $HL^2(L, L)_{(-1)} = 0$ we arrive into an equivalent statement.

Let $\mathfrak{g}$ be a finite dimensional semisimple complex Lie algebra and I be a finite dimensional $\mathfrak{g}$-module. Denote by $i.g$ the action of $g \in \mathfrak{g}$ on $i \in I$.

Proposition 5.6. *Let $\phi \in \mathrm{Hom}(I \otimes \mathfrak{g}, \mathfrak{g})$. Then $HL^2(L, L)_{(-1)} = 0$ if and only if*

$$\phi(i, [g_1, g_2]) = \phi(i.g_1, g_2) - \phi(i.g_2, g_1) \qquad (*)$$

for all $g_1, g_2 \in \mathfrak{g}, i \in I$ yields existence of a $d \in \mathrm{Hom}(I, \mathfrak{g})$ such that $\phi(i, g) = d(i.g)$.

Proof. Recall that $BL^2(L, L)_{(-1)}$ consists of $\partial d =: \psi = \psi_{01}^0 + \psi_{11}^1 + \psi_{10}^0$, where $d \in \mathrm{Hom}(L_1, L_0)$ and $\psi_{ij}^k \in H_{ij}^k$ are given by

$$\begin{cases} \psi_{01}^0(x_0, y_1) = & [x_0, d(y_1)] \\ \psi_{11}^1(x_1, y_1) = & [x_1, d(y_1)] \\ \psi_{10}^0(x_1, y_0) = & [d(x_1), y_0] - d([x_1, y_0]) \end{cases}$$

Consider $\varphi \in ZL^2(L,L)_{(-1)}$. By Proposition 5.5 its components $\varphi = \varphi_{11}^1 + \varphi_{01}^0 + \varphi_{10}^0$ are completely determined by system of equations 3 and a map $\phi \in \mathrm{Hom}(I \otimes \mathfrak{g}, \mathfrak{g})$ that satisfies $(*)$.

Our conjecture holds, if and only $\varphi = \psi$. Consider $\varphi_{01}^0 = \psi_{01}^0$. Then $\varphi_{01}^0(g_0,[i,g]) = [g_0,\phi(i,g)] = [g_0,d([i,g])]$ which yields $[g_0,\phi(i,g) - d([i,g])] = 0$ for any $g_0, g \in \mathfrak{g}, i \in I$. Due to semi-simplicity it follows that $\phi(i,g) = d([i,g])$. Note that by construction $[i,g] = i.g$ Conversely, if $\phi(i,g) = d([i,g])$ holds then one can check easily that $\varphi = \psi$. □

Let us consider $HL^2(L,L)_{(0)}$ space. Verification if the later one is zero is not known. However, we present an equivalent conjecture in the next proposition.

Proposition 5.7. *Let L be a semisimple Leibniz algebra over $\mathbb{C}$. Statement $HL^2(L,L)_{(0)} = 0$ is valid if and only if for any map $\psi \in \mathrm{Hom}(I \otimes \mathfrak{g}, I)$ that satisfies*

$$[\psi(i,g_1),g_2] - [\psi(i,g_2),g_1] - \psi(i,[g_1,g_2]) + \psi(i.g_1,g_2) - \psi(i.g_2,g_1) = 0,$$

it follows that there exist $g_0 \in \mathfrak{g}$ and $d \in \mathrm{Hom}(I,I)$ such that

$$\psi(i,g) = i.[g_0,g] + d(i).g - d(i.g).$$

Proof. By Theorem 5.4 the space $ZL^2(L,L)_{(0)}$ consists of $\varphi = \varphi_{00}^0 + \varphi_{10}^1$, where $\varphi_{00}^0 \in H_{00}^0$ and $\varphi_{10}^1 \in H_{10}^1$. We have the following defining equalities for φ :

$$\partial\varphi_{00}^0(x_1,y_0,z_0) = [x_1,\varphi_{00}^0(y_0,z_0)];$$

$\partial\varphi_{00}^0(x_0,y_0,z_0) = [x_0,\partial\varphi_{00}^0(y_0,z_0)] - [\partial\varphi_{00}^0(x_0,y_0),z_0] + [\partial\varphi_{00}^0(x_0,z_0),y_0] + \partial\varphi_{00}^0(x_0,[y_0,z_0]) - \partial\varphi_{00}^0([x_0,y_0],z_0) + \partial\varphi_{00}^0([x_0,z_0],y_0);$

$\partial\varphi_{10}^1(x_1,y_0,z_0) = -[\varphi_{10}^1(x_1,y_0),z_0] + [\varphi_{10}^1(x_1,z_0),y_0] + \varphi_{10}^1(x_1,[y_0,z_0]) - \varphi_{10}^1([x_1,y_0],z_0) + \varphi_{10}^1([x_1,z_0],y_0)$, where elements with subscripts equal to 1 belong to $L_1 := I$, and $L_0 := \mathfrak{g}$ otherwise.

Therefore, for $\varphi = \varphi_{00}^0 + \varphi_{10}^1 \in ZL^2(L,L)_{(0)}$ it is necessary and sufficient that

(1) $[x_1,\varphi_{00}^0(y_0,z_0)] - [\varphi_{10}^1(x_1,y_0),z_0] + [\varphi_{10}^1(x_1,z_0),y_0] + \varphi_{10}^1(x_1,[y_0,z_0]) - \varphi_{10}^1([x_1,y_0],z_0) + \varphi_{10}^1([x_1,z_0],y_0) = 0,$

(2) $\varphi_{00}^0|_{H_{000}^0}$ is a Lie 2-cocycle.

Recall, that for $d = d_1 + d_0 \in H_1^1 \oplus H_0^0$ where $d_0 \in H_0^0$, $d_1 \in H_1^1$ we have $\partial(H_0^0) \subseteq H_{00}^0 \oplus H_{10}^1$ and $\partial(H_1^1) \subseteq H_{10}^1$ with

$$\partial d(x_1,y_0) = \partial d_0(x_1,y_0) + \partial d_1(x_1,y_0) = [x_1,d_0(y_0)] + [d_1(x_1),y_0] - d_1([x_1,y_0]),$$

$$\partial d(x_0,y_0) = [d_0(x_0),y_0] + [x_0,d_0(y_0)] - d_0([x_0,y_0]).$$

Since $H^2(L_0,L_0) = 0$ we have $\varphi_{00}^0 = \partial d_0|_{L_0 \otimes L_0}$ for some $d_0 \in H_0^0$.

Using Leibniz identity one can check the following equality

$$[x_1, \varphi_{00}^0(y_0, z_0)] = -\partial d(x_1, [y_0, z_0]) + \partial d([x_1, y_0], z_0) \\ - \partial d([x_1, z_0], y_0) + [\partial d([x_1, y_0]), z_0] - [\partial d([x_1, z_0]), y_0],$$

where $d = d_0 + d_1$ and $d_1 \in \mathrm{Hom}(L_1, L_1)$ is arbitrary.

Let us denote by $\psi = \varphi_{10}^1 - \partial d|_{L_1 \otimes L_0}$. Then $\psi \in \mathrm{Hom}(L_1 \otimes L_0, L_1)$ and condition (1) is equivalent to

$$[\psi(x_1, y_0), z_0] - [\psi(x_1, z_0), y_0] - \psi(x_1, [y_0, z_0]) + \psi([x_1, y_0], z_0) - \psi([x_1, z_0], y_0) = 0.$$

Observe that $\varphi_{00}^0 + \varphi_{10}^1 = \partial d_0 + \partial d_1 + \psi$ and we are done if there exist $\bar{d}_0 \in H_0^0$ and $\bar{d}_1 \in H_1^1$ such that

$$\partial \bar{d}_0|_{L_0 \otimes L_0} + \partial(\bar{d}_0 + \bar{d}_1)|_{L_1 \otimes L_0} = \partial d_0|_{L_0 \otimes L_0} + \partial(d_0 + d_1)|_{L_1 \otimes L_0} + \psi.$$

Therefore, for $\varphi_{00}^0 + \varphi_{10}^1 \in ZL^2(L, L)_{(0)}$ to be a 2-coboundary it is necessary and sufficient to the following conditions to take place:

(i) $\partial \bar{d}_0|_{L_0 \otimes L_0} = \partial d_0|_{L_0 \otimes L_0}$.

(ii) $\partial(\bar{d}_0 + \bar{d}_1)|_{L_1 \otimes L_0} = \partial(d_0 + d_1)|_{L_1 \otimes L_0} + \psi$.

Now (i) implies that $\bar{d}_0 - d_0 \in Z^1(\mathfrak{g}, \mathfrak{g}) = \mathrm{Der}(\mathfrak{g})$. Since $\mathfrak{g}$ is semisimple it is known that $\mathrm{Der}(\mathfrak{g}) = \mathcal{I}\mathrm{nnder}(\mathfrak{g})$. Therefore, there exists $g_0 \in \mathfrak{g}$ such that $(\bar{d}_0 - d_0)(g) = [g_0, g]$. Putting this into (ii) yields

$$\psi(i, g) = [i, [g_0, g]] + [(\bar{d}_1 - d_1)(i), g] - (\bar{d}_1 - d_1)([i, g]).$$

Re-denoting $d = \bar{d}_1 - d_1$ finishes the proof. □

Summarizing, results of this section we have the following conjectures for semi-simple finite-dimensional Lie algebra $\mathfrak{g}$ and its right module I that are equivalent to $HL^2(L, L)_{(-1)} = 0$ and $HL^2(L, L)_{(0)} = 0$, correspondingly.

Conjecture 1. Let $\phi \in \mathrm{Hom}(I \otimes \mathfrak{g}, \mathfrak{g})$ satisfy

$$\phi(i, [g_1, g_2]) = \phi(i.g_1, g_2) - \phi(i.g_2, g_1).$$

Then there exists $d \in \mathrm{Hom}(I, \mathfrak{g})$ such that $\phi(i, g) = d(i.g)$.

Conjecture 2. Let $\psi \in \mathrm{Hom}(I \otimes \mathfrak{g}, I)$ satisfy

$$[\psi(i, g_1), g_2] - [\psi(i, g_2), g_1] - \psi(i, [g_1, g_2]) + \psi(i.g_1, g_2) - \psi(i.g_2, g_1) = 0.$$

Then there exist $g_0 \in \mathfrak{g}$ and $d \in \mathrm{Hom}(I, I)$ such that

$$\psi(i, g) = i.[g_0, g] + d(i).g - d(i.g).$$

Journal of Physics: Conference Series **697** (2016) 012008 doi:10.1088/1742-6596/697/1/012008

6. Verification of $HL^2(L,L)_{(-1)} = 0$ for some algebras

Let $\mathfrak{g}$ be a finite dimensional semisimple complex Lie algebra and I be finite dimensional right $\mathfrak{g}$-module. Denote by $i.g$ the action of $g \in \mathfrak{g}$ on $i \in I$.

For $\phi \in \mathrm{Hom}(I \otimes \mathfrak{g}, \mathfrak{g})$ let us introduce a map $\Phi_\phi \in \mathrm{Hom}(I \otimes \mathfrak{g} \otimes \mathfrak{g}, \mathfrak{g})$ defined by

$$\Phi_\phi(i, g_1, g_2) = \phi(i, [g_1, g_2]) - \phi(i.g_1, g_2) + \phi(i.g_2, g_1).$$

Note that $\Phi_\phi(i, g_1, g_2) = -\Phi_\phi(i, g_2, g_1)$ and $\Phi_\phi(i, g, g) = 0$.

Propositions 5.6 and 5.7 claim:

- $HL^2(L,L)_{(-1)} = 0 \iff [\ \Phi_\phi = 0$ implies $\phi(i,g) = d(i.g)$ for some $d \in \mathrm{Hom}(I, \mathfrak{g})$;]
- $HL^2(L,L)_{(0)} = 0 \iff [\ \Phi_\phi = [\psi(i, g_1), g_2] - [\psi(i, g_2), g_1]$ implies that there exist $g_0 \in \mathfrak{g}$ and $d \in \mathrm{Hom}(I, I)$ such that $\psi(i,g) = i.[g_0, g] + d(i).g - d(i.g)$.]

In [2] it was verified that $HL^2(L,L)_{(-1)} = 0$ for simple Leibniz algebra with liezation $\mathfrak{sl}_2$. Below we present more general result when L is not necessarily simple Leibniz algebra with liezation $\mathfrak{sl}_2$.

Let $I = \{x_0, x_1, \dots, x_m\}$ be an irreducible right $\mathfrak{sl}_2$-module. The action is very well-known to be as follows:

$$\begin{array}{ll} x_k.e = -k(m+1-k)x_{k-1}, & k = 1, \dots, m. \\ x_k.f = x_{k+1}, & k = 0, \dots, m-1, \\ x_k.h = (m-2k)x_k & k = 0, \dots, m, \end{array}$$

and Lie algebra multiplication on $\mathfrak{sl}_2 = \{e, f, h\}$ to be

$$\begin{array}{lll} [e,h] = 2e, & [h,f] = 2f, & [e,f] = h, \\ [h,e] = -2e & [f,h] = -2f, & [f,e] = -h. \end{array}$$

Theorem 6.1. *Let L be a finite dimensional Leibniz algebra with liezation $\mathfrak{sl}_2$. Then $HL^2(L,L)_{(-1)} = 0$.*

Proof. First assume that I is irreducible $\mathfrak{sl}_2$-module. As mentioned above, we have the basis $\{x_0, x_1, \dots, x_m\}$ of I .

Define a map $d : I \to \mathfrak{g}$ by $d(x_k) = \phi(x_{k-1}, f)$ for all $1 \le k \le m$ and $d(x_0) = \frac{1}{m}\phi(x_0, h)$.

From $\Phi_\phi(x_m, f, h) = 0$ one obtains $\phi(x_m, f) = 0$ which is in accordance with $d(x_m.f) = 0$. Hence, we have $\phi(x_k, f) = d(x_k.f)$ for all $0 \le k \le m$.

Condition $\Phi_\phi(x_k, f, h) = 0$ for $0 \le k \le m-1$ simplifies to

$$-2\phi(x_k, f) = \phi(x_k, [f,h]) = \phi(x_k.f, h) - \phi(x_k.h, f) = \phi(x_{k+1}, h) - (m-2k)\phi(x_k, f)$$

which yields

$$\phi(x_{k+1}, h) = (m - 2(k+1))\phi(x_k, f) = (m - 2(k+1))d(x_{k+1}) = d(x_{k+1}.h).$$

Together with $\phi(x_0, h) = md(x_0) = d(x_0.h)$ we obtain $\phi(x_k, h) = d(x_k.h)$ for all $0 \le k \le m$.

Using $\Phi_\phi(x_k, e, f) = 0$ for $1 \le k \le m-1$ we have the following chain of equalities

$$\begin{aligned}(m-2k)d(x_k) = & \ d(x_k.h) = \phi(x_k, h) = \phi(x_k, [e, f]) \\ = & \ \phi(x_k.e, f) - \phi(x_k.f, e) = \\ = & \ -k(m+1-k)\phi(x_{k-1}, f) - \phi(x_{k+1}, e) \\ = & \ -k(m+1-k)d(x_k) - \phi(x_{k+1}, e).\end{aligned}$$

This results in $\phi(x_{k+1}, e) = -(k+1)(m-k)d(x_k) = d(x_{k+1}.e)$ for $1 \le k \le m-1$.

Now $\Phi_\phi(x_0, e, f) = 0$ yields $\phi(x_1, e) = -\phi(x_0, h) = -md(x_0) = d(x_1.e)$ and $\Phi_\phi(x_0, e, h) = 0$ results in $\phi(x_0, e) = 0$ which is the same as $d(x_0.e) = 0$. Thus we have $\phi(x_k, e) = d(x_k.e)$ for all $0 \le k \le m$.

Thus $\phi(i, g) = d(i.g)$ for all $i \in I, g \in \mathfrak{g}$ and by Theorem 5.6 in this case we obtain $HL^2(L, L)_{(-1)} = 0$.

Now assume that I is a finite dimensional $\mathfrak{sl}_2$-module. Then it is completely reducible. Let J be an irreducible submodule of I. Then for restriction of ϕ on $\mathrm{Hom}(J \otimes \mathfrak{g}, \mathfrak{g})$ by the previous construction we have a map $d_J \in \mathrm{Hom}(J, \mathfrak{g})$ such that $\phi(j, g) = d(j.g)$. Defining $d \in \mathrm{Hom}(I, \mathfrak{g})$ as a direct sum of d_J of all irreducible submodules we obtain the desired map. □

Consider a Lie algebra $\mathfrak{g} = \mathfrak{sl}_2 \oplus \mathfrak{sl}_2$ and let $I = I_1 \oplus I_2$ be a sum of two irreducible $\mathfrak{sl}_2$-modules of the same dimension (hence, isomorphic). As shown in work [4] the action of $\mathfrak{g}$ on I is as follows:

$$\begin{array}{ll} x_k.e_1 = -k(m+1-k)x_{k-1}, & k = 1, \dots, m. \\ x_k.f_1 = y_{k+1}, & k = 0, \dots, m-1, \\ x_k.h_1 = (m-2k)y_k, & k = 0, \dots, m, \\ y_k.e_1 = -k(m+1-k)y_{k-1}, & k = 1, \dots, m. \\ y_k.f_1 = y_{k+1}, & k = 0, \dots, m-1, \\ y_k.h_1 = (m-2k)y_k, & k = 0, \dots, m, \\ & \\ x_j.e_2 = y_j.h_2 = y_j & j = 0, \dots, m \\ x_j.h_2 = y_j.f_2 = -x_j & j = 0, \dots, m \end{array}$$

where $I_1 = \mathrm{Span}\{x_0, x_1, \dots, x_m\}$, $I_2 = \mathrm{Span}\{y_0, y_1, \dots, y_m\}$ and $\mathfrak{sl}_2^i = \langle e_i, f_i, h_i \rangle$ for $i = 1, 2$.

Let $L_1 = I_1 \oplus I_2 \oplus \mathfrak{sl}_2^1 \oplus \mathfrak{sl}_2^2$ be a Leibniz algebra with table of multiplication as above.

Proposition 6.2. *For Leibniz algebra L_1 we have $HL^2(L_1, L_1)_{(-1)} = 0$.*

Proof. Define a map $d : I \to \mathfrak{sl}_2^1 \oplus \mathfrak{sl}_2^2$ by $d(x_k) = \phi(x_{k-1}, f_1), d(y_k) = \phi(y_{k-1}.f_1)$ for all $1 \le k \le m$ and $d(x_0) = \frac{1}{m}\phi(x_0, h_1), d(y_0) = \frac{1}{m}\phi(y_0, h_1)$. Then by the proof of Theorem 6.1 we have $\phi(i, g) = d(i.g)$ for all $i \in I_1 \oplus I_2$ and $g \in \mathfrak{sl}_2^1$. Hence, we only need to show the desired equality when $g \in \mathfrak{sl}_2^2$.

Conditions $\Phi_\phi(x_k, f_2, h_2) = 0$ and $\Phi_\phi(y_k, e_2, h_2) = 0$ yield

$$\phi(x_k, f_2) = 0 = d(x_k.f_2), \ \ \phi(y_k, e_2) = 0 = d(y_k.e_2),$$

respectively, for $0 \leq k \leq m$.

Next conditions $\Phi_\phi(x_k, e_2, f_2) = 0$ and $\Phi_\phi(x_k, e_2, h_2) = 0$ gives us $\phi(x_k, h_2) = \phi(y_k, f_2)$ and $\phi(x_k, e_2) = \phi(y_k, h_2)$ for $0 \leq k \leq m$, respectively. Recall $x_k.h_2 = y_k.f_2$ and $x_k.e_2 = y_k.h_2$.

Using $\Phi_\phi(x_k, e_2, f_1) = 0$ one obtain

$$\phi(x_{k+1}, e_2) = \phi(y_k, f_1) = d(y_k.f_1) = d(y_{k+1}) = d(x_{k+1}.e_2)$$

for $0 \leq k \leq m-1$.

One derives equality $\phi(x_0, e_2) = \frac{1}{m}\phi(y_0, h_1) = d(y_0) = d(x_0.e_2)$ using $\Phi_\phi(x_0, e_2, h_1) = 0$.

Similarly, using $\Phi_\phi(x_k, f_1, h_2) = 0$ one obtain

$$\phi(x_{k+1}, h_2) = -\phi(x_k, f_1) = -d(x_k.f_1) = -d(x_{k+1}) = d(x_{k+1}.h_2)$$

for $0 \leq k \leq m-1$. As for the only undefined missing part one verifies $\phi(x_0, h_2) = -\frac{1}{m}\phi(x_0, h_1) = -d(x_0) = d(x_0.h_2)$ using $\Phi_\phi(x_0, h_1, h_2) = 0$.

Hence, we have proved that $\phi(i, g) = d(i.g)$ for all $i \in I_1 \oplus I_2, g \in \mathfrak{sl}_2^1 \oplus \mathfrak{sl}_2^2$ and by Theorem 5.6 in this case we obtain $HL^2(L_1, L_1)_{(-1)} = 0$. □

Acknowledgments

Authors were supported by Ministerio de Economa y Competitividad (Spain), grant MTM2013-43687-P (European FEDER support included). The third author was also supported by Xunta de Galicia, grant GRC2013-045 (European FEDER support included).

References

[1] Abdukassymova A S, Dzhumadil'daev A S 2000 Simple Leibniz algebras of rank 1 Abstract presented to the IX International Conference of the Representation Theory of Algebras pp 17–18

[2] Adashev J Q, Ladra M, Omirov B A 2015 On the second cohomology group of simple Leibniz algebras arXiv:1502.00609 pp 1–11

[3] Bloh A 1965 On a generalization of the concept of Lie algebra *Dokl. Akad. Nauk SSSR* **165** pp 471–473

[4] Camacho L M, Gómez-Vidal S, Omirov B A, Karimjanov I A 2014 Leibniz algebras whose semisimple part is related to sl_2, *Bull. Malays. Math. Sci. Soc.*(to appear)

[5] Kinyon M K, Weinstein A, 2001 Leibniz algebras, Courant algebroids, and multiplications on reductive homogeneous spaces, *Amer. J. Math.* **123**(3) pp 525–550

[6] Loday J-L 1992 Cyclic homology *Grundl. Math. Wiss. Bd.* **301** Springer-Verlag, Berlin

[7] Loday J-L 1993 Une version non commutative des algèbres de Lie: les algèbres de Leibniz *Enseign. Math. (2)* **39** pp 269–293

[8] Loday J-L, Pirashvili T 1993 Universal enveloping algebras of Leibniz algebras and (co)homology *Math. Ann.* **296** pp 139–158

[9] Ntolo P 2005 Homologie de Leibniz d'algèbres de Lie semi-simples, *C. R. Acad. Sci. Paris Sér. I Math.* **318**(8) pp 707–710

[10] Pirashvili T 1994 On Leibniz homology *Annales de l'Institute Fourier.* **44**(2) pp 401–411

[11] Rakhimov I S, Masutova K K, Omirov B A 2014 On derivations of semisimple Leibniz algebras arXiv:1412.4367 pp 1–9

[12] Turdibaev R M 2015 Cohomologies of semisimple Leibniz algebras with coefficients in the adjoint representation *Proc. of conference "Modern methods of mathematical physics and their application"* Tashkent pp 1–2

Algebra, Analysis and Quantum Probability IOP Publishing
Journal of Physics: Conference Series **697** (2016) 012009 doi:10.1088/1742-6596/697/1/012009

On Central Extensions of Associative Dialgebras

Isamiddin S.Rakhimov

Dept. of Math., Faculty of Science, and Institute for Mathematical Research (INSPEM), Universiti Putra Malaysia, Malaysia.

E-mail: risamiddin@gmail.com

Abstract. The concept of central extensions plays an important in constructing extensions of algebras. This technique has been successfully used in the classification problem of certain classes of algebras. In 1978 Skjelbred and Sund reduced the classification of nilpotent Lie algebras in a given dimension to the study of orbits under the action of automorphism group on the space of second degree cohomology of a smaller Lie algebra with coefficients in a trivial module. Then W. de Graaf applied the Skjelbred and Sund method to the classification problem of low-dimensional nilpotent Lie and associative algebras over some fields. The main purpose of this note is to establish elementary properties of central extensions of associative dialgebras and apply the above mentioned method to the classification of low dimensional nilpotent associative dialgebras.

1. Introduction

A dissociative algebra is a vector space with two bilinear binary associative operations $\vdash$, $\dashv$, satisfying certain conditions. Associative algebras are particular case of the diassociative algebras where the two operations coincide. The class of the associative dialgebras has been introduced by Loday in 1990 (see [6] and references therein). The main motivation of Loday to introduce this class of algebras was the search of an "obstruction" to the periodicity in algebraic K-theory. Later some their important relations with classical and non-commutative geometry, and physics have been discovered.

Definition 1.1. *Associative dialgebra (the term "diassociative algebra" also is used) D over a field $\mathbb{F}$ is an algebra equipped with two bilinear binary associative operations $\dashv$ and $\vdash$, called left and right products, respectively, satisfying the following axioms:*

$$\begin{aligned}(x \dashv y) \dashv z &= x \dashv (y \vdash z),\\ (x \vdash y) \dashv z &= x \vdash (y \dashv z),\\ (x \dashv y) \vdash z &= x \vdash (y \vdash z),\end{aligned} \tag{1.1}$$

for all $x, y, z \in D$.

A homomorphism from associative dialgebra $(D_1, \dashv_1, \vdash_1)$ to associative dialgebra $(D_2, \dashv_2, \vdash_2)$ is a linear map $f : D_1 \longrightarrow D_2$ satisfying

$$f(x \dashv_1 y) = f(x) \dashv_2 f(y) \text{ and } f(x \vdash_1 y) = f(x) \vdash_2 f(y), \text{ for all } x, y \in D_1.$$

The kernel and the image of a homomorphism is defined naturally. A bijective homomorphism is called isomorphism. Main problem in structural theory of algebras is the problem of classification. The classification means the description of the orbits under base change linear transformations and list representatives of the orbits.

The center of D is defined by

$$Z(D) = \{x \in D|\ x * y = y * x = 0 \text{ for all } y \in D, \text{ where } * = \dashv \text{ and } \vdash\}.$$

The natural extensions of the concepts of nilpotency and solvability of algebras to diassociative algebras case have been given in [1]. The automorphism group of D is denoted by $AutD$.

There is the following naive approach for classifying of algebra structures on a vector space which often is being used up to now. It runs as follows. Let us fix a basis of the underlying vector space, then according to the identities which the algebra satisfies we get a system of equations with respect to the structure constants of the algebra on this basis. Solving this system of equations we get a redundant, in general, list of algebras via the tables of multiplications. The second step is to make the obtained list irredundant. The irredundancy can be achieved by identifying those algebras which are obtained from others by a base change. This approach has been applied to get classifications of two and three-dimensional diassociative algebras over $\mathbb{C}$. The classification results can be found in [2], [3], [7] in low-dimensional cases and four-dimensional nilpotent case has been obtained in [1]. In this paper we propose a different approach to the classification problem of associative dialgebras based on central extensions and action of group automorphisms on the Grassmanian of subspaces of the second cohomological groups of algebra with a smaller dimension.

So far central extensions of groups and algebras mostly are extensively studied in Lie algebras case and the results successfully applied in various branches of physics. In the theory of Lie groups, Lie algebras and their representation theory, a Lie algebra extension M is an enlargement of a given Lie algebra L by another Lie algebra K. Extensions arise in several ways. There is the trivial extension obtained by taking a direct sum of two Lie algebras. Other types are the split extension and the central extension. Extensions may arise naturally, for instance, when forming a Lie algebra from projective group representations. It is proven that a finite-dimensional simple Lie algebra has only trivial central extensions. Central extensions are needed in physics, because the symmetry group of a quantized system usually is a central extension of the classical symmetry group, and in the same way the corresponding symmetry Lie algebra of the quantum system is, in general, a central extension of the classical symmetry algebra. Kac-Moody algebras have been conjectured to be a symmetry groups of a unified superstring theory. The centrally extended Lie algebras play a dominant role in quantum field theory, particularly in conformal field theory, string theory and in M-theory.

The central extensions also are applied in the process of pre-quantization, namely in the construction of prequantum bundles in geometric quantization.

Starting with a polynomial loop algebra over finite-dimensional simple Lie algebra and performing two extensions, a central extension and an extension by a derivation, one obtains a Lie algebra which is isomorphic with an untwisted affine Kac-Moody algebra. Using the centrally extended loop algebra one may construct a current algebra in two spacetime dimensions. The Virasoro algebra is the universal central extension of the Witt algebra.

Due to the Lie correspondence, the theory, and consequently the history of Lie algebra extensions, is tightly linked to the theory and history of group extensions. A systematic study of group extensions has been given by the Austrian mathematician O. Schreier in 1923. Lie algebra extensions are most interesting and useful for infinite-dimensional Lie algebras. In 1967, Victor Kac and Robert Moody independently generalized the notion of classical Lie algebras, resulting in a new theory of infinite-dimensional Lie algebras, now called Kac-Moody algebras. They generalize the finite-dimensional simple Lie algebras and can often concretely be constructed as extensions.

The organization of the paper is as follows. Section 1 is an introduction followed by Section 2 where we introduce the concept of central extension of associative dialgebras. Then we define 2-cocyles and 2-boundaries of associative dialgebras with coefficients in a trivial module in Section 3. In Section 4 we establish a close relationship between the central extensions and cocycles and in Section 5 we suggest an algorithm on application of the central extension approach to the classification problem of associative dialgebras.

2. Central Extensions of Associative dialgebras

In this section we introduce the concept of central extension of associative dialgebras.

Definition 2.1. *Let D_1, D_2 and D_3 be associative dialgebras over a field $\mathbb{F}$. The associative dialgebra D_2 is called the **extension** of D_3 by D_1 if there are associative dialgebra homomorphisms $\alpha : D_1 \longrightarrow D_2$ and $\beta : D_2 \longrightarrow D_3$ such that the following sequence*

$$0 \longrightarrow D_1 \xrightarrow{\alpha} D_2 \xrightarrow{\beta} D_3 \longrightarrow 0$$

is exact.

Definition 2.2. *An extension is called **trivial** if there exists an ideal I of D_2 complementary to $ker\ \beta$, i.e.,*

$$D_2 = ker\ \beta \oplus I \quad (\text{the direct sum of algebras}).$$

It may happen that there exist several extensions of D_3 by D_1. To classify extensions the notion of equivalent extensions is defined.

Definition 2.3. *Two sequences*

$$0 \longrightarrow D_1 \xrightarrow{\alpha} D_2 \xrightarrow{\beta} D_3 \longrightarrow 0$$

and

$$0 \longrightarrow D_1 \xrightarrow{\alpha'} D_2' \xrightarrow{\beta'} D_3 \longrightarrow 0$$

*are called **equivalent extensions** if there exists a associative dialgebra isomorphism $f : D_2 \longrightarrow D_2'$ such that*

$$f \circ \alpha = \alpha' \text{ and } \beta' \circ f = \beta.$$

The equivalence of extensions is an equivalent relation.

Definition 2.4. *An extension*

$$0 \longrightarrow D_1 \xrightarrow{\alpha} D_2 \xrightarrow{\beta} D_3 \longrightarrow 0 \tag{2.1}$$

*is called **central** if the kernel of β is contained in the center $Z(D_2)$ of D_2, i.e., $ker\ \beta \subset Z(D)$.*

From the definition above it is easy to see that in a central extension the algebra D_1 must be abelian. A central extension of an associative dialgebra D_2 by an abelian associative dialgebra D_1 can be obtained with the help of 2-cocycles on D_3.

3. Cocycles on Diassociative algebras

This section introduces the concept of 2-cocycle for diassociative algebras and gives some of simple but important properties of the 2-cocycles.

Journal of Physics: Conference Series **697** (2016) 012009 doi:10.1088/1742-6596/697/1/012009

Definition 3.1. *Let D be an associative dialgebra over a field $\mathbb{F}$ and V be a vector space over the same field. A pair $\Theta = (\theta_1, \theta_2)$ of bilinear maps $\theta_1 : D \times D \longrightarrow V$ and $\theta_2 : D \times D \longrightarrow V$ is called a* 2-**cocycle** *on D with values in V if θ_1 and θ_2 satisfy the conditions*

$$\begin{aligned} \theta_1(x \dashv y, z) &= \theta_1(x, y \dashv z) &&= \theta_1(x, y \vdash z), \\ \theta_2(x \vdash y, z) &= \theta_2(x, y \vdash z) &&= \theta_2(x \dashv y, z), \\ \theta_1(x \vdash y, z) &= \theta_2(x, y \dashv z), \end{aligned} \tag{3.1}$$

for all $x, y, z \in D$.

The set of all 2-cocycles on D with values in V is denoted by $Z^2(D, V)$. One easily sees that $Z^2(D, V)$ is a vector space if one defines the vector space operations as follows:

$$(\Theta_1 \oplus \Theta_2)(x, y) := \Theta_1(x, y) + \Theta_2(x, y) \text{ and } (\lambda \odot \Theta)(x) := \lambda\Theta(x).$$

It is easy to see that a linear combination of 2-cocycles again is a 2-cocycle. A special types of 2-cocycles given by the following lemma are called 2-**coboundaries**.

Lemma 3.1. *Let $\nu : D \longrightarrow V$ be a linear map, and define $\varphi_1(x, y) = \nu(x \dashv y)$ and $\varphi_2(x, y) = \nu(x \vdash y)$. Then $\Phi = (\varphi_1, \varphi_2)$ is a cocycle.*

Proof. Let us check the axioms (3.1) one by one.

$$\begin{aligned} \varphi_1(x \dashv y, z) &= \nu((x \dashv y) \dashv z) \\ &= \nu(x \dashv (y \dashv z)) \quad = \varphi_1(x, y \dashv z). \end{aligned}$$

$$\begin{aligned} \varphi_1(x, y \dashv z) &= \nu(x \dashv (y \dashv z)) \\ &= \nu(x \dashv (y \vdash z)) \quad = \varphi_1(x, y \vdash z). \end{aligned}$$

$$\begin{aligned} \varphi_2(x \vdash y, z) &= \nu((x \vdash y) \vdash z) \\ &= \nu(x \vdash (y \vdash z)) \quad = \varphi_2(x, y \vdash z). \end{aligned}$$

$$\begin{aligned} \varphi_2(x, y \vdash z) &= \nu(x \vdash (y \vdash z)) \\ &= \nu((x \vdash y) \vdash z) \\ &= \nu((x \dashv y) \vdash z) \quad = \varphi_2(x \dashv y, z). \end{aligned}$$

$$\begin{aligned} \varphi_1(x \vdash y, z) &= \nu((x \vdash y) \dashv z) \\ &= \nu(x \vdash (y \dashv z)) = \varphi_2(x, y \dashv z). \end{aligned}$$

□

The set of all coboundaries is denoted by $B^2(D, V)$. Clearly, $B^2(D, V)$ is a subgroup of $Z^2(D, V)$. The group $H^2(D, V) = Z^2(D, V)/B^2(D, V)$ is said to be a second cohomology group of D with values in V. Two cocycles Θ_1 and Θ_2 are said to be **cohomologous** cocycles if $\Theta_1 - \Theta_2$ is a coboundary. If we view V as a trivial D-bimodule, then $H^2(D, V)$ is an analogue of the second Hochschild-cohomology space.

Theorem 3.1. *Let D be an associative dialgebra, V a vector space,*

$$\theta_1 : D \times D \longrightarrow V \text{ and } \theta_2 : D \times D \longrightarrow V$$

be bilinear maps. Set $D_\Theta = D \oplus V$, where $\Theta = (\theta_1, \theta_2)$. For $x, y \in D$, $v, w \in V$ we define

$$(x+v) \dashv (y+w) = x \dashv y + \theta_1(x,y) \text{ and } (x+v) \vdash (y+w) = x \vdash y + \theta_2(x,y).$$

Then $(D_\Theta, \dashv, \vdash)$ is an associative dialgebra if and only if $\Theta = (\theta_1, \theta_2)$ is a 2-cocycle.

Proof. Let D_Θ be the diassociative defined above. We show that the functions θ_1, θ_2 are 2-cocycles. Indeed, according to axioms (1.1) we have

$$((x+v) \dashv (y+w)) \dashv (z+t) = (x+v) \dashv ((y+w) \dashv (z+t)) = (x+v) \dashv ((y+w) \vdash (z+t)).$$

$$((x+v) \vdash (y+w)) \vdash (z+t) = (x+v) \vdash ((y+w) \vdash (z+t)) = ((x+v) \dashv (y+w)) \vdash (z+t)).$$

$$((x+v) \vdash (y+w)) \dashv (z+t)) = (x+v) \vdash ((y+w)) \dashv (z+t)).$$

Since

$$\begin{aligned} ((x+v) \dashv (y+w)) \dashv (z+t) &= ((x \dashv y) + \theta_1(x,y)) \dashv (z+t) \\ &= ((x \dashv y) \dashv z) + \theta_1(x \dashv y, z). \end{aligned} \tag{3.2}$$

$$\begin{aligned} (x+v) \dashv ((y+w)) \dashv (z+t)) &= (x+v) \dashv ((y \dashv z) + \theta_1(y,z)) \\ &= (x \dashv (y \dashv z)) + \theta_1(x, y \dashv z). \end{aligned} \tag{3.3}$$

$$\begin{aligned} (x+v) \dashv ((y+w) \vdash (z+t)) &= (x+v) \dashv ((y \vdash z) + \theta_2(y,z) \\ &= (x \dashv (y \vdash z)) + \theta_1(x, y \vdash z). \end{aligned} \tag{3.4}$$

$$\begin{aligned} ((x+v) \vdash (y+w)) \vdash (z+t) &= ((x \vdash y) + \theta_2(x,y)) \vdash (z+t) \\ &= ((x \vdash y) \vdash z) + \theta_2(x \vdash y, z). \end{aligned} \tag{3.5}$$

$$\begin{aligned} (x+v) \vdash ((y+w)) \vdash (z+t)) &= (x+v) \vdash ((y \vdash z) + \theta_2(y,z)) \\ &= (x \vdash (y \vdash z)) + \theta_2(x, y \vdash z). \end{aligned} \tag{3.6}$$

$$\begin{aligned} ((x+v) \dashv (y+w)) \vdash (z+t) &= (x \dashv y + \theta_1(x,y)) \vdash (z+t) \\ &= (x \dashv y) \vdash z) + \theta_2(x \dashv y, z). \end{aligned} \tag{3.7}$$

$$\begin{aligned} ((x+v) \vdash (y+w)) \dashv (z+t)) &= (x \vdash y + \theta_2(x,y) \dashv (z+t) \\ &= ((x \vdash y) \dashv z)) + \theta_1(x \vdash y, z). \end{aligned} \tag{3.8}$$

$$\begin{aligned} (x+v) \vdash ((y+w)) \dashv (z+t)) &= (x+v) \vdash ((y \dashv z) + \theta_1(y,z)) \\ &= (x \vdash (y \dashv z) + \theta_2(x, y \dashv z). \end{aligned} \tag{3.9}$$

Comparing (3.2), (3.3) and (3.4) we get

$$\theta_1(x \dashv y, z) = \theta_1(x, y \dashv z) = \theta_1(x, y \vdash z).$$

Comparing (3.5), (3.6) and (3.7) we obtain

$$\theta_2(x \vdash y, z) = \theta_2(x, y \vdash z) = \theta_2(x \dashv y, z).$$

Comparing (3.8) and (3.9) we derive

$$\theta_1(x \vdash y, z) = \theta_2(x, y \dashv z).$$

□

Lemma 3.2. *Let Θ be a cocycle and Φ a coboundary. Then $D_\Theta \cong D_{\Theta+\Phi}$.*

Proof. The isomorphism

$$f : D_\Theta \longrightarrow D_{\Theta+\Phi},$$

is given by $f(x+v) = x+\nu(x)+v$. First of all we note that f is a bijective linear transformation. The linear transformation f obeys the algebraic operations $\dashv$ and $\vdash$ as well. Indeed,
$f((x+v) \dashv (y+w)) = f(x \dashv y + \theta_1(x,y))$

$$\begin{aligned}
&= f(x \dashv y) + f(\theta_1(x,y)) \\
&= x \dashv y + \nu(x \dashv y) + \theta_1(x,y) \\
&= x \dashv y + \varphi_1(x,y) + \theta_1(x,y) \\
&= x \dashv y + (\varphi_1 + \theta_1)(x,y) \\
&= ((x + (\varphi_1 + \theta_1)(x) + v)) \dashv (y + (\varphi_1 + \theta_1)(y) + w) \\
&= f(x+v) \dashv f(y+w).
\end{aligned}$$

For $\vdash$ the proof is carried out similarly. □

The proof of the following corollary is straightforward.

Corollary 3.1. *Let Θ_1, Θ_2 be cohomologous 2-cocycles on a diassociative algebra D and D_1, D_2 be the central extensions constructed with these 2-cocycles, respectively. Then the central extensions D_1 and D_2 are equivalent extensions. Particularly, a central extension defined by a coboundary is equivalent with a trivial central extension.*

Let V be a k-dimensional vector space with a basis $e_1, e_2, ..., e_k$ and $\Theta = (\theta_1, \theta_2) \in Z^2(D,V)$. Then we have $Z^2(D,V) \cong Z^2(D,\mathbb{F})^k$, $B^2(D,V) \cong B^2(D,\mathbb{F})^k$ and

$$\Theta(x,y) = (\theta_1(x,y), \theta_2(x,y)) = \left(\sum_{i=1}^{k} \theta_1^i(x,y)e_i, \sum_{i=1}^{k} \theta_2^i(x,y)e_i\right).$$

Lemma 3.3. $\Theta = (\theta_1, \theta_2) \in Z^2(D,V)$ *if and only if* $\Theta^i = (\theta_1^i, \theta_2^i) \in Z^2(D,\mathbb{F})$.

Proof. The proof is the direct verification of axioms (3.1). □

4. Central Extensions and Cocycles.

There is a close relationship between the central extensions and cocycles which is established in this section.

Theorem 4.1. *There exists one to one correspondence between elements of $H^2(D,V)$ and nonequivalents central extensions of the diassociative algebra D by V.*

Proof. The fact that for a 2-cocycle Θ the algebra D_Θ is a central extension of D has been proven early.

Let us prove the converse, i.e., suppose that D_Θ is a central extension of D :

$$0 \longrightarrow V \overset{\alpha}{\to} D_\Theta \overset{\beta}{\to} D \longrightarrow 0.$$

We construct a 2-cocycle which generates this central extension. Considering the extension with the property $im\ \alpha = ker\ \beta \subset Z(D)$ we show that one can find bilinear maps $\theta_1 : D_2 \times D_2 \longrightarrow D_1$ and $\theta_2 : D_2 \times D_2 \longrightarrow D_1$ which satisfy (3.1) for all $x, y, z \in D$ $(V = D_1)$.

In case D_1 is a one-dimensional associative dialgebra it can be identified with $\mathbb{F}$. To obtain maps satisfying equations (3.1) we consider a linear map $s : D_3 \longrightarrow D_2$ satisfying $\beta \circ s = id_{D_3}$. A map with this property is called a *section* of D_2. With the help of a section one can define a bilinear maps $\vartheta_i : D_3 \times D_3 \longrightarrow D_2$ $(i = 1, 2)$ by taking for all $x, y \in D_2$

$$\vartheta_1(x, y) := s(x \dashv y) - s(x) \dashv s(y), \tag{4.1}$$

$$\vartheta_2(x, y) := s(x \vdash y) - s(x) \vdash s(y). \tag{4.2}$$

Notice that ϑ_i, $i = 1, 2$ is identically zero if s is an associative dialgebra homomorphism. Moreover, since β is an associative dialgebra homomorphism one sees that $\beta \circ \vartheta_i = 0$, $i = 1, 2$. Hence, for all $x, y \in D_3$ we have

$$\vartheta_i(x, y) \in ker\ \beta \subset Z(D_2). \tag{4.3}$$

Using this property and the associative dialgebra axioms one readily verifies that $\Omega = (\vartheta_1, \vartheta_2)$ satisfies

$$\begin{array}{lll} \vartheta_1(x \dashv y, z) & = \vartheta_1(x, y \dashv z) & = \vartheta_1(x, y \vdash z), \\ \vartheta_2(x \vdash y, z) & = \vartheta_2(x, y \vdash z) & = \vartheta_2(x \dashv y, z), \\ \vartheta_1(x \vdash y, z) & = \vartheta_2(x, y \dashv z). & \end{array} \tag{4.4}$$

In the final step to obtain the 2-cocycle $\Theta = (\theta_1, \theta_2)$ we use the injectivity of the map $\alpha : D_1 \longrightarrow D_2$. The map $\Theta = (\theta_1, \theta_2)$, where $\theta_i : D_3 \times D_3 \longrightarrow D_1$, $i = 1, 2$ are defined by $\theta_i := \alpha^{-1} \circ \vartheta_i$, $i = 1, 2$, is the required cocycle. □

The algebra D_Θ is a (dim V-dimensional) central extension of D by V since we have the exact sequence of associative dialgebras

$$0 \longrightarrow V \longrightarrow D_\Theta \longrightarrow D \longrightarrow 0.$$

Let us now transfer the description of the orbits with respect to the "transport of structure" action of $GL(D_\Theta)$ $(dim\ D_\Theta = n)$ on the variety of dialgebras $Dias_n$ to the description of orbits of action of the $AutD$ on the Grassmanian of $H^2(D, \mathbb{F})$. The action of $\sigma \in AutD$ on $\Theta \in Z^2(D, V)$ is defined as follows:

$$\sigma(\Theta(x, y)) = (\sigma(\theta_1(x, y)), \sigma(\theta_2(x, y))), \text{ where } \sigma(\theta_i(x, y)) = \theta_i(\sigma(x), \sigma(y)),\ i = 1, 2,$$

for $x, y \in D$. Thus $AutD$ operates on $Z^2(D, V)$. It is easy to see that $B^2(D, V)$ is stabilized by this action, so that there is an induced action of $AutD$ on $H^2(D, V)$.

Theorem 4.2. *Let*

$$\Theta(x,y) = (\theta_1(x,y), \theta_2(x,y)) = \left(\sum_{i=1}^{k} \theta_1^i(x,y)e_i, \sum_{i=1}^{k} \theta_2^i(x,y)e_i\right)$$

and

$$\Omega(x,y) = (\omega_1(x,y), \omega_2(x,y)) = \left(\sum_{i=1}^{k} \omega_1^i(x,y)e_i, \sum_{i=1}^{k} \omega_2^i(x,y)e_i\right).$$

Then $D_\Theta \cong D_\Omega$ if and only if there exists $\varphi \in Aut(D)$ such that $\varphi(\omega_j^i)$ span the same subspace of $H^2(D,\mathbb{F})$ as the θ_j^i, $j = 1,2$ and $i = 1,2,...,k$.

Proof. As vector spaces $D_\Theta = D \oplus V$ and $D_\Omega = D \oplus V$. Let $\sigma : D_\Theta \longrightarrow D_\Omega$ be an isomorphism. Since V is the center of both dialgebras, we have $\sigma(V) = V$. So σ induces an isomorphism of $D_\Theta/V = D$ to $D_\Omega/V = D$, i.e., it generates an automorphim φ of D. Let $D = Span\{x_1,...,x_n\}$. Then we write $\sigma(x_i) = \varphi(x_i) + v_i$, where $v_i \in V$, and $\sigma(e_i) = \sum\limits_{j=1}^{s} a_{ji}e_j$. Also write

$$x_i \dashv x_j = \sum_{k=1}^{n} \gamma_{ij}^k x_k, \quad x_i \vdash x_j = \sum_{k=1}^{n} \delta_{ij}^k x_k,$$

and $v_i = \sum\limits_{l=1}^{s} \beta_{il}^l e_l$. Then the relations

$$\sigma(x_i \dashv_\Theta x_j) = \sigma(x_i) \dashv_\Omega \sigma(x_j) \text{ and } \sigma(x_i \vdash_\Theta x_j) = \sigma(x_i) \vdash_\Omega \sigma(x_j)$$

amount to

$$\omega_r^l(\varphi(x_i), \varphi(x_j)) = \sum_{k=1}^{s} a_{lk}\theta_r^k(x_i,x_j) + \sum_{k=1}^{n} \gamma_{ij}^k \beta_{kl}, \text{ for } 1 \le l \le s \text{ and } r = 1,2. \quad (4.5)$$

$$\omega_r^l(\varphi(x_i), \varphi(x_j)) = \sum_{k=1}^{s} a_{lk}\theta_r^k(x_i,x_j) + \sum_{k=1}^{n} \delta_{ij}^k \beta_{kl}, \text{ for } 1 \le l \le s \text{ and } r = 1,2. \quad (4.6)$$

Now define the linear function $f_l : D \longrightarrow \mathbb{F}$ by $f_l(x_k) = \beta_{kl}$. Then

$$f_l^1(x_i \dashv x_j]) = \sum_{k=1}^{n} \gamma_{ij}^k \beta_{kl} \text{ and } f_l^2(x_i \vdash x_j]) = \sum_{k=1}^{n} \delta_{ij}^k \beta_{kl}.$$

From (4.5) and (4.6) it is obvious that modulo $B^2(D,\mathbb{F})$, $\varphi(\omega_r^i)$ and θ_r^i span the same space.

Let now suppose that there exists an automorphism $\varphi \in D$ such that the cocycles $\varphi(\omega_j^i)$ and θ_j^i, $j = 1,2$ and $i = 1,2,...,k$ span the same space in $Z^2(D,\mathbb{F})$, modulo $B^2(D,\mathbb{F})$. Then there are linear functions $f_l : D \longrightarrow \mathbb{F}$ and $a_{lk} \in \mathbb{F}$ so that

$$\omega_r^l(\varphi(x_i), \varphi(x_j)) = \sum_{k=1}^{s} a_{lk}\theta_r^k(x_i,x_j) + f_l(x_i \dashv x_j) \text{ for } 1 \le l \le s \text{ and } r = 1,2. \quad (4.7)$$

$$\omega_r^l(\varphi(x_i), \varphi(x_j)) = \sum_{k=1}^{s} a_{lk}\theta_r^k(x_i,x_j) + f_l(x_i \vdash x_j) \text{ for } 1 \le l \le s \text{ and } r = 1,2. \quad (4.8)$$

If we take $\beta_{kl} = f(x_k)$ then (4.5) and (4.6) hold. Now define $\sigma : D_\Theta \longrightarrow D_\Omega$ as follows

$$\sigma(x_i) = \varphi(x_i) + \sum_{l=1}^{s} \beta_{il}^l e_l, \quad \sigma(e_i) = \sum_{j=1}^{s} a_{ji}e_j.$$

Then the σ is the required isomorphism. □

5. Application

Let D' be an associative dialgebra and $Z(D')$ be its center which we suppose to be nonzero. Set $V = Z(D')$ and $D = D'/Z(D')$. Then there is a $\Theta \in H^2(D, V)$ such that $D' = D_\Theta$. We conclude that any associative dialgebra with a nontrivial center can be obtained as a central extension of an associative dialgebra of smaller dimension. So in particular, all nilpotent dialgebras can be constructed by this way.

Procedure: Let D be an associative dialgebra of dimension $n - s$. The procedure outputs all nilpotent algebras D' of dimension n such that $D'/Z(D') = D$. It runs as follows:

(1) Determine $Z^2(D, \mathbb{F})$, $B^2(D, \mathbb{F})$ and $H^2(D, \mathbb{F})$.

(2) Determine the orbits of $Aut(D)$ on s-dimensional subspaces of $H^2(D, \mathbb{F})$.

(3) For each of the orbits let Θ be the cocycle corresponding to a representative of it, and construct D_Θ.

6. Acknowledgements

The research was supported by Grant 01-02-14-1591FR MOHE, Malaysia.

References

[1] Basri W, Rakhimov I S and Rikhsiboev I M 2015 *J. Generalized Lie Theor. Appl,* **9**
[2] Basri W, Rakhimov I S and Rikhsiboev I M 2007 *Proc. 3rd Conf. Research and Educ. in Math. (ICREM3), UPM, Malaysia* 164–170.
[3] Gonzàlez CM 2013 *Commun. in Algebra,* **41** 1903–1912.
[4] de Graaf W A 2007 *J. Algebra* **309** 640–653.
[5] de Graaf W A 2010 *arXiv: 1009.5339*
[6] Loday JL, Frabetti A, Chapoton F and Goichot F 2001 *Lecture Notes in Math.* **1763** (Springer- Berlin)
[7] Rikhsiboev I M, Rakhimov I S and Basri W 2010 *Malays. J. Math. Sci.* **4** (2) 241–254.
[8] Skjelbred T and Sund T 1978 *C. R. Acad. Sci. Paris Ser. A-B,* **286** (5)

$b-$ Bistochastic Quadratic Stochastic Operators and Their Properties

Farrukh Mukhamedov[1], **Ahmad Fadillah Embong**[2]

[1] Department of Computational & Theoretical Sciences, Faculty of Science, International Islamic University Malaysia, P.O. Box, 25200, Kuantan, Pahang, Malaysia

[2] Department of Computational & Theoretical Sciences, Faculty of Science, International Islamic University Malaysia, P.O. Box, 25200, Kuantan, Pahang, Malaysia

E-mail: farrukh_m@iium.edu.my, ahmadfadillah.90@gmail.com

Abstract. In the present paper, we consider a class of quadratic stochastic operators (q.s.o.) called b-bistochastic q.s.o. We study several descriptive properties of b- bistochastic q.s.o. It turns out that, upper triangular stochastic matrix defines a linear b-stochastic operator. This allowed us to find some sufficient conditions on cubic stochastic matrix to be a $b-$bistochastic q.s.o.

1. Introduction

The history of quadratic stochastic operators (q.s.o.) can be traced back to Bernstein's work [1] where such kind of operators appeared from the problems of population genetics (see also [7]). Such kind of operators describe time evolution of variety species in biology are represented by so-called Lotka-Volterra(LV) systems [20]. Nowadays, scientists are interested in these operators, since they have a lot of applications especially in modelings in many different fields such as biology [5, 15], physics [17, 19], economics and mathematics [7, 15, 19].

Let us recall how q.s.o. appears in biology. The time evolution of species in biology can be comprehended by the following situation. Let $I = \{1, 2, \ldots, n\}$ be the n type of species (or traits) in a population and we denote $x^{(0)} = (x_1^{(0)}, \ldots, x_n^{(0)})$ to be the probability distribution of the species in an early state of that population. By $P_{ij,k}$ we denote the probability of an individual in the ith species and j^{th} species to cross-fertilize and produce an individual from k^{th} species (trait). Given $x^{(0)} = (x_1^{(0)}, \ldots, x_n^{(0)})$, we can find the probability distribution of the first generation, $x^{(1)} = (x_1^{(1)}, \ldots, x_n^{(1)})$ by using a total probability, i.e.

$$x_k^{(1)} = \sum_{i,j=1}^{n} P_{ij,k} x_i^{(0)} x_j^{(0)}, \quad k \in \{1, \ldots, n\}.$$

This relation defines an operator which is denoted by V and it is called *quadratic stochastic operator (q.s.o.)*. In other words, each q.s.o. describes the sequence of generations in terms of probabilities distribution, if the values of $P_{ij,k}$ and the distribution of the current generation are given. The main problem in the nonlinear operator theory is to study the behavior of nonlinear operators. Presently, there is only a small number of studies on dynamical phenomena on higher dimensional systems, even though they are very important (see for example, [6, 11, 12, 13, 18]). In case of q.s.o., the difficulty of the problem depends on the given cubic matrix $(P_{ijk})_{i,j,k=1}^{m}$. In [3, 10], it has given along self-contained exposition of the recent achievements and open problems in the theory of the q.s.o.

In [16] a new majorization was introduced, and it opened a path for the study to generalize the theory of majorization by Hardy, Littlewood and Polya [4]. The new majorization has an advantage as compared to the classical one, since it can be defined as a partial order on sequences. While the classical one is not an antisymmetric order (because any sequence is majorized by any2 of its permutations), it is only defined as a preorder on sequence [16]. Most of the works in the mentioned paper were devoted to the investigation of majorized linear operators (see [4, 16]. Therefore, it is natural to study nonlinear majorized operators.

In what follows, to differentiate between the terms majorization and classical majorization that was popularized by Hardy et al.[4], we call majorization as $b-$order (which is denoted as $\leq^b$) while classical majorization as majorization (which is denoted as $\prec$) only. In [2] it was introduced and studied q.s.o. with a property $V(\mathbf{x}) \prec \mathbf{x}$ for all $\mathbf{x} \in S^{n-1}$. Such an operator is called *bistochastic*. In [9], it was proposed to a definition of bistochastic q.s.o. in terms of b- order. Namely, a q.s.o. is called $b-$*bistochastic* if $V(\mathbf{x}) \leq^b \mathbf{x}$ for all $\mathbf{x}$ taken from the $n-1$-dimensional simplex. In this paper we continue our previous investigations on b-bistochastic operators. We study several descriptive properties of b- bistochastic q.s.o. It turns out that, upper triangular stochastic matrix defines a linear b-stochastic operator. This allowed us to find some sufficient conditions on cubic stochastic matrix to be a $b-$bistochastic q.s.o.

2. Preliminaries

In this section we recall necessary definitions and facts about b-bistochastic operators.

Throughout this paper we consider the simplex

$$S^{n-1} = \left\{ \mathbf{x} = (x_1, x_2, ..., x_n) \in \mathbb{R}^n \quad |x_i \geq 0, \quad \sum_{i=1}^{n} x_i = 1 \right\}. \tag{2.1}$$

For each $k \in \{1, \ldots, n-1\}$ we define functional $\mathcal{U}_k : \mathbb{R}^n \to \mathbb{R}$ by

$$\mathcal{U}_k(x_1, \ldots, x_n) = \sum_{i=1}^{k} x_i. \tag{2.2}$$

Let $\mathbf{x}, \mathbf{y} \in S^{n-1}$. We say that $\mathbf{x}$ is *b-ordered* or *b-majorized* by $\mathbf{y}$ ($\mathbf{x} \leq^b \mathbf{y}$) if and only if $\mathcal{U}_k(\mathbf{x}) \leq \mathcal{U}_k(\mathbf{y})$, for all $k \in \{1, \ldots, n-1\}$.

The introduced relation is partial order i.e. it satisfies the following conditions:

(i) For any $\mathbf{x} \in S^{n-1}$ one has $\mathbf{x} \leq^b \mathbf{x}$,

(ii) If $\mathbf{x} \leq^b \mathbf{y}$ and $\mathbf{y} \leq^b \mathbf{x}$ then $\mathbf{x} = \mathbf{y}$,

(iii) If $\mathbf{x} \leq^b \mathbf{y}$, and $\mathbf{y} \leq^b \mathbf{z}$ then $\mathbf{x} \leq^b \mathbf{z}$.

Using the defined order, one can define the classical majorization [8]. First, recall that for any $\mathbf{x} = (x_1, x_2, \ldots, x_n) \in S^{n-1}$, by $\mathbf{x}_{[\downarrow]} = (x_{[1]}, x_{[2]}, \ldots, x_{[n]})$ one denotes the *rearrangement* of $\mathbf{x}$ in non-increasing order, i.e. $x_{[1]} \geq x_{[2]} \geq \cdots \geq x_{[n]}$. Take $\mathbf{x}, \mathbf{y} \in S^{n-1}$, then it is said that an element $\mathbf{x}$ *is majorized by* $\mathbf{y}$ and denoted $\mathbf{x} \prec \mathbf{y}$ if $\mathbf{x}_{[\downarrow]} \leq^b \mathbf{y}_{[\downarrow]}$. We refer the reader to [8] for more information regarding to this topic.

Recall that any operator V with $V(S^{n-1}) \subset S^{n-1}$ is called *stochastic*.

Definition 2.1. *A stochastic operator* V *is called* $b-$*bistochastic if one has* $V(\mathbf{x}) \leq^b \mathbf{x}$ *for all* $\mathbf{x} \in S^{n-1}$.

Note that, the simplest nonlinear operators are quadratic ones. Therefore, we restrict ourselves to such kind of operators. Namely, a stochastic operator $V : S^{n-1} \to S^{n-1}$ is called *quadratic stochastic operator (q.s.o.)* if V has the following form:

$$V(\mathbf{x})_k = \sum_{i,j=1}^{n} P_{ij,k} x_i x_j, \quad k = 1, 2, \dots, n, \quad \mathbf{x} = (x_1, x_2, \dots, x_n) \in S^{n-1}, \tag{2.3}$$

where $\{P_{ij,k}\}$ are the heredity coefficients with the following properties:

$$P_{ij,k} \geq 0, \quad P_{ij,k} = P_{ji,k}, \quad \sum_{k=1}^{n} P_{ij,k} = 1, \quad i, j, k = 1, 2, \dots, n\,. \tag{2.4}$$

Remark 2.2. *If a q.s.o. V satisfies $V(\mathbf{x}) \prec \mathbf{x}$ for all $\mathbf{x} \in S^{n-1}$, then it is called bistochastic [2]. In our definition, we are taking $b-$order instead of the majorization. Note that if one takes absolute continuity instead of the $b-$order, then $b-$bistochastic operator reduces to Volterra q.s.o. [14].*

Let V be a q.s.o., then one can define an associated matrix $\mathbb{T}_n(\mathbf{x}) = [T_{ik}(\mathbf{x})]_{i,k=1}^{n}$, $\mathbf{x} \in S^{n-1}$ by

$$T_{ik}(\mathbf{x}) = \sum_{j=1}^{n} P_{ij,k} x_j, \ \mathbf{x} = (x_i) \in S^{n-1}, \tag{2.5}$$

where $\{P_{ij,k}\}$ are the heredity coefficients. One can see that $\mathbb{T}_n(\mathbf{x})$ is a stochastic matrix. Moreover, one has $V(\mathbf{x}) = \mathbf{x}\mathbb{T}_n(\mathbf{x})$, and $\mathbb{T}_n(\mathbf{x}) = \sum\limits_{l=1}^{n} x_l \mathbb{T}_n(\mathbf{e}_l)$. Hence, each q.s.o. V can be uniquely defined by stochastic matrices, i.e.

$$V = \{\mathbb{T}_n(\mathbf{e}_1), \mathbb{T}_n(\mathbf{e}_2), \dots, \mathbb{T}_n(\mathbf{e}_n)\}\,. \tag{2.6}$$

3. Description of b-bistochastic q.s.o.

In this section, we are going to provide some general properties of $b-$bistochastic q.s.o. In [9], we have proved the following fact.

Theorem 3.1. *[9] Let V be a $b-$bistochastic q.s.o. defined on S^{n-1}, then the following statements hold:*

(i) $\sum\limits_{m=1}^{k} \sum\limits_{i,j=1}^{n} P_{ij,m} \leq kn, \ k \in \{1, \dots, n\}$

(ii) $P_{ij,k} = 0$ *for all* $i, j \in \{k+1, \dots, n\}$ *where* $k \in \{1, \dots, n-1\}$

(iii) $P_{nn,n} = 1$

(iv) for every $\mathbf{x} \in S^{n-1}$ *one has*

$$V(\mathbf{x})_k = \sum_{l=1}^{k} P_{ll,k} x_l^2 + 2 \sum_{l=1}^{k} \sum_{j=l+1}^{n} P_{lj,k} x_l x_j \quad \text{where} \quad k = \overline{1, n-1}$$

$$V(\mathbf{x})_n = x_n^2 + \sum_{l=1}^{n-1} P_{ll,n} x_l^2 + 2 \sum_{l=1}^{n-1} \sum_{j=l+1}^{n} P_{lj,n} x_l x_j.$$

(v) $P_{lj,l} \leq \frac{1}{2}$ *for all* $j \geq l+1, \ l \in \{1, \dots, n-1\}$.

This properties are only necessary conditions for a $b-$bistochastic q.s.o.. Indeed, consider the following example.

Example 3.2. *Let $V : S^3 \to S^3$ be a q.s.o given by the following heredity coefficients*

$$\left\{\begin{bmatrix} 0 & 1 & 0 & 0 \\ 0 & 1 & 0 & 0 \\ 0 & 1 & 0 & 0 \\ 0 & 0 & P_{4,1,3} & 1-P_{4,1,3} \end{bmatrix}, \begin{bmatrix} 0 & 1 & 0 & 0 \\ 0 & 1 & 0 & 0 \\ 0 & 1/2 & P_{3,2,3} & 1/2-P_{3,2,3} \\ 0 & 1/2 & P_{4,2,3} & 1/2-P_{4,2,3} \end{bmatrix},\right.$$

$$\left.\begin{bmatrix} 0 & 1 & 0 & 0 \\ 0 & 1/2 & P_{3,2,3} & 1/2-P_{3,2,3} \\ 0 & 0 & P_{3,3,3} & 1-P_{3,3,3} \\ 0 & 0 & P_{4,3,3} & 1-P_{4,3,3} \end{bmatrix}, \begin{bmatrix} 0 & 0 & P_{4,1,3} & 1-P_{4,1,3} \\ 0 & 1/2 & P_{4,2,3} & 1/2-P_{4,2,3} \\ 0 & 0 & P_{4,3,3} & 1-P_{4,3,3} \\ 0 & 0 & 0 & 1 \end{bmatrix},\right\}$$

where $2P_{4,1,3} + 2P_{3,2,3} + 2P_{4,2,3} + P_{3,3,3} + 2P_{4,3,3} \leq 4$. Note that $P_{ij,k} = P_{ji,k}$. One can see that such an operator is not a $b-$bistochastic q.s.o., since for $\mathbf{x} = (0.1, 0.1, 0.8, 0)$ we have

$$\begin{aligned} V(\mathbf{x})_1 + V(\mathbf{x})_2 &= P_{1,1,1}{x_1}^2 + 2\,P_{1,2,1}x_1x_2 + 2\,P_{1,3,1}x_1x_3 + 2\,P_{1,4,1}x_1\,(1-x_1-x_2-x_3) \\ &\quad + P_{1,1,2}{x_1}^2 + 2\,P_{1,2,2}x_1x_2 + 2\,P_{1,3,2}x_1x_3 + 2\,P_{1,4,2}x_1\,(1-x_1-x_2-x_3) \\ &\quad + P_{2,2,2}{x_2}^2 + 2\,P_{2,3,2}x_2x_3 + 2\,P_{2,4,2}x_2\,(1-x_1-x_2-x_3) \\ &= 0 + 0.28 \geq 0.1 + 0.1. \end{aligned}$$

Moreover, under the condition $2P_{4,1,3} + 2P_{3,2,3} + 2P_{4,2,3} + P_{3,3,3} + 2P_{4,3,3} \leq 4$, the property (i) in Theorem 3.1 holds. The other properties also satisfy accordingly due to the construction of the q.s.o.

In what follow, we want to recall a fascinating result on $b-$bistochastic linear operators. Let T be a linear stochastic operator $T : S^{n-1} \to S^{n-1}$ such that

$$T(\mathbf{x})_k = \sum_{i=1}^{n} t_{ik}x_i \text{ where } t_{ik} \geq 0, \sum_{k=1}^{n} t_{ik} = 1, \quad \mathbf{x} = (x_1, \dots, x_n) \in S^{n-1}. \tag{3.1}$$

In [16], it was showed the simplified form of linear $b-$bistochastic operators. Namely,

Theorem 3.3. *[16] Let T be a linear stochastic operator defined on S^{n-1}. Then T is a $b-$bistochastic if and only if $\mathbb{T}$ is an upper triangular stochastic matrix, i.e.*

$$\mathbb{T} = \begin{bmatrix} t_{11} & t_{12} & t_{13} & \dots & t_{1n} \\ 0 & t_{22} & t_{23} & \dots & t_{2n} \\ 0 & 0 & t_{33} & \dots & t_{3n} \\ \vdots & \vdots & \vdots & \ddots & \vdots \\ 0 & 0 & 0 & \dots & 1 \end{bmatrix}.$$

From this theorem, it is interesting to know the following question: if we take $\{\mathbb{T}_n(\mathbf{e}_k)\}$ (in the representation (2.6)) all the stochastic matrices to be upper triangular, then would V be a b-stochastic q.s.o.?

For each $j \in \{1, \dots, n\}$ let $\mathbb{T}_n(\mathbf{e}_j)$ be a stochastic matrix given be

$$\mathbb{T}_n(\mathbf{e}_j) = \left(a_{ik}^{(j)}\right)_{i,k=1}^{n}; \quad j = 1, 2, \dots, n. \tag{3.2}$$

We define a quadratic operator $V : S^{n-1} \to S^{n-1}$ by

$$V(\mathbf{x})_k = \sum_{i,j}^{n} a_{ik}^{(j)} x_i x_j; \quad k = \overline{1,n}, \quad \mathbf{x} = (x_1, x_2, \dots, x_n) \in S^{n-1}. \tag{3.3}$$

Let us assume that, $\mathbb{T}_n(\mathbf{e}_j)$ is taking upper triangular stochastic form i.e.

$$\mathbb{T}_n(\mathbf{e}_j) = \begin{bmatrix} a_{11}^{(j)} & a_{12}^{(j)} & a_{13}^{(j)} & \dots & a_{1n}^{(j)} \\ 0 & a_{22}^{(j)} & a_{23}^{(j)} & \dots & a_{2n}^{(j)} \\ 0 & 0 & a_{33}^{(j)} & \dots & a_{3n}^{(j)} \\ \vdots & \vdots & \vdots & \ddots & \vdots \\ 0 & 0 & 0 & \dots & 1 \end{bmatrix} \tag{3.4}$$

for every $j = 1, 2, \dots, n$.

Theorem 3.4. *Let $\{\mathbb{T}_n(\mathbf{e}_j)\}$, $j = 1, 2, \dots, n$ be a collection of upper triangular stochastic matrices and V be the associated q.s.o. defined by (3.3). Then V is a b-bistochastic operator.*

Proof. Let $\mathbf{x} = (x_1, x_2, \dots, x_n)$. Using the fact

$$V(\mathbf{x})_k = \sum_{i,j=1}^{n} a_{ik}^{(j)} x_i x_j = \sum_{i=1}^{k}\sum_{j=1}^{n} a_{ik}^{(j)} x_i x_j$$

one gets

$$\begin{aligned} \sum_{l=1}^{k} V(\mathbf{x})_l &= \sum_{j=1}^{n} a_{11}^{(j)} x_1 x_j + \sum_{i=1}^{2}\sum_{j=1}^{n} a_{i2}^{(j)} x_i x_j + \dots + \sum_{i=1}^{k}\sum_{j=1}^{n} a_{ik}^{(j)} x_i x_j \\ &= x_1 \sum_{j=1}^{n} a_{11}^{(j)} x_j + x_1 \sum_{j=1}^{n} a_{12}^{(j)} x_j + x_2 \sum_{j=1}^{n} a_{22}^{(j)} x_j + \dots + \\ &\quad x_1 \sum_{j=1}^{n} a_{1k}^{(j)} x_j + x_2 \sum_{j=1}^{n} a_{2k}^{(j)} x_j + \dots + x_k \sum_{j=1}^{n} a_{kk}^{(j)} x_j \\ &= x_1 \left(\sum_{l=1}^{k}\sum_{j=1}^{n} a_{1l}^{(j)} x_j \right) + x_2 \left(\sum_{l=2}^{k}\sum_{j=1}^{n} a_{2l}^{(j)} x_j \right) + \dots + x_k \sum_{j=1}^{n} a_{kk}^{(j)} x_j \\ &= \sum_{i=1}^{k} x_i \left(\sum_{j=1}^{n}\sum_{l=i}^{k} a_{il}^{(j)} x_j \right). \end{aligned}$$

We know that $\sum_{l=i}^{k} a_{il}^{(j)} \leq 1$ for each i, hence

$$\sum_{l=1}^{k} V(x)_l \leq \sum_{i=1}^{k} x_i \left(\sum_{j=1}^{n} x_j \right) = \sum_{i=1}^{k} x_i$$

This completes the prove. □

Remark 3.5. *It is worth to note, in general the necessary and sufficient conditions of b−bistochastic quadratic operators may not need to be upper triangular stochastic matrices. Let us choose the following quadratic operators (indeed it is a q.s.o.):*

$$\left\{ \begin{bmatrix} 0 & 0 & 1 \\ \frac{1}{2} & 0 & \frac{1}{2} \\ 0 & 0 & 1 \end{bmatrix}, \begin{bmatrix} \frac{1}{2} & 0 & \frac{1}{2} \\ 0 & 0 & 1 \\ 0 & 0 & 1 \end{bmatrix}, \begin{bmatrix} 0 & 0 & 1 \\ 0 & 0 & 1 \\ 0 & 0 & 1 \end{bmatrix} \right\}. \tag{3.5}$$

Let $\mathbf{x} = (x_1, x_2, x_3) \in S^2$*, then one has* $V(\mathbf{x})_1 = x_1x_2$ *and* $V(\mathbf{x})_2 = 0$*. In addition, the statement* $x_2 \leq 1$ *and* $x_1x_2 \leq x_1$ *is always true for any* $\mathbf{x} = (x_1, x_2, x_3) \in S^2$*. Therefore, one finds that*

$$V(\mathbf{x})_1 = x_1x_2 \leq x_1, \quad V(\mathbf{x})_1 + V(\mathbf{x})_2 = x_1x_2 \leq x_1 + x_2.$$

which shows q.s.o. given by (3.5) *is a b−bistochastic q.s.o..*

Furthermore, due to Theorem 3.1, we introduce a stochastic cubic matrix such that

$$\left\{ \begin{bmatrix} a_{11}^{(1)} & a_{12}^{(1)} & a_{13}^{(1)} & \cdots & a_{1n}^{(1)} \\ a_{21}^{(1)} & a_{22}^{(1)} & a_{23}^{(1)} & \cdots & a_{2n}^{(1)} \\ a_{31}^{(1)} & a_{32}^{(1)} & a_{33}^{(1)} & \cdots & a_{3n}^{(1)} \\ \vdots & \vdots & \vdots & \ddots & \vdots \\ a_{n1}^{(1)} & a_{n2}^{(1)} & a_{n3}^{(1)} & \cdots & a_{nn}^{(1)} \end{bmatrix}, \begin{bmatrix} a_{11}^{(2)} & a_{12}^{(2)} & a_{13}^{(2)} & \cdots & a_{1n}^{(2)} \\ 0 & a_{22}^{(2)} & a_{23}^{(2)} & \cdots & a_{2n}^{(2)} \\ 0 & a_{32}^{(2)} & a_{33}^{(2)} & \cdots & a_{3n}^{(2)} \\ \vdots & \vdots & \vdots & \ddots & \vdots \\ 0 & a_{n2}^{(2)} & a_{n3}^{(2)} & \cdots & a_{nn}^{(2)} \end{bmatrix}, \right.$$

$$\left. \begin{bmatrix} a_{11}^{(3)} & a_{12}^{(3)} & a_{13}^{(3)} & \cdots & a_{1n}^{(3)} \\ 0 & a_{22}^{(3)} & a_{23}^{(3)} & \cdots & a_{2n}^{(2)} \\ 0 & 0 & a_{33}^{(3)} & \cdots & a_{3n}^{(3)} \\ \vdots & \vdots & \vdots & \ddots & \vdots \\ 0 & 0 & a_{n3}^{(3)} & \cdots & a_{nn}^{(3)} \end{bmatrix}, \dots, \begin{bmatrix} a_{11}^{(n)} & a_{12}^{(n)} & a_{13}^{(n)} & \cdots & a_{1n}^{(n)} \\ 0 & a_{22}^{(n)} & a_{23}^{(n)} & \cdots & a_{2n}^{(n)} \\ 0 & 0 & a_{33}^{(n)} & \cdots & a_{3n}^{(n)} \\ \vdots & \vdots & \vdots & \ddots & \vdots \\ 0 & 0 & 0 & \cdots & 1 \end{bmatrix} \right\}$$

In short,

$$\mathbb{T}_n(\mathbf{e}_j) = \left(a_{ik}^{(j)}\right)_{i,k=1}^{n}, \quad j = 1, 2, \dots n$$

where

$$a_{i,k}^{(j)} = 0; \ i, j = \overline{k+1, n}, \quad k = \overline{1, n-1} \tag{3.6}$$

Note that, under the constraints given by (3.6), we shall use Theorem 3.1 which reduces the components of vector $V(x)_k$ to

$$V(\mathbf{x})_k = \sum_{l=1}^{k} a_{l,k}^{(l)} x_l^2 + \sum_{l=1}^{k} \sum_{j=l+1}^{n} \left(a_{l,k}^{(j)} + a_{j,k}^{(l)}\right) x_l x_j \quad \text{where } k = \overline{1, n-1} \tag{3.7}$$

$$V(\mathbf{x})_n = x_n^2 + \sum_{l=1}^{n-1} a_{l,n}^{(l)} x_l^2 + \sum_{l=1}^{n-1} \sum_{j=l+1}^{n} \left(a_{l,n}^{(j)} + a_{j,n}^{(l)}\right) x_l x_j \tag{3.8}$$

Theorem 3.6. *Let* V *be a quadratic operator given by* (3.3) *and let the corresponding collection* $\{\mathbb{T}_n(\mathbf{e}_j)\}$ *satisfy the constraints* (3.6)*. If*

$$\sum_{l=1}^{n-1} \left(a_{i,l}^{(j)} + a_{j,l}^{(i)}\right) \leq 1 \quad \text{for every } j = \overline{i+1, n} \quad \text{where } i = \overline{1, n-1}, \tag{3.9}$$

then V *is a b-bistochastic operator.*

Proof. Clearly from (3.7) one gets

$$\begin{aligned}
V(\mathbf{x})_1 &= a_{1,1}^{(1)}x_1^2 + \sum_{j=2}^{n}\left(a_{1,1}^{(j)} + a_{j,1}^{(1)}\right)x_1x_j \\
V(\mathbf{x})_2 &= a_{1,2}^{(1)}x_1^2 + a_{2,2}^{(1)}x_2^2 + \sum_{j=2}^{n}\left(a_{1,2}^{(j)} + a_{j,2}^{(1)}\right)x_1x_j + \sum_{j=3}^{n}\left(a_{2,2}^{(j)} + a_{j,2}^{(2)}\right)x_2x_j \\
&\vdots \\
V(\mathbf{x})_k &= a_{1,k}^{(1)}x_1^2 + \cdots + a_{k,k}^{(k)}x_k^2 + \sum_{j=2}^{n}\left(a_{1,k}^{(j)} + a_{j,k}^{(1)}\right)x_1x_j + \cdots + \\
&+ \sum_{j=k+1}^{n}\left(a_{k,k}^{(j)} + a_{j,k}^{(k)}\right)x_kx_j
\end{aligned}$$

Therefore, one finds that

$$\begin{aligned}
\sum_{l=1}^{k} V(\mathbf{x})_l &= x_1\left(x_1\left(\sum_{l=1}^{k} a_{1,l}^{(1)}\right) + \sum_{j=2}^{n}\left(\sum_{l=1}^{k}\left(a_{1,l}^{(j)} + a_{j,l}^{(1)}\right)\right)x_j\right) + \\
&\quad x_2\left(x_2\left(\sum_{l=2}^{k} a_{2,l}^{(2)}\right) + \sum_{j=3}^{n}\left(\sum_{l=2}^{k}\left(a_{2,l}^{(j)} + a_{j,l}^{(2)}\right)\right)x_j\right) + \cdots + \\
&\quad x_k\left(x_k\left(a_{k,k}^{(k)}\right) + \sum_{j=k+1}^{n}\left(a_{k,k}^{(j)} + a_{j,k}^{(k)}\right)x_j\right),
\end{aligned}$$

for any $k = 1, 2, \ldots, n-1$.

Due to stochasticity, then we know $\sum_{l=i}^{k} a_{i,l}^{(i)} \le 1$ for every $i = \overline{1,k}$. Moreover, based on the conditions (3.9), we have

$$x_t\left(x_t\left(\sum_{l=t}^{k} a_{t,l}^{(t)}\right) + \sum_{j=t+1}^{n}\left(\sum_{l=t}^{k}\left(a_{t,l}^{(j)} + a_{j,l}^{(t)}\right)\right)x_j\right) \le x_t\left(x_t + \sum_{j=2}^{n} x_j\right)$$

for any $t \in \{1, \ldots, k\}$. By virtue of the last inequality, it will imply that

$$\begin{aligned}
\sum_{l=1}^{k} V(\mathbf{x})_l &\le x_1\left(x_1 + \sum_{j=2}^{n} x_j\right) + x_2\left(x_2 + \sum_{j=3}^{n} x_j\right) + \cdots + x_k\left(x_k + \sum_{j=k+1}^{n} x_j\right) \\
&\le \sum_{l=1}^{k} x_l
\end{aligned}$$

This completes the proof. □

Remind that, the reverse is not true. For instance, a quadratic operator (again it is a q.s.o.)

$$\left\{\begin{bmatrix} 0 & 0 & 1 \\ \frac{1}{2} & \frac{1}{2} & 0 \\ 0 & 0 & 1 \end{bmatrix}, \begin{bmatrix} \frac{1}{2} & \frac{1}{2} & 0 \\ 0 & 0 & 1 \\ 0 & 0 & 1 \end{bmatrix}, \begin{bmatrix} 0 & 0 & 1 \\ 0 & 0 & 1 \\ 0 & 0 & 1 \end{bmatrix}\right\}, \tag{3.10}$$

which is a $b-$bistochastic operator i.e.

$$V(\mathbf{x})_1 = x_1x_2 \leq x_1 \text{ and } V(\mathbf{x})_1 + V(\mathbf{x})_2 = x_1x_2 + x_1x_2 \leq x_1 + x_2.$$

One can see that if $i = 1$ and $j = 2$, then $2a_{1,1}^{(2)} + 2a_{2,2}^{(1)} = 2 > 1$.

References

[1] Bernstein S 1942 *Ann. Math. Statistics* **13** 53–61
[2] Ganikhodzhaev R N 1993 *Russ. Math. Surv.* **48** 244–246.
[3] Ganikhodzhaev R, Mukhamedov F and Rozikov U 2011 *Infin. Dimens. Anal. Quant. Prob. Relat. Top.* **14** 279–335
[4] Hardy G H, Littlewood JE and Polya G 1952 *Inequalities* (Cambridge: Cambridge Univ. Press)
[5] Hofbauer J and Sigmund K 1998 *Evolutionary Games and Population Dynamics* (Cambridge: Cambridge Univ. Press)
[6] Kesten H 1972 *Adv. Appl. Probab.* **2** 1–82
[7] Lyubich Yu I 1992 *Mathematical structures in population genetics* (Berlin: Springer)
[8] Marshall A W, Olkin I and Arnold B.C 2011 *Inequalities: Theory of Majorization and its Applications* (Berlin: Springer)
[9] Mukhamedov F and Embong A F 2015 *J. Inequal. Appl.* **2015** 226
[10] Mukhamedov F and Ganikhodjaev N 2015 *Lect. Notes Math.* **2133** (Berlin: Springer)
[11] Mukhamedov F, Qaralleh I and Rozali W N F A W 2014 *Sains Malaysiana* **43** 1275–1281
[12] Mukhamedov F, Saburov M and Qaralleh I 2013 *Abst. Appl. Anal.* **2013** 942038
[13] Mukhamedov F and Supar N A 2015 *Bull. Malays. Math. Sci. Soc.* **38** 1281–1296
[14] Mukhamedov F and Taha M H 2015 *AIP Conf. Proc.* **1660** 050057
[15] Narendra S G, Samaresh C M and Elliott W M 1971 *Rev. Mod. Phys.* **43** 231–276
[16] Parker D S and Ram P (1996) *Greed and majorization* (US: University of California)
[17] Plank M and Losert V 1995 *J. Math. Phys.* **36** 3520-3543
[18] Rozikov U A and Zhamilov U U 2009 *Sb. Math.* **200** 1339–1351
[19] Takeuchi Y 1996 *Global dynamicsal properties of Lotka-Volterra systems* (Singapore: World Scientific)
[20] Volterra V 1927 *ssociation Franc. Lyon* **1926** 96-98

Dynamical systems generated by quadratic homeomorphisms of 2D simplex [1]

Rasul Ganikhodzhaev[1], **Abror Pirnapasov**[2],

[1,2] Faculty of Mechanics and Mathematics, National University of Uzbekistan, 100174, Tashkent, Uzbekistan

E-mail: rganikhodzhaev@gmail.com, pirnapasov_imc@mail.ru

Abstract. In this paper we analyze a general form of homeomorphisms of the two-dimensional simplex and study the asymptotical behavior of their trajectories.

1. Introduction

The notion of quadratic stochastic operators (QSO) was introduced by Bernstein [2], and the theory of QSO has been developed for more than 90 years. In recent years, there has been a considerable growth of interest in this theory due to its numerous applications to problems in mathematics, biology, and physics ([4],[6],[11]).

The quadratic stochastic operator (QSO) is a mapping of the simplex

$$S^{m-1} = \left\{ x = (x_1, ..., x_m) \in \mathbf{R}^m : x_i \geq 0, \sum_{i=1}^{m} x_i = 1 \right\}$$

into itself, of the form

$$V : x'_k = \sum_{i,j=1}^{m} P_{ij,k} x_i x_j, \quad k = 1, ..., m, \tag{1.1}$$

where $P_{ij,k}$ are coefficients of heredity and

$$P_{ij,k} \geq 0, \quad P_{ij,k} = P_{ji,k}, \quad \sum_{k=1}^{m} P_{ij,k} = 1, (i,j,k = 1, ..., m). \tag{1.2}$$

Thus, each QSO V can be uniquely defined by a cubic matrix $P = (P_{ij,k})_{i,j,k=1}^{m}$ with conditions (1.2). Note that each element $x \in S^{m-1}$ is a probability distribution on $E = \{1, ..., m\}$.

Let $x^0 = (x_1^0, ..., x_m^0)$ be a probability distribution (where $x_i^0 = P(i)$ is the probability of i, $i = 1, 2, ..., m$) of species in the initial generation, and $P_{ij,k}$ is the probability that individuals in the $i-$th and $j-$th species interbred to produce an individual k.

In this paper, we consider a model of free population, i.e. there is no difference of sex and in any generation, the parents ij are independent, i.e. $P(i,j) = P(i)P(j) = x_i x_j$. Then the

[1] To the memory of our teacher T. A. Sarymsakov on the occasion of his 100th birthday.

probability distribution $x' = (x'_1, ..., x'_m)$ (the state) of the species in the first generation can be found by the total probability

$$x'_k = \sum_{i,j=1}^{m} P(k|i,j)P(i)P(j) = \sum_{i,j=1}^{m} P_{ij,k}x_i^0x_j^0, \quad k = 1, ..., m. \tag{1.3}$$

This means that the correspondence $x^0 \to x'$ defines a map V called the evolution operator. The population evolves by starting from an arbitrary state x^0 , then passing to the state $x' = V(x^0)$ (in the next generation), after to the state $x'' = V(V(x^0))$, and so on. Thus, the states of the population are described by the following discrete-time dynamical system

$$x^{(0)}, x^{(1)} = V(x^{(0)}), x^{(2)} = V^2(x^{(0)}), ... \tag{1.4}$$

where $V^n(x) = V(V^{n-1}(x))$ denotes the n times iteration of V to x.

For a given $x^{(0)} \in S^{m-1}$ the trajectory (orbit) $x^{(n)}, n = 0, 1, 2, ...$ of $x^{(0)}$ under action of QSO (1.1) is defined by $x^{(n+1)} = V(x^{(n)})$, where $n = 0, 1, 2, ...$

One of the main problems in mathematical biology consists in the study of the asymptotical behavior of the trajectories. The difficulty of the problem depends on the given matrix P. An asymptotic behavior of the QSO even on the small dimensional simplex is complicated. In order to solve this problem, many researchers always introduced a certain class of QSO and studied their behavior (see for example, [8, 12, 13, 16, 17]).In this paper we shall consider several particular cases of P for which the above-mentioned problem is (particularly) solved.

A QSO given by (1.1),(1,2) is called Volterra if one has

$$P_{ij,k} = 0, \quad \text{if } k \in \{i, j\}, \forall i, j, k \in E. \tag{1.5}$$

The biological treatment of condition (1.5) is clear: the offspring repeats the genotype of one of its parents. In [5], the general form of Volterra QSO is given by

$$x'_k = x_k \left(1 + \sum_{i=1}^{m} a_{ki}x_i\right), \quad k \in E, \tag{1.6}$$

where

$$a_{ki} = 2P_{ik,k} - 1, \quad \text{for } i \neq k \text{ and } a_{ii} = 0, i \in E. \tag{1.7}$$

Moreover, $a_{ki} = -a_{ik}$ and $|a_{ki}| \leq 1$.

Denote by $A = (a_{ij})_{i,j=1}^m$ the skew-symmetric matrix with entries (1.7). Note that the operator (1.6) is a discretization of the Volterra model which describes interacting, competing species in population. Such a model has attracted considerable attention in the fields of biology, ecology, mathematics . It is known from [5], that Volterra operators are quadratic homeomorphism of the simplex.

Theorem 1.1. *[5] Let V be a quadratic homeomorphism of the simplex. Then there is a Volterra opetator V_0 and a permutation π of $1, 2, ..., n$ such that $V = V_0 \circ \pi$.*

Corollary 1.2. *[5] The set of all quadratic homeomorphisms of the simplex S^{m-1} can be geometrically presented as the union of $m!$ disjoint cubes of dimension $\frac{m(m-1)}{2}$.*

Remark 1.3. *In [1, 15] quadratic homeomorphisms of the simplex are characterized as orthogonal preserving QSO.*

2. Main result

In this section, we consider the case $m = 3$ and study the dynamical systems of quadratic homeomorphisms of the two-dimensional simplex. In [15] these quadratic homeomorphisms classified into three disjoint classes. Namely, let π be a permutation of the set $\{1, 2, 3\}$. One has the following three possibilities $\pi = id$, $\pi^2 = id$ and $\pi^3 = id$. The cases $\pi = id$ and $\pi^3 = id$, respectively, define two disjoint classes, which were studied in [8], [7], [17]. The remaining class corresponds to $\pi^2 = id$.

In this paper, we study the mentioned class. It is easy to see that there are three possibilities

$$\pi_3 = \begin{pmatrix} 1 & 2 & 3 \\ 2 & 1 & 3 \end{pmatrix}, \quad \pi_4 = \begin{pmatrix} 1 & 2 & 3 \\ 3 & 2 & 1 \end{pmatrix} \quad \text{and} \quad \pi_5 = \begin{pmatrix} 1 & 2 & 3 \\ 1 & 3 & 2 \end{pmatrix}.$$

Let us consider the case π_3 (the cases π_4 and π_5 can be considered in a similar way). By $T_{\pi_3} : S^2 \to S^2$ we denote the corresponding linear transformation defined by $T_{\pi_3}(x_1, x_2, x_3) = (x_2, x_1, x_3)$.

In [6] the dynamics of the following operator

$$V : \begin{cases} x_1' = x_1(1 + ax_2 - bx_3) \\ x_2' = x_2(1 + cx_3 - ax_1) \\ x_3' = x_3(1 + bx_1 - cx_2) \end{cases}$$

has been studied.

Let us define

$$V_{\pi_3} = V \circ T_{\pi_3} : \begin{cases} x_1' = x_2(1 + ax_1 - bx_3) \\ x_2' = x_1(1 + cx_3 - ax_2) \\ x_3' = x_3(1 + bx_2 - cx_1) \end{cases}, \tag{2.1}$$

where $-1 \le a, b, c \le 1, abc \neq 0$.

Remark 2.1. *From the followings*

$$V_{\pi_3}^2 : \begin{cases} x_1' = x_1(1 + cx_3 - ax_2)(1 + ax_2 - bx_3 - a^2x_1x_2 - b^2x_1x_3 + (ac - bc)x_2x_3) \\ x_2' = x_2(1 + ax_1 - bx_3)(1 + cx_3 - ax_1 - a^2x_1x_2 - (bc - ac)x_1x_3 - c^2x_2x_3) \\ x_3' = x_3(1 + bx_2 - cx_1)(1 + bx_1 - cx_2 - (ab + ac)x_1x_2 + bcx_1x_3 + bcx_2x_3) \end{cases} \tag{2.2}$$

and

$$V^2 : \begin{cases} x_1' = x_1(1 + ax_2 - bx_3)(1 + ax_2 - bx_3 - a^2x_1x_2 - b^2x_1x_3 + (ac + bc)x_2x_3) \\ x_2' = x_2(1 + cx_3 - ax_1)(1 + cx_3 - ax_1 - a^2x_1x_2 - (bc + ab)x_1x_3 - c^2x_2x_3) \\ x_3' = x_3(1 + bx_1 - cx_2)(1 + bx_1 - cx_2 - c^2x_2x_3 - b^2x_1x_3 + (ab + ac)x_1x_2) \end{cases}$$

we obtain that the operator $V_{\pi_3}^2$ is not equal to operator V^2.

Recall that a point $x \in S^{m-1}$ is called a *fixed point* of V if $V(x) = x$. The set of all fixed points of V is denoted by $Fix(V)$.

Let $D_xV(a) = (\frac{\partial V_i}{\partial x_j})(a)$ be the Jacobian of V at the point $a \in S^2$. A fixed point a is called *hyperbolic* if the Jacobian $D_xV(a)$ at a has no eigenvalues on the unit circle. A hyperbolic fixed point a is called: *attracting* if all the eigenvalues of the Jacobian $D_xV(a)$ are in the unit ball; *repelling* if all the eigenvalues of the Jacobian $D_xV(a)$ are outside the unit ball; a *saddle* otherwise.

We denote one dimensional face $\Gamma_{i,j} = \{x = (x_1, ..., x_m) \in S^{m-1} : x_i + x_j = 1\}$ and $int\Gamma_{i,j} = \{x = (x_1, ..., x_m) \in S^{m-1} : x_i + x_j = 1,\ x_i x_j \neq 0\}$.

We have

$$D_x V_{\pi_3} = \begin{pmatrix} ax_2 & 1 + ax_1 - bx_3 & -bx_2 \\ 1 + cx_3 - ax_2 & -ax_1 & cx_1 \\ -cx_3 & bx_3 & 1 + bx_2 - cx_1 \end{pmatrix}.$$

Definition 2.2. *[3] If $f : I \to I$ is a C^3 map and $Df(x) \neq 0$, the Schwarzian derivative of f at x is*

$$(Sf)(x) = \frac{D^3 f(x)}{Df(x)} - \frac{3}{2}\left(\frac{D^2 f(x)}{Df(x)}\right)^2, \tag{2.3}$$

where $Df(x)$ is derivation of $f(x)$.

Denote $e_1 = (1, 0, 0)$, $e_2 = (0, 1, 0)$ and $e_3 = (0, 0, 1)$.

Definition 2.3. *A curve γ is called an invariant curve, if $x \in \gamma$ implies that $V^i(x) \in \gamma$, for all $i = 1, 2, ...$.*

It is evident that the face Γ_{12} is invariant set with respect to V_{π_3}.

For the sake of simplicity, we denote $W = V_{\pi_3}^2$. First we study dynamics of W on the invariant face Γ_{12}.

Theorem 2.4. *The operator W has a unique fixed point $x^3 = (x_1^3, x_2^3, 0)$ belonging to $int\Gamma_{12}$. If $x \in \{x = (x_1, x_2, 0) : x_1 \in [0, x_1^3)\}$ then $\lim_{n\to\infty} W^{(n)}(x) = e_1$ and if $x \in \{x = (x_1, x_2, 0) : x_1 \in (x_1^3, 1]\}$ then $\lim_{n\to\infty} W^{(n)}(x) = e_2$.*

Proof. Since Γ_{12} is invariant subset of the operator (2.1), then the restriction of (2.1) on Γ_{12} has the following form

$$V_{\pi_3}(x_0) : \begin{cases} x_1' = x_2(1 + ax_1), \\ x_2' = x_1(1 - ax_2). \end{cases} \tag{2.4}$$

Due to $x_1 + x_2 = 1$ one has $x_1' = x_2(1 + ax_1) = (1 - x_1)(1 + ax_1)$.

Using notation $x = x_1$, we have $x' = g(x)$, where $g(x) = (1 - x)(1 + ax)$, $x \in [0, 1]$.

Simple algebra gives that

$$x_1^3 = \begin{cases} \frac{a-2+\sqrt{a^2+4}}{2a} & if \quad a > 0 \\ \frac{a-2-\sqrt{a^2+4}}{2a} & if \quad a < 0 \end{cases}$$

is a fixed point of $g(x)$ and $|Dg(x_1^3)| > 1$, i.e. x_1^3 is a repelling fixed point of $g(x)$. One can see that $D(g(x)) < 0$ for any $x \in (0, 1)$ and $g(x)$ decreasing function in $[0, 1]$.

It is easy to check that $g^2(0) = 0$ and $g^2(1) = 1$, besides $x = 0$ and $x = 1$ are attracting fixed points of $g^2(x)$. From (2.3) one has that $(Sg)(x) < 0$ and $(Sg^{(2)})(x) < 0$.

From Singer's Theorem [18] we infer that if $x \in \{x = (x_1, x_2, 0) : x_1 \in [0, x_1^3)\}$ then $\lim_{n\to\infty} W^{(n)}(x) = e_1$ and if $x \in \{x = (x_1, x_2, 0) : x_1 \in (x_1^3, 1]\}$ then $\lim_{n\to\infty} W^{(n)}(x) = e_2$. □

Note that the operator W has the following form

$$x_i' = x_i(1 + f_i(x)), \quad i = 1, 2, 3$$

where $f_1(x),\ f_2(x),\ f_3(x)$ are some polynomials.

Note that this kind of operators are called *Lotka-Volterra* operators. Some properties of such kind operators have been studied in [14].

Denote

$$P = \left\{x \in S^2 : f_i(x) \geq 0,\ i = 1, 2, 3\right\}, \quad Q = \left\{x \in S^2 : f_i(x) \leq 0,\ i = 1, 2, 3\right\}.$$

Theorem 2.5. *[9] For any $x \in P$ and $y \in Q$ there exists an invariant curve γ connecting these two points.*

Next theorem gives full dynamics of the operator W in $S^2\backslash\Gamma_{12}$.

Theorem 2.6. *The following statements hold:*

(i) If $(1-b)(1+c) < 1$ and $(1+b)(1-c) > 1$, then for every $x \in S^2\backslash\Gamma_{12}$ one has $\lim_{n\to\infty} W^{(n)}(x) = e_3$;

(ii) If $(1-b)(1+c) > 1$ and $(1+b)(1-c) < 1$, then there exist an invariant curve γ which passes from e_3 and the fixed point x^3. This curve divides the simplex into two open parts such that e_1 and e_2 vertices are situated in different parts, respectively. If x belongs to the part of simplex where the vertex e_1 is located, then $\lim_{n\to\infty} W^{(n)}(x) = e_1$, if $x \in \gamma$ then $\lim_{n\to\infty} W^{(n)}(x) = x^3$, otherwise one has $\lim_{n\to\infty} W^{(n)}(x) = e_2$.

(iii) If $(1-b)(1+c) < 1$ and $(1+b)(1-c) < 1$ then there are $x^1 \in int\Gamma_{23}$, $x^2 \in int\Gamma_{13}$ and $x^3 \in int\Gamma_{12}$ fixed points of W. There are three invariant curves which pass over the fixed points. These curves divide the simplex into three open parts and vertices of the simplex are situated in each parts of the simplex, respectively. If x belongs to a containing the vertex part, then the trajectory of x converges to the corresponding vertex.

Remark 2.7. $(1-b)(1+c)(1+b)(1-c) = (1-b^2)(1-c^2) < 1$, *because* $(1-b)(1+c) \geq 1$ *and* $(1+b)(1-c) \geq 1$ *inequalities does not hold.*

Proof. From (2.2) we have

$$W = \begin{cases} x_1' = x_1(1 + f_1(x)) \\ x_2' = x_2(1 + f_2(x)) \\ x_3' = x_3(1 + f_3(x)) \end{cases}$$

where f_1, f_2, f_3 are some polynomials.

One can see that e_1 and e_2 are fixed points of W. The eigenvalues of D_xW at points e_1 and e_2 are equal to $|\lambda_1| = 1$, $|\lambda_2| = 1 - a^2 < 1$, $|\lambda_3| = (1-c)(1+b)$, and eigenvalues of D_xW at e_3 are $|\lambda_1| = |\lambda_2| = \sqrt{(1-b)(1+c)}$ and $|\lambda_3| = 1$.

Let $x \in \Gamma_{13}$ then the restriction of W to Γ_{13} has the following form

$$W(x) : \begin{cases} x_1' = f(k(x_1)), \\ x_3' = 1 - f(k(x_1)) \end{cases}$$

where $f(x) = (1-x)(1-bx)$ and $k(x) = (1-x)(1-cx)$.

Using notion $x = x_1$ we have $x' = g_1(x)$, where $g_1(x) = f(k(x))$, $x \in [0, 1]$. From (2.3) one find $(Sg_1)(x) < 0$.

It is easy to check that

$$\begin{cases} |Dg_1(0)| < 1 \; and \; |Dg_1(1)| > 1 \quad if \;\; (1-b)(1+c) < 1 \; and \; (1+b)(1-c) > 1 \\ |Dg_1(0)| < 1 \; and \; |Dg_1(1)| > 1 \quad if \;\; (1-b)(1+c) > 1 \; and \; (1+b)(1-c) < 1 \\ |Dg_1(0)| < 1 \; and \; |Dg_1(1)| > 1 \quad if \;\; (1-b)(1+c) < 1 \; and \; (1+b)(1-c) < 1 \end{cases}. \tag{2.5}$$

(i) From $(1-b)(1+c) < 1$ we have $|\lambda_1| = |\lambda_2| < 1$. Therefore e_3 is an attracting fixed point of W.

From (2.5), we infer that $x = 0$ is attracting, and $x = 1$ is repelling points of $g_1(x)$.

From Singer's theorem for $x \in (0,1)$, it follows $\lim\limits_{n\to\infty} g_1(x) = 0$. Therefore if $x \in int\Gamma_{13}$ we have $\lim\limits_{n\to\infty} W^{(n)}(x) = e_3$.

Similarly, for Γ_{23} one can prove that $\lim\limits_{n\to\infty} W^{(n)}(x) = e_3$.

From Lemma 2.2 of [10] we obtain

$$\lim_{n\to\infty} W^{(n)}(x) = e_3$$

for all $x \in S^2 \backslash \Gamma_{12}$.

(ii) Let $(1-b)(1+c) > 1$ and $(1+b)(1-c) < 1$. Then e_1 and e_2 are attracting fixed points, and e_3 is repelling fixed point for W.

Note that $(x^3, e_3) \in P \times Q$. There is an invariant curve γ which passes over two given fixed points.

It's easy to see that x^3 is a saddle point for operator W. From Grobman-Hartman's theorem it follows that from x^3 one passes only two invariant curves: the fist curve is Γ_{12} and the second one is γ. So γ is a unique invariant curve.

From (2.5) we find that $x = 1$ is attracting and $x = 0$ is repelling points of $g_1(x)$.

From Singer's Theorem we infer that if $x \in (0,1)$, then $\lim\limits_{n\to\infty} g_1(x) = 1$. Therefore, if $x \in \Gamma_{13}$, then $\lim\limits_{n\to\infty} W^{(n)}(x) = e_1$. Similarly, for Γ_{23} one can prove that $\lim\limits_{n\to\infty} W^{(n)}(x) = e_2$.

If follows from Lemma 2.2 of [10] that if x belongs to the part of simplex where the vertex e_1 is located, then $\lim\limits_{n\to\infty} W^{(n)}(x) = e_1$, if $x \in \gamma$ then $\lim\limits_{n\to\infty} W^{(n)}(x) = x^3$. Otherwise, we obtain $\lim\limits_{n\to\infty} W^{(n)}(x) = e_2$.

(iii) From $(1-b)(1+c) < 1$ and $(1+b)(1-c) < 1$, one has that e_1, e_2 and e_3 are attracting fixed points of W.

It follows from (2.5) that $x = 1$ and $x = 0$ are attracting points of $g_1(x)$.

The Singer's Theorem yields that there exists another fixed point $x_1^2 \in (0,1)$ of $g_1(x)$.

Moreover, if $x \in [0, x_1^2)$, then $\lim\limits_{n\to\infty} g_1(x) = 0$, and if $x \in (x_1^2, 1]$, then $\lim\limits_{n\to\infty} g_1(x) = 1$. Therefore, if $x \in \{x = (x_1, 0, x_3) : x_1 \in [0, x_1^2)\}$ then $\lim\limits_{n\to\infty} W^{(n)}(x) = e_1$. If $x \in \{x = (x_1, 0, x_3) : x_1 \in (x_1^2, 1]\}$ then $\lim\limits_{n\to\infty} W^{(n)}(x) = e_3$.

Similarly, for Γ_{23} one can prove that there exists a fixed point $x^1 = (0, x_2^1, x_3^1)$. Moreover, if $x \in \{x = (x_1, 0, x_3) : x_1 \in [0, x_2^1)\}$ then $\lim\limits_{n\to\infty} W^{(n)}(x) = e_2$, and if $x \in \{x = (x_1, 0, x_3) : x_1 \in (x_2^1, 1]\}$ then $\lim\limits_{n\to\infty} W^{(n)}(x) = e_3$.

For Γ_{12}, the claim of theorem was proved in the proof of Theorem 2.5.

In the case (iii) there is a fixed point

$$x_0 = \left(\frac{b-c+bc}{c(a+b+c)}, \frac{b-c+bc}{b(a+b+c)}, 1 - \frac{b-c+bc}{c(a+b+c)} \frac{b-c+bc}{b(a+b+c)} \right) \in intS^2$$

It is easy to check that x_0 is repelling fixed point.

Note that (x^1, x_0), (x^2, x_0), (x^3, x_0) are in $P \times Q$. Consequently there are invariant curves each of which passes over two given fixed points. Since x_0 is a repelling point, the points x^1, x^2 and x^3 are saddle points, these invariant curves divide the simplex into three parts and the boundaries of the parts are invariant in the corresponding part.

Let us consider the part containing the vertex e_1. Since e_1 is an attracting point by Lemma 2.2 of [10] one has that for an arbitrary initial point takes from the part where e_1 is located, then the trajectory of W converges to e_1. This completes the proof. □

Remark 2.8. *If we know trajectory of operator W than we can know the trajectory of the operator V_{π_3}.*

Acknowledgements
The second author thanks to Professor U. Rozikov and Dr. U. Jamilov for useful discussions.We thank the referee for careful reading of the manuscript and for useful suggestions.

References
[1] Akin H and Mukhamedov F 2015 *AIP Conf. Proc.* **1676** 020008.
[2] Bernstein S 1924 *Uchn. Zapiski NI Kaf. Ukr. Otd. Mat.* 83-115.
[3] Devaney R 2003 *An introduction to chaotic Dynamical system.* (Westview press).
[4] Lyubich Yu 1992 *Mathematical structures in population genetics.* (Berlin: Springer).
[5] Ganikhodzhaev R and Eshmamatova D 2006 *Vladikavkaz Math. Jour.* **8** 12-28.
[6] Ganikhodzhaev R, Mukhamedov F and Rozikov U 2011 *Inf. Dim. Analysis, Quantum Probab. Related Topics* **2** 279-335.
[7] Ganikhodzhaev R and Pirnapasov A 2016 *Contem. Math. AMS Proc.* (to appear)
[8] Ganikhodjaev N and Zanin D 2004 *Russ Math Surv* **3** 571-572.
[9] Hartman P 1964 *Ordinary differential equations.* (New York-London-Sydney: John Wiley).
[10] Franks J and Handel M 2009 *Geometry and Topology* **13** 87-98.
[11] Mukhamedov F and Ganikhodjaev N 2015 *Lect. Notes. Math.* **2133** (Berlin: Springer).
[12] Mukhamedov F, Qaralleh I and Rozali W N F A W 2014 *Sains Malaysiana* **43** 1275–1281.
[13] Mukhamedov F, Saburov M and Qaralleh I 2013 *Abst. Appl. Anal.* Article ID 942038.
[14] Mukhamedov F and Saburov M 2015 *Bull. Malays. Math. Sci. Soc.* **37** 59-64.
[15] Mukhamedov F and Taha M H M 2015 *AIP Conf. Proc.* **1660** 050057.
[16] Rozikov U A and Zada A 2010 *Inter. Journal Biomath.* **3** 143–159.
[17] Rozikov U and Zhamilov U 2009 *Sbornik. Math.* **9** 81-94.
[18] Singer D. 1968 *SIAM J. Appl. Math.* **35** 260-267.

On Quadratic Stochastic Operators Having Three Fixed Points

Mansoor Saburov[1], **Nur Atikah Yusof**[2]

[1,2]Department of Computational & Theoretical Sciences, Faculty of Science, International Islamic University Malaysia, P.O. Box, 25200, Kuantan, Pahang, MALAYSIA

E-mail: msaburov@iium.edu.my, msaburov@gmail.com, y.nuratikah@gmail.com

Abstract. We knew that a trajectory of a linear stochastic operator associated with a positive square stochastic matrix starting from any initial point from the simplex converges to a unique fixed point. However, in general, the similar result for a quadratic stochastic operator associated with a positive cubic stochastic matrix does not hold true. In this paper, we provide an example for the quadratic stochastic operator with positive coefficients in which its trajectory may converge to different fixed points depending on initial points.

1. Introduction

Let $\|\mathbf{x}\|_1 = \sum\limits_{k=1}^{m} |x_k|$ be a norm of a vector $\mathbf{x} = (x_1, \cdots, x_m) \in \mathbb{R}^m$. We say that $\mathbf{x} \geq 0$ (resp. $\mathbf{x} > 0$) if $x_k \geq 0$ (resp. $x_k > 0$) for all $k = \overline{1,m}$. Let $\mathbb{S}^{m-1} = \{\mathbf{x} \in \mathbb{R}^m : \|\mathbf{x}\|_1 = 1, \ \mathbf{x} \geq 0\}$ be the $(m-1)-$dimensional standard simplex. An element of the simplex $\mathbb{S}^{m-1}$ is called *a stochastic vector*. Recall that a square matrix $\mathbb{P} = (p_{ij})_{i,j=1}^m$ is called *stochastic* if every row is a stochastic vector. A square stochastic matrix $\mathbb{P} = (p_{ij})_{i,j=1}^m$ is called *positive* if $p_{ij} > 0, \ \forall \ i,j = \overline{1,m}$. The classical Perron-Frobenius theorem states that any positive square stochastic matrix has a unique fixed point in the simplex. However, in general, the similar result for higher dimensional hyper-matrices (tensors) does not hold true. In this paper, we discuss this problem for cubic stochastic matrices.

A cubic matrix $\mathcal{P} = (p_{ijk})_{i,j,k=1}^m$ is called *stochastic* if $\sum\limits_{k=1}^{m} p_{ijk} = 1, \ p_{ijk} \geq 0, \ \forall i,j,k = \overline{1,m}$. Every cubic stochastic matrix is associated with a quadratic stochastic operator $\mathcal{Q} : \mathbb{S}^{m-1} \to \mathbb{S}^{m-1}$ as follows

$$(\mathcal{Q}(\mathbf{x}))_k = \sum_{i,j=1}^{m} x_i x_j p_{ijk}, \quad \forall \ k = \overline{1,m}. \tag{1.1}$$

By being the simplest nonlinear mapping, a quadratic stochastic operator has an incredible application in population genetics [1, 2, 3, 6, 8], control systems [18, 19]. In population genetics, the quadratic stochastic operator describes a distribution of the next generation of the system if the current distribution is given [8, 21]. In this sense, the quadratic stochastic operator is a primary source for investigations of evolution of population genetics. The detailed exposure of the theory of quadratic stochastic operators is presented in [4, 5],[10]-[17].

A cubic stochastic matrix $\mathcal{P} = (p_{ijk})_{i,j,k=1}^m$ is said to be *positive* (written $\mathcal{P} > 0$) if $p_{ijk} > 0$, $\forall i,j,k = \overline{1,m}$. A quadratic stochastic operator associated with a positive cubic stochastic matrix is called *positive*. Let $\mathbf{Fix}(\mathcal{Q}) = \{\mathbf{x} \in \mathbb{S}^{m-1} : \mathcal{Q}(\mathbf{x}) = \mathbf{x}\}$ be a fixed point set. Due to Brouwer's theorem, $\mathbf{Fix}(\mathcal{Q}) \neq \emptyset$. Meanwhile, a fixed point of the quadratic operator is an equilibrium for the system. In contrast to the linear case, the fixed point set of the quadratic operator is sophisticated. In general, if $\mathcal{P} > 0$ then it is not necessary to be true that $|\mathbf{Fix}(\mathcal{Q})| = 1$. The first attempt to give an example for such kind of quadratic operators was done by A. A. Krapivin [7] and Yu. I. Lyubich [8]. However, it turns out that their examples are wrong. In fact, we shall show that Krapivin's example as well as Lyubich's example has a unique fixed point in the simplex. At the same time, we shall also provide an example for the quadratic operator associated with the positive cubic stochastic matrix which has three fixed points in the simplex.

2. Krapivin's Example

A. A. Krapivin has considered the following quadratic operator $V_\varepsilon : \mathbb{S}^2 \to \mathbb{S}^2$, $V_\varepsilon(\mathbf{x}) = \mathbf{x}' = (x_1', x_2', x_3')$ in his paper [7]

$$V_\varepsilon : \begin{cases} x_1' = (1-4\varepsilon)x_1^2 + 2\varepsilon x_2^2 + 10\varepsilon x_3^2 + 4\varepsilon x_1x_2 + (1+4\varepsilon)x_1x_3 + 8\varepsilon x_2x_3 \\ x_2' = 2\varepsilon x_1^2 + (1-3\varepsilon)x_2^2 + \varepsilon x_3^2 + (\frac{1}{2}+2\varepsilon)x_1x_2 + 2\varepsilon x_1x_3 + (1+8\varepsilon)x_2x_3 \\ x_3' = 2\varepsilon x_1^2 + \varepsilon x_2^2 + (1-11\varepsilon)x_3^2 + (\frac{3}{2}-6\varepsilon)x_1x_2 + (1-6\varepsilon)x_1x_3 + (1-16\varepsilon)x_2x_3 \end{cases}$$

A.A. Krapivin claimed [7] that the quadratic operator $V_\varepsilon : \mathbb{S}^2 \to \mathbb{S}^2$ has two fixed points on the line segment $L = \left\{\left(\frac{t}{2}, \frac{1-t}{2}, \frac{1}{2}\right)\right\}_{0\le t\le 1}$ where $0 < \varepsilon < \frac{1}{100}$. However, this claim is wrong.

Proposition 2.1. *The quadratic operator V_ε does not have any fixed point on L.*

Proof. We search for a fixed point $\mathbf{x}_0 = (\frac{t}{2}, \frac{1-t}{2}, \frac{1}{2})$ on the line segment L. We should have that $V_\varepsilon(\mathbf{x}_0) = \mathbf{x}_0$. After some algebraic calculations, we obtain the following system of equations

$$\begin{cases} (1-6\varepsilon)t^2 - (1+4\varepsilon)t + 20\varepsilon = 0 \\ (1-6\varepsilon)t^2 - (1-4\varepsilon)t + 12\varepsilon = 0 \\ (1-6\varepsilon)t^2 - \left(1+\frac{4}{3}\varepsilon\right)t + \frac{52}{3}\varepsilon = 0 \end{cases} . \qquad (2.1)$$

If we take a difference of the first two equations of the system (2.1) we then get that $8\varepsilon t = 8\varepsilon$ or $t = 1$. However, if we substitute $t = 1$ into the third equation in the system (2.1) we then have that $10\varepsilon = 0$ which contradicts to the condition $0 < \varepsilon < \frac{1}{100}$. Therefore, the system (2.1) does not have any solution. This completes the proof. □

Now, we are aiming to prove that $|\mathbf{Fix}(V_\varepsilon)| = 1$.
It is clear that $V_\varepsilon(\mathbb{S}^2) \subset \mathrm{int}\mathbb{S}^2 = \{\mathbf{x} \in \mathbb{S}^2 : x_1x_2x_3 > 0\}$. Hence, $\mathbf{Fix}(V_\varepsilon) \subset \mathrm{int}\mathbb{S}^2$.
In order to find all fixed points, we have to solve the system of equations

$$\begin{cases} x_1 = (1-4\varepsilon)x_1^2 + 2\varepsilon x_2^2 + 10\varepsilon x_3^2 + 4\varepsilon x_1x_2 + (1+4\varepsilon)x_1x_3 + 8\varepsilon x_2x_3 \\ x_2 = 2\varepsilon x_1^2 + (1-3\varepsilon)x_2^2 + \varepsilon x_3^2 + (\frac{1}{2}+2\varepsilon)x_1x_2 + 2\varepsilon x_1x_3 + (1+8\varepsilon)x_2x_3 \\ x_3 = 2\varepsilon x_1^2 + \varepsilon x_2^2 + (1-11\varepsilon)x_3^2 + (\frac{3}{2}-6\varepsilon)x_1x_2 + (1-6\varepsilon)x_1x_3 + (1-16\varepsilon)x_2x_3 \end{cases}$$

Proposition 2.2. *One has that $\xi_1 \neq \eta_1$, $\xi_2 \neq \eta_2$, $\xi_3 \neq \eta_3$ for $\xi, \eta \in$ **Fix**(V_ε) and $\xi \neq \eta$.*

Proof. Let ξ, η be two distinct solutions of the system given above (if any). Since $x_3 = 1-x_1-x_2$, we can rewrite the system in terms of x_1 and x_2 as follows

$$\begin{cases} 2\varepsilon x_1^2 + 4\varepsilon x_2^2 + (12\varepsilon - 1)x_1x_2 - 16\varepsilon x_1 - 12\varepsilon x_2 + 10\varepsilon = 0 \\ \varepsilon x_1^2 - 10\varepsilon x_2^2 - (\frac{1}{2} + 6\varepsilon)x_1x_2 + 6\varepsilon x_2 + \varepsilon = 0 \end{cases} \tag{2.2}$$

Case $\xi_1 \neq \eta_1$ Our aim is to show that $\xi_1 \neq \eta_1$. We suppose the contrary that is $\xi_1 = \eta_1$. Since $\xi \neq \eta$, we must have that $\xi_2 \neq \eta_2$. This means that for $x_1 = \xi_1 = \eta_1$, the following two quadratic equations (with respect to x_2)

$$4\varepsilon x_2^2 + ((12\varepsilon - 1)\xi_1 - 12\varepsilon)x_2 + (2\xi_1^2 - 16\xi_1 + 10)\varepsilon = 0,$$
$$-10\varepsilon x_2^2 + \left[-\left(\frac{1}{2} + 6\varepsilon\right)\xi_1 + 6\varepsilon\right]x_2 + (\xi_1^2 + 1)\varepsilon = 0$$

must have two distinct common roots ξ_2 and η_2. Consequently, we should have that

$$\frac{4\varepsilon}{-10\varepsilon} = \frac{(12\varepsilon - 1)\xi_1 - 12\varepsilon}{-\left(\frac{1}{2} + 6\varepsilon\right)\xi_1 + 6\varepsilon} = \frac{2\xi_1^2 - 16\xi_1 + 10}{\xi_1^2 + 1}.$$

It follows from the first equality that $\xi_1 = \frac{8\varepsilon}{8\varepsilon - 1}$. Since $0 < \varepsilon < \frac{1}{100}$, we get that $\xi_1 < 0$ which is a contradiction.

Case $\xi_2 \neq \eta_2$ Our aim is to show that $\xi_2 \neq \eta_2$. We again suppose the contrary that is $\xi_2 = \eta_2$. Since $\xi \neq \eta$, we must have that $\xi_1 \neq \eta_1$. It follows from (2.2) that for $x_2 = \xi_2 = \eta_2$, the following two quadratic equations (with respect to x_1)

$$2\varepsilon x_1^2 + ((12\varepsilon - 1)\xi_2 - 16\varepsilon)x_1 + (4\xi_2^2 - 12\xi_2 + 10)\varepsilon = 0,$$
$$\varepsilon x_1^2 - \left(\frac{1}{2} + 6\varepsilon\right)\xi_2 x_1 + (-10\xi_2^2 + 6\xi_2 + 1)\varepsilon = 0$$

must have two distinct common roots ξ_1 and η_1. Consequently, we should have that

$$\frac{2\varepsilon}{\varepsilon} = \frac{(12\varepsilon - 1)\xi_2 - 16\varepsilon}{-\left(\frac{1}{2} + 6\varepsilon\right)\xi_2} = \frac{4\xi_2^2 - 12\xi_2 + 10}{-10\xi_2^2 + 6\xi_2 + 1}.$$

It follows from the first equality that $\xi_2 = \frac{2}{3}$. By substituting $\xi_2 = \frac{2}{3}$ into the last fraction, we get that $\frac{4\xi_2^2 - 12\xi_2 + 10}{-10\xi_2^2 + 6\xi_2 + 1} = \frac{34}{5} \neq 2 = \frac{2\varepsilon}{\varepsilon}$. This is again contradiction.

Case $\xi_3 \neq \eta_3$ Our aim is to show that $\xi_3 \neq \eta_3$. We again suppose the contrary that is $\xi_3 = \eta_3$. Since $\xi \neq \eta$, we must have that $\xi_1 \neq \eta_1$. In this case, we can rewrite the system of equations in terms of x_1 and x_3 as follows

$$\begin{cases} (1 - 6\varepsilon)x_1^2 + 4\varepsilon x_3^2 + (1 - 4\varepsilon)x_1x_3 - x_1 + 4\varepsilon x_3 + 2\varepsilon = 0 \\ (9\varepsilon - \frac{3}{2})x_1^2 + 6\varepsilon x_3^2 + (18\varepsilon - \frac{3}{2})x_1x_3 + (\frac{3}{2} - 8\varepsilon)x_1 - 18\varepsilon x_3 + \varepsilon = 0 \end{cases}.$$

This yields that for $x_3 = \xi_3 = \eta_3$, the following two quadratic equations (with respect to x_1)

$$(1 - 6\varepsilon)x_1^2 + ((1 - 4\varepsilon)\xi_3 - 1)x_1 + (4\xi_3^2 + 4\xi_3 + 2)\varepsilon = 0,$$
$$(9\varepsilon - \frac{3}{2})x_1^2 + [(18\varepsilon - \frac{3}{2})\xi_3 + (\frac{3}{2} - 8\varepsilon)]x_1 + (6\xi_3^2 - 18\xi_3 + 1)\varepsilon = 0$$

must have two distinct common roots ξ_1 and η_1. Consequently, we should have that

$$\frac{1-6\varepsilon}{9\varepsilon-\frac{3}{2}} = \frac{(1-4\varepsilon)\xi_3-1}{(18\varepsilon-\frac{3}{2})\xi_3+\frac{3}{2}-8\varepsilon} = \frac{4\xi_3^2+4\xi_3+2}{6\xi_3^2-18\xi_3+1}.$$

It follows from the first equality that $\xi_3 = \frac{2}{3}$. By substituting $\xi_3 = \frac{2}{3}$ into the last fraction, we get that $\frac{4\xi_3^2+4\xi_3+2}{6\xi_3^2-18\xi_3+1} = -\frac{58}{75} \neq -\frac{2}{3} = \frac{1-6\varepsilon}{9\varepsilon-\frac{3}{2}}$. This is again contradiction.

This completes the proof. □

Theorem 2.3. *Let $V_\varepsilon : \mathbb{S}^2 \to \mathbb{S}^2$ be the quadratic operator given above. Then for sufficiently small ε, one has that $\boldsymbol{Fix}(V_\varepsilon) = \left\{\left(\frac{3a_0^2-3a_0+1}{2-3a_0},\ a_0,\ \frac{1-2a_0}{2-3a_0}\right)\right\}$ where a_0 is the unique positive root in $\left(0, \frac{1}{2}\right)$ of the following quartic equation*

$$(9-54\varepsilon)a^4 + (132\varepsilon-15)a^3 + (9-68\varepsilon)a^2 - (12\varepsilon+2)a + 10\varepsilon = 0.$$

Proof. Since $x_3 = 1 - x_1 - x_2$, it is enough to find all solutions (x_1, x_2) of the following system of equations

$$\begin{cases} 2\varepsilon x_1^2 + 4\varepsilon x_2^2 + (12\varepsilon-1)x_1x_2 - 16\varepsilon x_1 - 12\varepsilon x_2 + 10\varepsilon = 0 \\ \varepsilon x_1^2 - 10\varepsilon x_2^2 - (\frac{1}{2}+6\varepsilon)x_1x_2 + 6\varepsilon x_2 + \varepsilon = 0 \end{cases} \tag{2.3}$$

which satisfy the conditions $0 < x_1, x_2 < 1$ and $0 < x_1 + x_2 < 1$.

Let $x_1 = x$ be a variable and $x_2 = a$ be a parameter. Then the system (2.3) takes the following form

$$x^2 + \frac{(12\varepsilon-1)a-16\varepsilon}{2\varepsilon}x + (2a^2-6a+5) = 0, \tag{2.4}$$

$$x^2 + \frac{a(-1-12\varepsilon)}{2\varepsilon}x + (-10a^2+6a+1) = 0. \tag{2.5}$$

Due to Proposition 2.2, these two quadratic equations cannot have two common roots. Hence, the system (2.3) has a solution (x_1, x_2) with $0 < x_1, x_2 < 1,\ 0 < x_1 + x_2 < 1$ if and only if two quadratic equations (2.4) and (2.5) must have a unique common root in $(0, 1)$ for $a \in (0, 1)$. We know (see [20]) that two quadratic equations (2.4) and (2.5) have a unique common root if and only if their resultant is equal to zero, i.e.,

$$(9-54\varepsilon)a^4 + (132\varepsilon-15)a^3 + (9-68\varepsilon)a^2 + (-12\varepsilon-2)a + 10\varepsilon = 0. \tag{2.6}$$

In this case, $x = \frac{3a^2-3a+1}{2-3a}$ is the unique common root of two quadratic equations (2.4) and (2.5). In order to have conditions $0 < x_1, x_2 < 1,\ 0 < x_1 + x_2 < 1$, we have to solve the following system of inequalities

$$\begin{cases} 0 < \frac{3a^2-3a+1}{2-3a} < 1 \\ 0 < a < 1 \\ 0 < a + \frac{3a^2-3a+1}{2-3a} < 1. \end{cases} \tag{2.7}$$

The solution of the system (2.7) is $a \in (0, \frac{1}{2})$. Therefore, the total number of solutions (x_1, x_2), $0 < x_1, x_2 < 1,\ 0 < x_1 + x_2 < 1$ of the system (2.3) is the same as the total number of roots of the quartic equation (2.6) in the interval $(0, \frac{1}{2})$. Moreover, there is one-to-one correspondence between a root $a_0 \in (0, \frac{1}{2})$ of the quartic equation (2.6) and a fixed point $\left(\frac{3a_0^2-3a_0+1}{2-3a_0},\ a_0,\ \frac{1-2a_0}{2-3a_0}\right)$ of the quadratic operator $V_\varepsilon : \mathbb{S}^2 \to \mathbb{S}^2$.

Now, we want to show that the quartic equation (2.6) has a unique root in the interval $(0, \frac{1}{2})$ for sufficiently small ε. To do so, we have to apply the Sturm theorem for the quartic equation (2.6) in $(0, \frac{1}{2})$ (see [20]).

Let $p(a) = (9 - 54\varepsilon)a^4 + (132\varepsilon - 15)a^3 + (9 - 68\varepsilon)a^2 + (-12\varepsilon - 2)a + 10\varepsilon$ be a quartic polynomial. Let $\{p_0(a), p_1(a), p_2(a), p_3(a), p_4(a)\}$ be a Sturm sequence of the quartic polynomial $p(a)$. Let $\sigma(\xi)$ be the number of sign changes (ignoring zero terms) in the sequence $p_0(\xi), p_1(\xi), p_2(\xi), p_3(\xi), p_4(\xi)$. Then due to the Sturm theorem, the number of roots of the quartic polynomial $p(x)$ in the interval $(0, \frac{1}{2})$ is equal to $\sigma(0) - \sigma(\frac{1}{2})$. Simple calculations show that for sufficiently small ε, one has that

$$\begin{aligned}
p_0(0) &= 10\varepsilon > 0, & p_0(\frac{1}{2}) &= -\frac{1}{16} + \frac{1}{8}\varepsilon < 0,\\
p_1(0) &= -(12\varepsilon + 2) < 0, & p_1(\frac{1}{2}) &= \frac{1}{4} - 8\varepsilon > 0,\\
p_2(0) &\simeq -\frac{5}{24(6\varepsilon - 1)} > 0, & p_2(\frac{1}{2}) &\simeq -\frac{13}{6\varepsilon - 1} > 0,\\
p_3(0) &\simeq \frac{9}{(3 - 344\varepsilon)^2} > 0, & p_3(\frac{1}{2}) &\simeq \frac{9}{(3 - 344\varepsilon)^2} > 0,\\
p_4(0) &\simeq \frac{3}{6\varepsilon - 1} < 0, & p_4(\frac{1}{2}) &\simeq \frac{3}{6\varepsilon - 1} < 0.
\end{aligned}$$

Therefore, we get that $\sigma(0) - \sigma(\frac{1}{2}) = 3 - 2 = 1$. Consequently, this means that the quartic equation (2.6) has a unique root a_0 in the interval $(0, \frac{1}{2})$, or equivalently, the quadratic operator $V_\varepsilon : \mathbb{S}^2 \to \mathbb{S}^2$ has a unique fixed point $\left(\frac{3a_0^2-3a_0+1}{2-3a_0},\ a_0,\ \frac{1-2a_0}{2-3a_0}\right)$ in the simplex $\mathbb{S}^2$. This completes the proof. □

3. Lyubich's Example

Yu. I. Lyubich has considered (see [8], page 296) the following quadratic operator $W_\varepsilon : \mathbb{S}^2 \to \mathbb{S}^2$, $W_\varepsilon(\mathbf{x}) = \mathbf{x}' = (x_1', x_2', x_3')$

$$W_\varepsilon : \begin{cases} x_1' = (1 - 4\varepsilon)x_1^2 + 2\varepsilon x_2^2 + 10\varepsilon x_3^2 + 4\varepsilon x_1x_2 + (1 + 4\varepsilon)x_1x_3 + 8\varepsilon x_2x_3 \\ x_2' = 2\varepsilon x_1^2 + (1 - 3\varepsilon)x_2^2 + \varepsilon x_3^2 + (\frac{1}{2} + 2\varepsilon)x_1x_2 + 2\varepsilon x_1x_3 + (1 - 12\varepsilon)x_2x_3 \\ x_3' = 2\varepsilon x_1^2 + \varepsilon x_2^2 + (1 - 11\varepsilon)x_3^2 + (\frac{3}{2} - 6\varepsilon)x_1x_2 + (1 - 6\varepsilon)x_1x_3 + (1 + 4\varepsilon)x_2x_3 \end{cases}$$

where $0 < \varepsilon < \frac{1}{12}$. In commentaries and references section, Yu.I. Lyubich wrote that the quadratic operator $W_\varepsilon : \mathbb{S}^2 \to \mathbb{S}^2$ was constructed by A.A. Krapivin in [7]. However, Krapivin's example $V_\varepsilon : \mathbb{S}^2 \to \mathbb{S}^2$ considered in the previous section is slightly different from the quadratic operator $W_\varepsilon : \mathbb{S}^2 \to \mathbb{S}^2$ given in Lyubich's book [8]. Yu. I. Lyubich claimed that if $0 < \varepsilon < \frac{9-5\sqrt{2}}{124}$ then the quadratic operator $W_\varepsilon : \mathbb{S}^2 \to \mathbb{S}^2$ has three fixed points in the simplex $\mathbb{S}^2$. However, this is wrong. Namely, the quadratic operator W_ε has a unique fixed point for any $0 < \varepsilon < \frac{1}{12}$. For the sake of argument, we shall present its proof by repeating the same method used in Krapivin's example.

It is clear that $W_\varepsilon(\mathbb{S}^2) \subset \mathrm{int}\mathbb{S}^2$. Hence, $\mathbf{Fix}(W_\varepsilon) \subset \mathrm{int}\mathbb{S}^2$.

In order to find all fixed points, we have to solve the system of equations

$$\begin{cases} x_1 = (1 - 4\varepsilon)x_1^2 + 2\varepsilon x_2^2 + 10\varepsilon x_3^2 + 4\varepsilon x_1x_2 + (1 + 4\varepsilon)x_1x_3 + 8\varepsilon x_2x_3 \\ x_2 = 2\varepsilon x_1^2 + (1 - 3\varepsilon)x_2^2 + \varepsilon x_3^2 + (\frac{1}{2} + 2\varepsilon)x_1x_2 + 2\varepsilon x_1x_3 + (1 - 12\varepsilon)x_2x_3 \\ x_3 = 2\varepsilon x_1^2 + \varepsilon x_2^2 + (1 - 11\varepsilon)x_3^2 + (\frac{3}{2} - 6\varepsilon)x_1x_2 + (1 - 6\varepsilon)x_1x_3 + (1 + 4\varepsilon)x_2x_3 \end{cases}$$

Proposition 3.1. *One has that $\xi_1 \neq \eta_1, \xi_2 \neq \eta_2, \xi_3 \neq \eta_3$ for $\xi, \eta \in \boldsymbol{Fix}(W_\varepsilon)$ and $\xi \neq \eta$.*

Proof. Let ξ, η be two distinct solutions of the system given above (if any). Since $x_3 = 1 - x_1 - x_2$, we can rewrite the system of equations in terms of x_1 and x_2 as

$$\begin{cases} 2\varepsilon x_1^2 + 4\varepsilon x_2^2 + (12\varepsilon - 1)x_1x_2 - 16\varepsilon x_1 - 12\varepsilon x_2 + 10\varepsilon = 0 \\ \varepsilon x_1^2 + 10\varepsilon x_2^2 + (14\varepsilon - \frac{1}{2})x_1x_2 - 14\varepsilon x_2 + \varepsilon = 0 \end{cases} \tag{3.1}$$

Case $\xi_1 \neq \eta_1$ Our aim is to show that $\xi_1 \neq \eta_1$. We suppose the contrary that is $\xi_1 = \eta_1$. Since $\xi \neq \eta$, we must have that $\xi_2 \neq \eta_2$. This means that for $x_1 = \xi_1 = \eta_1$, the following two quadratic equations (with respect to x_2)

$$4\varepsilon x_2^2 + ((12\varepsilon - 1)\xi_1 - 12\varepsilon)x_2 + (2\xi_1^2 - 16\xi_1 + 10)\varepsilon = 0,$$
$$10\varepsilon x_2^2 + ((14\varepsilon - \frac{1}{2})\xi_1 - 14\varepsilon)x_2 + (\xi_1^2 + 1)\varepsilon = 0$$

must have two distinct common roots ξ_2 and η_2. Consequently, we should have that

$$\frac{4\varepsilon}{10\varepsilon} = \frac{(12\varepsilon - 1)\xi_1 - 12\varepsilon}{(14\varepsilon - \frac{1}{2})\xi_1 - 14\varepsilon} = \frac{2\xi_1^2 - 16\xi_1 + 10}{\xi_1^2 + 1}.$$

It follows from the first equality that $\xi_1 = \frac{8\varepsilon}{8\varepsilon - 1}$. Since $0 < \varepsilon < \frac{1}{12}$, we get that $\xi_1 < 0$ which is a contradiction.

Case $\xi_2 \neq \eta_2$ Our aim is to show that $\xi_2 \neq \eta_2$. We again suppose the contrary that is $\xi_2 = \eta_2$. Since $\xi \neq \eta$, we must have that $\xi_1 \neq \eta_1$. It follows from (3.1) that for $x_2 = \xi_2 = \eta_2$, the following two quadratic equations (with respect to x_1)

$$2\varepsilon x_1^2 + ((12\varepsilon - 1)\xi_2 - 16\varepsilon)x_1 + (4\xi_2^2 - 12\xi_2 + 10)\varepsilon = 0,$$
$$\varepsilon x_1^2 + (14\varepsilon - \frac{1}{2})\xi_2 x_1 + (10\xi_2^2 - 14\xi_2 + 1)\varepsilon = 0$$

must have two distinct common roots ξ_1 and η_1. Consequently, we should have that

$$\frac{2\varepsilon}{\varepsilon} = \frac{(12\varepsilon - 1)\xi_2 - 16\varepsilon}{(14\varepsilon - \frac{1}{2})\xi_2} = \frac{4\xi_2^2 - 12\xi_2 + 10}{10\xi_2^2 - 14\xi_2 + 1}.$$

It follows from the first equality that $\xi_2 = -1$ which is a contradiction.

Case $\xi_3 \neq \eta_3$ Our aim is to show that $\xi_3 \neq \eta_3$. We again suppose the contrary that is $\xi_3 = \eta_3$. Since $\xi \neq \eta$, we must have that $\xi_1 \neq \eta_1$. In this case, we can rewrite the system of equations in terms of x_1 and x_3 as follows

$$\begin{cases} (1 - 6\varepsilon)x_1^2 + 4\varepsilon x_3^2 + (1 - 4\varepsilon)x_1x_3 - x_1 + 4\varepsilon x_3 + 2\varepsilon = 0 \\ (9\varepsilon - \frac{3}{2})x_1^2 - 14\varepsilon x_3^2 - (\frac{3}{2} + 2\varepsilon)x_1x_3 + (\frac{3}{2} - 8\varepsilon)x_1 + 2\varepsilon x_3 + \varepsilon = 0 \end{cases}$$

This yields that for $x_3 = \xi_3 = \eta_3$, the following two quadratic equations (with respect to x_1)

$$(1 - 6\varepsilon)x_1^2 + ((1 - 4\varepsilon)\xi_3 - 1)x_1 + (4\xi_3^2 + 4\xi_3 + 2)\varepsilon = 0,$$
$$(9\varepsilon - \frac{3}{2})x_1^2 + [-(\frac{3}{2} + 2\varepsilon)\xi_3 + (\frac{3}{2} - 8\varepsilon)]x_1 + [-14\xi_3^2 + 2\xi_3 + 1]\varepsilon = 0$$

must have two distinct common roots ξ_1 and η_1. Consequently, we should have that

$$\frac{1 - 6\varepsilon}{9\varepsilon - \frac{3}{2}} = \frac{(1 - 4\varepsilon)\xi_3 - 1}{-(\frac{3}{2} + 2\varepsilon)\xi_3 + \frac{3}{2} - 8\varepsilon} = \frac{4\xi_3^2 + 4\xi_3 + 2}{-14\xi_3^2 + 2\xi_3 + 1}.$$

It follows from the first equality that $\xi_3 = -1$ which is a contradiction.

This completes the proof □

Theorem 3.2. *Let* $W_\varepsilon : \mathbb{S}^2 \to \mathbb{S}^2$ *be the quadratic operator given above. Then for any* $0 < \varepsilon < \frac{1}{12}$, *one has that* $\textbf{Fix}(W_\varepsilon) = \left\{\left(\frac{1+2a_0(1-a_0)}{2(1+a_0)},\ a_0,\ \frac{1-2a_0}{2(1+a_0)}\right)\right\}$ *where* a_0 *is the unique positive root in* $(0, \frac{1}{2})$ *of the following quartic equation*

$$(2-12\varepsilon)a^4 + 16\varepsilon a^3 + (16\varepsilon - 3)a^2 - (16\varepsilon + 1)a + 5\varepsilon = 0.$$

Proof. Since $x_3 = 1 - x_1 - x_2$, it is enough to find all solutions (x_1, x_2) of the following system of equations

$$\begin{cases} 2\varepsilon x_1^2 + 4\varepsilon x_2^2 + (12\varepsilon - 1)x_1x_2 - 16\varepsilon x_1 - 12\varepsilon x_2 + 10\varepsilon = 0 \\ \varepsilon x_1^2 + 10\varepsilon x_2^2 + (14\varepsilon - \frac{1}{2})x_1x_2 - 14\varepsilon x_2 + \varepsilon = 0 \end{cases} \tag{3.2}$$

which satisfy the conditions $0 < x_1, x_2 < 1$ and $0 < x_1 + x_2 < 1$.

Let $x_1 = x$ be a variable and $x_2 = a$ be a parameter. Then the system (3.2) takes the following form

$$x^2 + \frac{(12\varepsilon - 1)a - 16\varepsilon}{2\varepsilon}x + (2a^2 - 6a + 5) = 0, \tag{3.3}$$

$$x^2 + \frac{a(14\varepsilon - \frac{1}{2})}{\varepsilon}x + (10a^2 - 14a + 1) = 0. \tag{3.4}$$

Due to Proposition 3.1, these two quadratic equations cannot have two common roots. Hence, the system (3.2) has a solution (x_1, x_2) with $0 < x_1, x_2 < 1,\ 0 < x_1 + x_2 < 1$ if and only if two quadratic equations (3.3) and (3.4) must have a unique common root in $(0, 1)$ for $a \in (0, 1)$. We know (see [20]) that two quadratic equations (3.3) and (3.4) have a unique common root if and only if their resultant is equal to zero, i.e.,

$$(2-12\varepsilon)a^4 + 16\varepsilon a^3 + (16\varepsilon - 3)a^2 - (16\varepsilon + 1)a + 5\varepsilon = 0. \tag{3.5}$$

In this case, $x = \frac{1+2a(1-a)}{2(1+a)}$ is the unique common root of two quadratic equations (3.3) and (3.4). In order to have conditions $0 < x_1, x_2 < 1,\ 0 < x_1 + x_2 < 1$, we have to solve the following system of inequalities

$$\begin{cases} 0 < \frac{1+2a(1-a)}{2(1+a)} < 1 \\ 0 < a < 1 \\ 0 < a + \frac{1+2a(1-a)}{2(1+a)} < 1. \end{cases} \tag{3.6}$$

The solution of the system (3.6) is $a \in (0, \frac{1}{2})$. Therefore, the total number of solutions (x_1, x_2), $0 < x_1, x_2 < 1,\ 0 < x_1 + x_2 < 1$ of the system (3.2) is the same as the total number of roots of the quartic equation (3.5) in the interval $(0, \frac{1}{2})$. Moreover, there is one-to-one correspondence between a root $a_0 \in (0, \frac{1}{2})$ of the quartic equation (3.5) and a fixed point $\left(\frac{1+2a_0(1-a_0)}{2(1+a_0)},\ a_0,\ \frac{1-2a_0}{2(1+a_0)}\right)$ of the quadratic operator $W_\varepsilon : \mathbb{S}^2 \to \mathbb{S}^2$.

We are aiming to study the number of positive roots of the quartic equation (3.5) in the interval $(0, \frac{1}{2})$. Let $f(a) = (2-12\varepsilon)a^4 + 16\varepsilon a^3 + (16\varepsilon - 3)a^2 - (16\varepsilon + 1)a + 5\varepsilon$. Since $0 < \varepsilon < \frac{1}{12}$, it is easy to check that

$$f(0) = 5\varepsilon > 0,\ f\left(\frac{1}{2}\right) = \frac{18\varepsilon - 9}{8} < 0,\ f(2) = 18 - 27\varepsilon > 0.$$

This means that the quartic equation (3.5) has at least two positive roots. On the other hand, due to Descartes's theorem, the number of positive roots cannot be more than the number of sign changes between consecutive nonzero coefficients $2-12\varepsilon$, 16ε, $16\varepsilon-3$, $-(16\varepsilon+1)$, 5ε of the quartic equation (3.5) which is two.

Therefore, the quartic equation (3.5) has exactly two positive roots in which one of them belongs to $(0,\frac{1}{2})$ and another one belongs to $(\frac{1}{2},2)$. Hence, for any $0<\varepsilon<\frac{1}{12}$, there exists a unique positive root a_0 of the quartic equation (3.5) in the interval $(0,\frac{1}{2})$. Consequently, for any $0<\varepsilon<\frac{1}{12}$, the quadratic operator W_ε has a unique fixed point $\left(\frac{1+2a_0(1-a_0)}{2(1+a_0)},\ a_0,\ \frac{1-2a_0}{2(1+a_0)}\right)$.
This completes the proof □

4. Positive Quadratic Stochastic Operator Having Three Fixed Points

In this section, we provide an example for a quadratic operator with positive coefficients having three fixed points in the simplex $\mathbb{S}^2$.

Let $\mathbf{A}(0.1,0.2,0.7)$, $\mathbf{B}(0.4,0.3,0.3)$ and $\mathbf{C}(0.59,0.31,0.1)$ be points in the simplex. We define a positive quadratic operator $\mathcal{Q}_0:\mathbb{S}^2\to\mathbb{S}^2$, $\mathcal{Q}_0(\mathbf{x})=\mathbf{x}'=(x_1',x_2',x_3')$ as follows

$$\mathcal{Q}_0:\begin{cases} x_1'=\frac{232873}{319300}x_1^2+\frac{4717}{10300}x_2^2+\frac{207}{63860}x_3^2+\frac{7}{5}x_1x_2+\frac{3}{5}x_1x_3+\frac{1}{50}x_2x_3\\ x_2'=\frac{27}{100}x_1^2+\frac{1}{2}x_2^2+\frac{3}{20}x_3^2+\frac{470171}{814300}x_1x_2+\frac{378421}{407150}x_1x_3+\frac{158157}{814300}x_2x_3\\ x_3'=\frac{54}{79825}x_1^2+\frac{433}{10300}x_2^2+\frac{27037}{31930}x_3^2+\frac{18409}{814300}x_1x_2+\frac{191589}{407150}x_1x_3+\frac{1454157}{814300}x_2x_3 \end{cases}.$$

The straightforward calculation shows that $\mathbf{A},\mathbf{B},\mathbf{C}$ are fixed points of the quadratic operator $\mathcal{Q}_0:\mathbb{S}^2\to\mathbb{S}^2$.

We can define another positive quadratic operator $\mathcal{Q}_1:\mathbb{S}^2\to\mathbb{S}^2$, $\mathcal{Q}_1(\mathbf{x})=\mathbf{x}'=(x_1',x_2',x_3')$ as follows

$$\mathcal{Q}_1:\begin{cases} x_1'=\frac{17322871}{22351000}x_1^2+\frac{990257}{2163000}x_2^2+\frac{1559}{13410600}x_3^2+\frac{13}{10}x_1x_2+\frac{16}{25}x_1x_3+\frac{11}{500}x_2x_3\\ x_2'=\frac{224}{1000}x_1^2+\frac{488}{1000}x_2^2+\frac{125}{1000}x_3^2+\frac{703327}{1017875}x_1x_2+\frac{19461451}{24429000}x_1x_3+\frac{8271787}{24429000}x_2x_3\\ x_3'=\frac{4301}{4470200}x_1^2+\frac{117199}{2163000}x_2^2+\frac{2933179}{3352650}x_3^2+\frac{18371}{2035750}x_1x_2+\frac{13761989}{24429000}x_1x_3+\frac{1601951}{977160}x_2x_3 \end{cases}$$

The straightforward calculation shows that $\mathbf{A},\mathbf{B},\mathbf{C}$ are also fixed points of the quadratic operator $\mathcal{Q}_1:\mathbb{S}^2\to\mathbb{S}^2$.

In addition, if $\mathcal{Q}:\mathbb{S}^2\to\mathbb{S}^2$ is a positive quadratic stochastic operator on 2D simplex then $|\mathbf{Fix}(\mathcal{Q})|=1$ or 3 (see [8]). Therefore, the quadratic operators $\mathcal{Q}_0$ and $\mathcal{Q}_1$ have exactly three fixed points; $\mathbf{A},\mathbf{B}$ and $\mathbf{C}$.

Now, we can define a family of positive quadratic operators $\mathcal{Q}_\varepsilon:\mathbb{S}^2\to\mathbb{S}^2$ as $\mathcal{Q}_\varepsilon(\mathbf{x})=(1-\varepsilon)\mathcal{Q}_0(\mathbf{x})+\varepsilon\mathcal{Q}_1(\mathbf{x})$ for any $x\in\mathbb{S}^2$ and $0\leq\varepsilon\leq1$. It is clear that $\mathbf{A},\mathbf{B},\mathbf{C}$ are also fixed points of the family of positive quadratic operators $\mathcal{Q}_\varepsilon:\mathbb{S}^2\to\mathbb{S}^2$.

In the paper [9], it was conjectures that *if the set of stationary vectors of a second-order Markov chain contains $k-$interior points of the $(k-1)-$dimensional face of the simplex then every vector in the $(k-1)-$dimensional face is a stationary vector.*

However, this conjecture is wrong. The family of quadratic stochastic operators $\mathcal{Q}_\varepsilon:\mathbb{S}^2\to\mathbb{S}^2$ defined above are counterexamples to this conjecture.

Acknowledgments

This work was partially supported by the MOHE grant FRGS14-141-0382. The first author (M.S.) is grateful to the Junior Associate scheme of the Abdus Salam International Centre for Theoretical Physics, Trieste, Italy.

References

[1] Bernstein S 1942 *Annals Math. Statis.* **13** 53–61
[2] Ganikhodjaev N, Saburov M and Jamilov U 2013 *App. Math. Info. Sci.* **7**(5) 1721–1729
[3] Ganikhodjaev N, Saburov M and Nawi A.M. 2014 *The Scientific World Jour.* **2014** 1–11
[4] Ganikhodzhaev R, Mukhamedov F and Rozikov U 2011 *Inf. Dim. Anal. Quan. Prob. Rel. Top.* **14**(2) 279–335
[5] Ganikhodjaev R, Saburov M and Saburov M 2013 *AIP Conf. Proc.* **1557** 108–111
[6] Kesten H 1970 *Adv. App. Prob.* **2** 1–82
[7] Krapivin A A 1975 *Theor. Funct Funct Anal Appl* **25** 62–67
[8] Lyubich Yu I 1992 *Mathematical structures in population genetics* (Berlin:Springer–Verlag)
[9] Li C-K and Zhang S 2015 *Lin. Alg. Appl.* **473** 114–125
[10] Mukhamedov F and Ganikhodjaev N 2015 *Lect. Notes Math.* **2133** (Berlin: Springer)
[11] Mukhamedov F, Akin H and Temir S 2005 *J. Math. Anal. Appl.* **310** 533–556
[12] Mukhamedov F and Saburov M 2010 *App. Math. Info. Sci.* **4** 47–62
[13] Mukhamedov F and Saburov M 2014 *Bull. Malay. Math. Sci. Soc* **37** 59–64
[14] Mukhamedov F, Saburov M and Qaralleh I 2013 *Abst. Appl. Anal.* **2013** 942038
[15] Saburov M 2013 *World Applied Science Jour.* **21** 94–97
[16] Saburov M 2015 *Math Notes* **97**(5-6) 759–763
[17] Saburov M 2015 *Ann. Funct. Anal.* **6**(4) 247–254
[18] Saburov M and Saburov Kh. 2014 *ScienceAsia* **40**(4) 306–312
[19] Saburov M and Saburov Kh. 2014 *Inter Journal of Cont Auto Systems* **12**(6) 1276–1282
[20] Sturmfels B 2002 *Solving systems of polynomial equations* (Texas)
[21] Ulam S 1960 *A collection of mathematical problems* (New-York, London)

Journal of Physics: Conference Series **697** (2016) 012013 doi:10.1088/1742-6596/697/1/012013

On a Family of Strictly Non-Volterra Quadratic Stochastic Operators

Uygun Jamilov

Institute of mathematics, National University of Uzbekistan, 29, Do'rmon Yo'li str., 100125, Tashkent, Uzbekistan

E-mail: jamilovu@yandex.ru

Abstract. In this paper we consider a class of strictly non-Volterra quadratic stochastic operators defined on the two-dimensional simplex. We show that such operators have a unique fixed point and the set of limit points is either a single point or an infinite set.

1. Introduction

Quadratic stochastic operators (QSOs) were first introduced by Bernstein in [2]. Such operators frequently arise in many models of mathematical genetics, namely, in the theory of heredity (see [5]-[8],[11, 12],[15]-[18],[21, 22]). Consider a biological population, that is, a community of organisms closed with respect to reproduction. Assume that each individual in this population belongs precisely to one of the species (genotype) $1, \dots, m$. The scale of species is such that the species of the parents i and j, unambiguously, determine the probability of every species k for the first generation of direct descendants. Denote this probability, called the heredity coefficient, by $p_{ij,k}$. It is then obvious that $p_{ij,k} \geq 0$ for all i, j, k and that

$$\sum_{k=1}^{m} p_{ij,k} = 1, \quad i, j, k = 1, \dots, m.$$

The state of the population can be described by the tuple $(x_1, x_2, \dots, x_m)$ of species probabilities, that is, x_k is the fraction of the species k in the total population. In the case of panmixia (random interbreeding) the parent pairs i and j arise for a fixed state $\mathbf{x} = (x_1, x_2, \dots, x_m)$ with probability $x_i x_j$. Hence, the total probability of the species k in the first generation of direct descendants is defined by

$$x'_k = \sum_{i,j=1}^{m} p_{ij,k} x_i x_j, \quad k = 1, \dots, m.$$

The association $\mathbf{x} \mapsto \mathbf{x}'$ defines an evolutionary quadratic operator. Thus evolution of a population can be studied as a dynamical system of a quadratic stochastic operator [15].

The paper is organized as follows. In Section 2 we recall definitions and well known results from the theory of Volterra and non-Volterra QSOs . In Section 3 we consider a new class of non-Volterra QSOs which are generated by convex combination of two strictly non-Volterra QSOs. Then we show that each QSO from this class has a unique fixed point. Moreover, we prove that the set of limit points is either a single point or an infinite set.

2. Preliminaries

Let

$$S^{m-1} = \{\mathbf{x} = (x_1, x_2, \ldots, x_m) \in \mathbb{R}^m : \text{ for any } i,\ x_i \geq 0, \text{ and } \sum_{i=1}^{m} x_i = 1\}$$

be the $(m-1)$-dimensional simplex. A map V of S^{m-1} into itself is called a *quadratic stochastic operator* (QSO) if

$$(V\mathbf{x})_k = \sum_{i,j=1}^{m} p_{ij,k} x_i x_j \tag{2.1}$$

for any $\mathbf{x} \in S^{m-1}$ and for all $k = 1, \ldots, m$, where

$$p_{ij,k} \geq 0, \quad p_{ij,k} = p_{ji,k} \text{ for all } i, j, k; \quad \sum_{k=1}^{m} p_{ij,k} = 1. \tag{2.2}$$

By $\{\mathbf{x}^{(n)} \in S^{m-1} : n = 0, 1, 2, \ldots\}$ we denote the the trajectory of V starting from the initial point $\mathbf{x} \in S^{m-1}$, where $\mathbf{x}^{(n+1)} = V(\mathbf{x}^{(n)})$ for all $n = 0, 1, 2, \ldots$, with $\mathbf{x}^{(0)} = \mathbf{x}$. Recall that a point $\mathbf{x} \in S^{m-1}$ is called a *fixed point* of V if $V(\mathbf{x}) = \mathbf{x}$.

Definition 2.1. *A QSO V is called regular if for any initial point $\mathbf{x} \in S^{m-1}$, the limit $\lim_{n\to\infty} V^n(\mathbf{x})$ exists.*

Note that the limit point of the trajectory $\{\mathbf{x}^{(n)}\}$ is a fixed point of V. Thus, the fixed points of a QSO describe limit or long run behavior of the trajectories for any initial point. The limit behavior of trajectories and fixed points play an important role in many applied problems (see [7, 8, 14, 15, 16]). The biological treatment of the regularity of a QSO is rather clear: in the long run the distribution of species in the next generation coincides with the distribution of species in the previous one, i.e. it is stable.

In statistical mechanics, an ergodic hypothesis proposes a connection between dynamics and statistics. In the classical theory the assumption was made that the average time spent in any region of phase space is proportional to the volume of the region in terms of the invariant measure, more generally, that time averages may be replaced by space averages. For nonlinear dynamical systems (2.1) Ulam [21] suggested an analogue of a measure-theoretic ergodicity, the following ergodic hypothesis:

Definition 2.2. *A QSO V is said to be ergodic if the limit*

$$\lim_{n\to\infty} \frac{1}{n} \sum_{k=0}^{n-1} V^k(\mathbf{x})$$

exists for any $\mathbf{x} \in S^{m-1}$.

On the basis of numerical calculations Ulam, in 1960 [21], conjectured that the ergodic theorem holds for any QSO. In 1977, Zakharevich [22] proved that this conjecture is false in general. Later, in [5], a sufficient condition of non-ergodicity for QSOs defined on S^2 was established. The biological treatment of non-ergodicity of a QSO is the following: in the long run the behavior of the distributions of species is chaotic, i.e. it is unpredictable. Note that a regular QSO is ergodic, but in general from ergodicity does not follow regularity.

Definition 2.3. *[13] Let (X, d) be a metric space. A continuous map $V : X \to X$ is called Li-Yorke chaotic if there exists an uncountable subset $S \subset X$ such that for every pair $(\mathbf{x}, \mathbf{y}) \in S \times S$ of distinct points, we have that*

$$\lim_{n\to\infty} \inf d(V^n(\mathbf{x}), V^n(\mathbf{y})) = 0$$

$$\lim_{n\to\infty} \sup d(V^n(\mathbf{x}), V^n(\mathbf{y})) > 0.$$

Note that the QSO which was studied by Zakharevich [22] exhibits the Li-Yorke chaos (see [20]).

Let π be a permutation of the set $\{1, 2, \ldots, m\}$. Then we can define a one-to-one transformation $T_\pi\colon S^{m-1} \to S^{m-1}$ as

$$T_\pi(x_1, x_2, \ldots, x_m) = (x_{\pi(1)}, x_{\pi(2)}, \ldots, x_{\pi(m)}). \tag{2.3}$$

It is evident that the transformation $T_\pi^{-1}VT_\pi$ is a QSO and the limit behavior of trajectories of the QSO $T_\pi^{-1}VT_\pi$ coincides with limit behavior of trajectories of the QSO V (see [11]).

The following notations will be used throughout this paper. We let ∂S^{m-1} denote the boundary of S^{m-1}, $\partial S^{m-1} = \{\mathbf{x} \in S^{m-1} : x_i = 0 \text{ for at least one } i \in \{1, 2, \ldots, m\}\}$; the interior of S^{m-1} be the set $\operatorname{int} S^{m-1} = \{\mathbf{x} \in S^{m-1} : x_1 x_2 \cdots x_m > 0\}$.

Let $\omega_V(\mathbf{x}^0)$ be the set of limit points of the trajectory $\{V^k(\mathbf{x}^0) \in S^{m-1} : k = 0, 1, 2, \ldots\}$. Using Lyapunov functions, one can handle on ω-limit points. Recall the definition of a Lyapunov function.

Definition 2.4. *A continuous function $\varphi\colon \operatorname{int} S^{m-1} \to R$ is called a Lyapunov function for a QSO V if $\varphi(V(\mathbf{x})) \geq \varphi(\mathbf{x})$ for all $\mathbf{x}$ (or $\varphi(V(\mathbf{x})) \leq \varphi(\mathbf{x})$ for all $\mathbf{x}$).*

A Lyapunov function is very helpful to describe an upper estimate of $\omega(\mathbf{x}^0)$. However there is no general recipe on how to find such Lyapunov functions. In [1, 5, 7] the authors presented examples of Lyapunov functions. A QSO given by (2.1) is called *Volterra* operator, if $p_{ij,k} = 0$, whenever $k \notin \{i, j\}$ for every $k, i, j \in \{1, \ldots, m\}$. Many properties of Lyapunov functions of Volterra operators have been investigated in [7]-[11],[17, 19]. In [23] it was defined an opposite class to Volterra operators. Such kind of operators are called *strictly non-Volterra* ones.

Now recall the definition of strictly non-Volterra operators.

Definition 2.5. *[23] A quadratic stochastic operator* (2.1),(2.2) *is said a strictly non-Volterra operator if*

$$p_{ij,k} = 0, \quad \text{for } \ k \in \{i, j\}, \quad i, j, k = 1, \ldots, m. \tag{2.4}$$

It is easy to see that for $m = 2$ conditions (2.2) and (2.4) cannot be satisfied simultaneously, hence strictly non-Volterra operators exist only when $m \geq 3$.

For $m = 3$ an arbitrary strictly non-Volterra operator has the following representation:

$$V : \begin{cases} x_1' = \alpha x_2^2 + c x_3^2 + 2x_2 x_3, \\ x_2' = a x_1^2 + d x_3^2 + 2x_1 x_3, \\ x_3' = b x_1^2 + \beta x_2^2 + 2x_1 x_2, \end{cases} \tag{2.5}$$

where

$$a, b, c, d, \alpha, \beta \geq 0, \quad \text{and} \quad a + b = c + d = \alpha + \beta = 1. \tag{2.6}$$

Theorem 2.6 ([23]). *For any parameters* (2.6) *the QSO* (2.5) *has a unique fixed point $\mathbf{x}^* \in S^2$.*

Let $D_\mathbf{x}V(\mathbf{x}^*) = (\partial V_i/\partial x_j)(\mathbf{x}^*)$ be a Jacobian of V at a point $\mathbf{x}^*$.

Definition 2.7. *[3] A fixed point $\mathbf{x}^*$ is called hyperbolic if its Jacobian $D_\mathbf{x}V(\mathbf{x}^*)$ has no eigenvalues on the unit circle.*

Definition 2.8. *[3] A hyperbolic fixed point $\mathbf{x}^*$ is called:*

i) attracting if all the eigenvalues of the Jacobian $D_\mathbf{x}V(\mathbf{x}^*)$ are less than 1 in absolute value;

ii) repelling if all the eigenvalues of the Jacobian $D_\mathbf{x}V(\mathbf{x}^*)$ are greater than 1 in absolute value;

iii) a saddle otherwise.

3. Main result

In this section, we consider the following two strictly non-Volterra QSOs on the two-dimensional simplex

$$V_1: \begin{cases} x_1' = x_3^2 + 2x_2x_3, \\ x_2' = x_1^2 + 2x_1x_3, \\ x_3' = x_2^2 + 2x_1x_2, \end{cases} \quad (3.1) \qquad V_2: \begin{cases} x_1' = x_2^2 + 2x_2x_3, \\ x_2' = x_3^2 + 2x_1x_3, \\ x_3' = x_1^2 + 2x_1x_2. \end{cases} \quad (3.2)$$

Remark 3.1. *It is known that the operators* (3.1), (3.2) *have unique fixed point* $C(1/3, 1/3, 1/3)$ *which is repelling and for any initial point* $\mathbf{x}^{(0)} \in S^2$ *one has* $\omega_{V_i}(\mathbf{x}^{(0)}) \subset \partial S^2$, $i = 1, 2$ *and* $\omega_{V_i}(\mathbf{x}^{(0)})$, $i = 1, 2$ *is infinite set, moreover they are non-ergodic transformations (see [23]).*

Note that in [9] a convex combination of a regular and non-ergodic QSOs was studied. It is shown that the properties of this operator depend on the combination parameter. Also some examples of invariant curves and the set of limit points of the trajectories of the convex combination operator are given. Now it seems natural to study dynamical systems of a QSO which is a convex combination of two non-ergodic QSOs.

In [10] the asymptotical behavior of the general form of homeomorphisms defined on the two-dimensional simplex was studied.

Let us consider the convex combination of given QSOs, that is convex linear combination of

$$V_\theta = \theta V_1 + (1 - \theta)V_2.$$

It is clear that V_θ has the form

$$V_\theta: \begin{cases} x_1' = \theta x_3^2 + (1 - \theta)x_2^2 + 2x_2x_3, \\ x_2' = \theta x_1^2 + (1 - \theta)x_3^2 + 2x_1x_3, \\ x_3' = \theta x_2^2 + (1 - \theta)x_1^2 + 2x_1x_2. \end{cases} \quad (3.3)$$

It is easy to see that the QSO (3.3) is strictly non-Volterra QSO and it coincides with operator (2.5) when $a = c = \beta = \theta$. One can verify that C is the fixed point of (3.3) and by Theorem 2.6 it is the unique fixed point.

Note that the operator (3.3) doesn't coincides with QSO which studied in [10].

Theorem 3.2. *If* $\theta = \frac{1}{2}$ *then* C *is non-hyperbolic; if* $\theta \neq \frac{1}{2}$ *then* C *is repelling;*

Proof. Using $x_3 = 1 - x_1 - x_2$ one can rewrite the QSO (3.3) as follows

$$\begin{cases} x_1' = \theta x_1^2 - x_2^2 - 2(1 - \theta)x_1x_2 - 2\theta x_1 + 2(1 - \theta)x_2 + \theta, \\ x_2' = -x_1^2 + (1 - \theta)x_2^2 - 2\theta x_1x_2 + 2\theta x_1 - 2(1 - \theta)x_2 + 1 - \theta. \end{cases} \quad (3.4)$$

where $(x_1, x_2) \in \{(x_1, x_2) : \ x_1, x_2 \geq 0, \ x_1 + x_2 \in [0, 1]\}$ and x_1, x_2 are the first two coordinates of a point lying in the simplex S^2. One can check that

$$-1 \pm \frac{\sqrt{3}}{3}|1 - 2\theta|i$$

are eigenvalues of Jacobian $D_{\mathbf{x}}V(\mathbf{x})$ of the the operator (3.4) at the point C and from it immediately implies the statements in the theorem. □

Consider the operator $W_\tau : S^2 \to S^2$ defined as follows

$$W_\tau: \begin{cases} x_1' = (1 - \tau)x_1^2 + \tau x_2^2 + 2x_1x_2, \\ x_2' = (1 - \tau)x_2^2 + \tau x_3^2 + 2x_2x_3, \\ x_3' = (1 - \tau)x_3^2 + \tau x_1^2 + 2x_1x_3. \end{cases} \quad (3.5)$$

Theorem 3.3. *[4] The following statements hold true.*

i) If $\tau = 1/2$ then $\varphi(\mathbf{x}) = |x_1 - x_2||x_1 - x_3||x_2 - x_3|$ is a Lyapunov function for the operator (3.5). Moreover for any $\mathbf{x}^{(0)} \in S^2$ one has that $\omega_{W_\tau}(\mathbf{x}^{(0)}) = \{C\}$;

ii) If $\tau \neq 1/2$ then for any $\mathbf{x}^{(0)} \in S^2 \setminus \{C\}$ one has that $\omega_{W_\tau}(\mathbf{x}^{(0)}) \subset \operatorname{int} S^2$ is an infinite compact subset.

Remark 3.4. *On the basis of numerical calculations shown that the operator (3.5) is a nonergodic transformation and it exhibits the Li-Yorke chaos[4].*

Consider the permutation

$$\pi = \begin{pmatrix} 1 & 2 & 3 \\ 2 & 3 & 1 \end{pmatrix}$$

and the corresponding map T_π defined by (2.3).

Proposition 3.5. *One has that*

i) $V_\theta(\mathbf{x}) = W_\theta(T_\pi(\mathbf{x}))$;

ii) $W_\theta^n(T_\pi(\mathbf{x})) = T_\pi(W_\theta^n(\mathbf{x}))$ for all $n \geq 1$;

iii) For any $k = 0, 1, \ldots$ the following hold true

$$V_\theta^n(\mathbf{x}) = \begin{cases} W_\theta^n(\mathbf{x}), & \text{if } n = 3k, \\ T_\pi(W_\theta^n(\mathbf{x})), & \text{if } n = 3k+1, \\ T_{\pi^2}(W_\theta^n(\mathbf{x})), & \text{if } n = 3k+2. \end{cases}$$

Proof. i) Using (2.3) one has

$$V_\theta(\mathbf{x})) = \begin{cases} x_1' = (1-\theta)x_2^2 + \theta x_3^2 + 2x_2x_3, \\ x_2' = (1-\theta)x_3^2 + \theta x_1^2 + 2x_1x_3, \\ x_3' = (1-\theta)x_1^2 + \theta x_2^2 + 2x_1x_2, \end{cases}$$

$$= \begin{cases} x_1' = (1-\theta)x_{\pi(1)}^2 + \theta x_{\pi(2)}^2 + 2x_{\pi(1)}x_{\pi(2)}, \\ x_2' = (1-\theta)x_{\pi(2)}^2 + \theta x_{\pi(3)}^2 + 2x_{\pi(2)}x_{\pi(3)}, \\ x_3' = (1-\theta)x_{\pi(2)}^2 + \theta x_{\pi(1)}^2 + 2x_{\pi(1)}x_{\pi(2)}, \end{cases} = W_\theta(T_\pi(\mathbf{x});$$

ii) For $n = 1$ easy to see that $W_\theta(T_\pi(\mathbf{x})) = V_\theta(\mathbf{x}) = T_\pi(W_\theta(\mathbf{x}))$. Suppose for $n = k > 1$ we have $W_\theta^k(T_\pi(\mathbf{x})) = T_\pi(W_\theta^k(\mathbf{x}))$. Then for $k+1$ we have $W_\theta^{k+1}(T_\pi(\mathbf{x})) = W_\theta(W_\theta^k(T_\pi(\mathbf{x}))) = W_\theta(T_\pi(W_\theta^k(\mathbf{x}))) = T_\pi(W_\theta(W_\theta^k(\mathbf{x}))) = T_\pi(W_\theta^{k+1}(\mathbf{x}))$;

iii) Using $\pi^3 = id$, where $id-$ is identity permutation and $i), ii)$ we obtain $V_\theta^n(\mathbf{x}) = \underbrace{T_\pi(W_\theta(T_\pi(W_\theta(\ldots T_\pi(W_\theta(\mathbf{x}))))))}_{n} = T_\pi^n(W_\theta^n(\mathbf{x})) = T_{\pi^n}(W_\theta^n(\mathbf{x}))$ which immediately implies the statement in the iii). □

Since the QSO (3.3) is a permutation of the operator (3.5) using the results of Proposition 3.5 and Theorem 3.3 and by means of same methods and techniques which were used in [23], we shall prove the following Theorem.

Theorem 3.6. *For the operator (3.3) the following statements hold true.*

i) If $\theta = 1/2$ then $\varphi(\mathbf{x}) = |x_1 - x_2||x_1 - x_3||x_2 - x_3|$ is a Lyapunov function. Besides for any $\mathbf{x}^{(0)} \in S^2$ the trajectory converges to the center of simplex;

ii) If $\theta \neq 1/2$ then for any $\mathbf{x}^{(0)} \in S^2 \setminus \{C\}$ one has that $\omega(\mathbf{x}^{(0)}) \subset \operatorname{int} S^2$ is an infinite compact subset.

Proof. i) Let $\theta = 1/2$ then since $\varphi(\mathbf{x}) = |x_1 - x_2||x_1 - x_3||x_2 - x_3|$ is a Lyapunov function for the operator W_θ one has $\varphi(W_{1/2}(\mathbf{x})) \leq \varphi(\mathbf{x})$. We have $\varphi(V_{1/2}(\mathbf{x})) = \varphi(W_{1/2}(T_\pi(\mathbf{x}))) \leq \varphi(T_\pi(\mathbf{x}))$. It is easy to see that $\varphi(T_\pi(\mathbf{x})) = \varphi(\mathbf{x})$. So, $\varphi(\mathbf{x})$ is a Lyapunov function for QSO $V_{1/2}$.

Let $\mathbf{x}^{(0)} \in S^2$ be an initial point. Using Proposition 3.5 and Theorem 3.3 one has

$$\lim_{n\to\infty} V_{1/2}^n(\mathbf{x}^{(0)}) = \lim_{n\to\infty} W_{1/2}^n(T_\pi(\mathbf{x}^{(0)})) = T_\pi\Big(\lim_{n\to\infty} W_{1/2}^n(\mathbf{x}^{(0)})\Big) = C,$$

that is the trajectory converges to the center of the simplex. The operator $V_{1/2}$ has the regularity property.

ii) Let $\theta \neq 1/2$ and $\mathbf{x}^{(0)} \in S^2 \setminus \{C\}$ be any initial point. Let $\mathbf{y} \in \omega_{W_\theta}(\mathbf{x}^{(0)})$ be any point. Then there exists a sequence $\{n_k\}, k = 1, 2, \ldots$ such that $\lim\limits_{k\to\infty} W_\theta^{n_k}(\mathbf{x}^{(0)}) = \mathbf{y}$. It is evident that at least one of the following subsequences is infinite:

$$\{n_k\}_{k=1,2,\ldots} \cap \{3t\}_{t=1,2,\ldots}, \quad \{n_k\}_{k=1,2,\ldots} \cap \{3t+1\}_{t=1,2,\ldots}, \quad \{n_k\}_{k=1,2,\ldots} \cap \{3t+2\}_{t=1,2,\ldots}.$$

We denote this subsequence by $\{n_{k_s}\}_{s=1,2,\ldots}$ then from Proposition 3.5 it follows

$$\lim_{s\to\infty} V_\theta^{n_{k_s}}(\mathbf{x}^{(0)}) = \mathbf{y},$$

that is $\mathbf{y} \in \omega_{V_\theta}(\mathbf{x}^{(0)})$.

Consequently, for each $\mathbf{x}^{(0)} \in S^2$ and each $\mathbf{y} \in \omega_{W_\theta}(\mathbf{x}^{(0)})$ we obtain that $|\omega_{V_\theta}(\mathbf{x}^{(0)}) \cap \{\mathbf{y}, T_\pi(\mathbf{y}), T_{\pi^2}(\mathbf{y})\}| \geq 1$, where $|\cdot|$ denotes the number of elements of a set. Hence cardinality of the set $\omega_{V_\theta}(\mathbf{x}^{(0)})$ at least equal to cardinality of the set $\omega_{W_\theta}(\mathbf{x}^{(0)})$. This by Theorem 3.3 ii) completes the proof. □

Similarly to from Remark 3.4 we have

Remark 3.7. *The operator* (3.3) *is a nonergodic transformation and it exhibits the Li-Yorke chaos.*

Remark 3.8. *The operator* (3.3) *at* $\theta = 1/2$ *is regular. The Lipschitz constant of an operator* V *is*

$$l(V) = \sup_{\mathbf{x}\neq\mathbf{y}} \frac{||V(\mathbf{x}) - V(\mathbf{y})||}{||\mathbf{x} - \mathbf{y}||}$$

where $||\cdot||$ *is some norm in* $\mathbb{R}^m$. *It is evident that if* $l(V) < 1$ *then the operator* V *will be strict contraction and consequently one has that* V *has a unique fixed point and the all trajectories of* V *converges to this fixed point exponentially rapidly. By Theorem 8.2.1 of [15] we have* $l(V) = \max\limits_{i_1,i_2,j} \sum\limits_{k=1}^{m} |p_{i_1j,k} - p_{i_2j,k}|$. *It is easy to check that* $l(V) = 2$ *for the operator* (2.5) *with* (2.6). *Thus our operator* (2.5) *is not contraction.*

Acknowledgments

The author wishes to thank professors U.A. Rozikov and F.M. Mukhamedov for useful suggestions.

References

[1] Akin E and Losert V 1984 *J. Math. Biol.* **20** 231–258
[2] Bernstein S 1942 *Ann. Math. Statistics* **13** 53–61
[3] Devaney R L 1989 *An introduction to chaotic dynamical systems* (New York: Addison-Westley)
[4] Ganikhodzhaev N N, Saburov M and Nawi A M 2014 *The Scientific World Journal* **2014** 1–11
[5] Ganikhodzhaev N N and Zanin D V 2004 *Russ. Math. Surv.* **59** 571–572

[6] Ganikhodzhaev N N and Zanin D V 2012 *arXiv:1205.3841* (in Russian)
[7] Ganikhodzhaev R N 1993 *Russ. Acad. Sci., Sb., Math.* **76** 489–506
[8] Ganikhodzhaev R N 1994 *Math. Notes* **56** 1125–1131
[9] Ganikhodzhaev R N 1989 *Dokl. Acad. Nauk UzSSR* **1** 3–5 (in Russian)
[10] Ganikhodzhaev R N and Pirnapasov A T *Cont. Math. AMS Proc.* (to appear)
[11] Ganikhodzhaev R N and Eshmamatova D B 2006 *Vladikavkaz. Mat. Zh.* **8** 12–28 (in Russian)
[12] Ganikhodzhaev R, Mukhamedov F and Rozikov U 2011 *Infin. Dimens. Anal. Quant. Prob. Relat. Top.* **14** 279–335
[13] Li T Y and Yorke J A 1975 *The American Mathematical Monthly* **82** 985–992
[14] Jenks R D 1969 *J. Differential Equations* **5** 497–514
[15] Lyubich Y I 1992 *Mathematical structures in population genetics* (Berlin: Springer)
[16] Mukhamedov F and Ganikhodjaev N 2015 *Lect. Notes Math.* **2133** (Berlin: Springer)
[17] Mukhamedov F, Akin H and Temir S 2005 *J. Math. Anal. Appl.* **310** 533–556
[18] Mukhamedov F, Qaralleh I and Rozali W N F A W 2014 *Sains Malaysiana* **43** 1275–1281.
[19] Mukhamedov F and Taha M H M 2015 *AIP Conf. Proc.* **1660** 050057.
[20] Saburov M 2012 In book: *Proc. Int. Conf. Adv. Sci. Tech., Contem. Math., Math. Phys. Appl., Malaysia, November 2012* 54–55
[21] Ulam S M 1960 *Interscience Tracts in Pure and Applied Mathematics* **8**
[22] Zakharevich M I 1978 *Russ. Math. Surv.* **33** 265–266
[23] Rozikov U A and Zhamilov U U 2009 *Sb. Math.* **200** 1339–1351

Journal of Physics: Conference Series **697** (2016) 012014 doi:10.1088/1742-6596/697/1/012014

Dynamics of Double Stochastic Operators

Mansoor Saburov

Department of Computational & Theoretical Sciences, Faculty of Science, International Islamic University Malaysia, P.O. Box, 25200, Kuantan, Pahang, MALAYSIA

E-mail: msaburov@iium.edu.my, msaburov@gmail.com

Abstract. A double stochastic operator is a generalization of a double stochastic matrix. In this paper, we study the dynamics of double stochastic operators. We give a criterion for a regularity of a double stochastic operator in terms of absences of its periodic points. We provide some examples to insure that, in general, a trajectory of a double stochastic operator may converge to any interior point of the simplex.

1. Introduction

A notion of a double stochastic operator (in short DSO) was firstly introduced in the paper [7] as a generalization of a double stochastic matrix in a class of nonlinear operators. For reasons of self-exposition, it is convenient to provide some necessary notations and notions in the theory of majorization (for a detail, see [1, 2, 13]).

Let $\|\mathbf{x}\|_1 = \sum\limits_{k=1}^{m} |x_k|$ be a norm of a vector $\mathbf{x} = (x_1, \cdots, x_m) \in \mathbb{R}^m$. We say that $\mathbf{x} \geq 0$ (resp. $\mathbf{x} > 0$) if $x_k \geq 0$ (resp. $x_k > 0$) for all $k = \overline{1,m}$. Let $S^{m-1} = \{\mathbf{x} \in \mathbb{R}^m : \|\mathbf{x}\|_1 = 1, \ \mathbf{x} \geq 0\}$ be the $(m-1)$−dimensional standard simplex. An element of the simplex S^{m-1} is called *a stochastic vector*. Recall that a square matrix $\mathbb{P} = (p_{ij})_{i,j=1}^{m}$ is called *stochastic* if every row is a stochastic vector. A square matrix $\mathbb{P} = (p_{ij})_{i,j=1}^{m}$ is said to be *double stochastic* if every row and column are stochastic vectors. For a given vector $\mathbf{x} = (x_1, \cdots, x_m) \in \mathbb{R}^m$, let $x_{[1]} \geq \cdots \geq x_{[m]}$ denote the components of $\mathbf{x}$ in a non-increasing order and $\mathbf{x}_\downarrow = (x_{[1]}, \cdots, x_{[m]})$. We set $\mathbb{R}^m_\downarrow = \{x \in \mathbb{R}^m : x_1 \geq \cdots \geq x_m\}$.

We say that $\mathbf{x}$ is majorized by $\mathbf{y}$ written $\mathbf{x} \prec \mathbf{y}$ if

$$\sum_{i=1}^{k} x_{[i]} \leq \sum_{i=1}^{k} y_{[i]}, \quad \forall\, k = \overline{1, m-1}, \qquad \sum_{i=1}^{m} x_{[i]} = \sum_{i=1}^{m} y_{[i]}.$$

The following results are the classical results in the theory of majorization [13].

Theorem 1.1. *Let $\mathbf{x}, \mathbf{y} \in \mathbb{R}^m$ be vectors. The following statements are equivalent:*

(i) The vector $\mathbf{x}$ is majorized by $\mathbf{y}$, i.e., $\mathbf{x} \prec \mathbf{y}$;

(ii) One has $D\mathbf{y} = \mathbf{x}$ for some double stochastic matrix D;

(iii) The vector $\mathbf{x}$ is in the convex hull of the $m!$ permutations of $\mathbf{y}$.

(iv) One has $\sum \phi(x_i) \leq \sum \phi(y_i)$ for all convex continuous functions ϕ;

We shall interchangeably use three equivalent statements $(i) - (iii)$ throughout this paper. A set of all $m!$ permutations of x and its convex hull are denoted, respectively, by $Per_{m!}(\mathbf{x})$ and $Conv(Per_{m!}(\mathbf{x}))$. Due to Theorem 1.1, the conditions (i) and (ii) are equivalent. Therefore, we can give an equivalent definition of a double stochastic matrix which is convenient in nonlinear settings: a matrix D is said to be *double stochastic* if one has that $D\mathbf{x} \prec \mathbf{x}$ for any $\mathbf{x} \in \mathbb{R}^m$.

Any mapping $V : S^{m-1} \to S^{m-1}$ is called *a stochastic operator.*

Definition 1.2 ([7]). *A stochastic operator $V : S^{m-1} \to S^{m-1}$ is said to be double stochastic if one has that $V(\mathbf{x}) \prec \mathbf{x}$ for any $\mathbf{x} \in S^{m-1}$.*

Proposition 1.3. *Let $V : S^{m-1} \to S^{m-1}$ be a stochastic operator. Then the following statements are equivalent:*

(i) V is a double stochastic operator;

(ii) One has that $V(Conv(Per_{m!}(\mathbf{x}))) \subset Conv(Per_{m!}(\mathbf{x}))$ for any $\mathbf{x} \in S^{m-1}$.

Proof. It follows from Theorem 1.1 that one has that $\mathbf{x} \prec \mathbf{y}$ if and only if $Conv(Per_{m!}(\mathbf{x})) \subset Conv(Per_{m!}(\mathbf{y}))$. Consequently, one has that $V(\mathbf{x}) \prec \mathbf{x}$ for any $\mathbf{x} \in S^{m-1}$ if and only if $V(Conv(Per_{m!}(\mathbf{x}))) \subset Conv(Per_{m!}(\mathbf{x}))$ for any $\mathbf{x} \in S^{m-1}$. □

Sometimes, the following geometric definition of a DSO is very useful in a practice.

Definition 1.4. *A stochastic operator $V : S^{m-1} \to S^{m-1}$ is said to be double stochastic if one has that $V(Conv(Per_{m!}(\mathbf{x}))) \subset Conv(Per_{m!}(\mathbf{x}))$ for any $\mathbf{x} \in S^{m-1}$.*

Throughout this paper, we shall consider a continuous DSO without mentioning "continuity".

A cubic matrix $\mathcal{P} = (p_{ijk})_{i,j,k=1}^m$ is called *stochastic* if $\sum_{k=1}^m p_{ijk} = 1,\ p_{ijk} \geq 0,\ \forall i,j,k = \overline{1,m}$. Every cubic stochastic matrix is associated with a quadratic stochastic operator $V : S^{m-1} \to S^{m-1}$ as follows

$$(V(\mathbf{x}))_k = \sum_{i,j=1}^m x_i x_j p_{ijk}, \quad \forall\ k = \overline{1,m}. \tag{1.1}$$

A Birkhoff theorem states that a set of extreme points of a set of double stochastic matrices coincides with a set of all permutations matrices. One of the main purposes of studying DSO was to solve Birkhoff's problem in a class of quadratic double stochastic operators. However, Birkhoff's problem remains open in the class of quadratic double stochastic operators [9, 10].

By being the simplest nonlinear mapping, a quadratic stochastic operator has an incredible application in population genetics[3, 5, 6, 11, 12], control systems [23, 24]. In population genetics, the quadratic stochastic operator describes a distribution of the next generation of the system if the current distribution is given [12, 26]. In this sense, the quadratic stochastic operator is a primary source for investigations of evolution of population genetics. The detailed exposure of the theory of quadratic stochastic operators is presented in [8, 10],[14]-[22].

Let $\mathbf{Fix}(V) = \{\mathbf{x} \in S^{m-1} : V(\mathbf{x}) = \mathbf{x}\}$ be a fixed point set. Due to Brouwer's theorem, $\mathbf{Fix}(V) \neq \emptyset$. In this paper, we study the dynamics of double stochastic operators. We give a criterion for a regularity of a double stochastic operator in terms of absences of its periodic points. This answers Problem 2.35 mentioned in the survey paper [8]. We provide some examples to insure that, in general, a trajectory of a double stochastic operator may converge to any interior point of the simplex.

2. Examples for Double Stochastic Operators

Let S^1 be 1D simplex, $e_1 = (1,0)$ and $e_2 = (0,1)$ be vertexes of the simplex S^1. Let $[a,b] = \{(1-\lambda)a + \lambda b\}_{0\leq\lambda\leq1}$ be a closed line segment connecting points $a,b \in S^1$. For any $a = (a_1;a_2) \in S^1$ we set $a' = (a_2;a_1) \in S^1$. Let us fix a point $a_\circ = (a_1^\circ;a_2^\circ) \in S^1$ where $a_1^\circ \leq a_2^\circ$. Throughout this paper, for any mapping V, we denote by $V^{\circ n} = \underbrace{V \circ \cdots \circ V}_{n}$.

Example 2.1. *We define an operator $V_{S^1} : S^1 \to S^1$ as follows:*

$$V_{S^1}(x) = \begin{cases} \dfrac{x + a_\circ}{2} & if \quad x \in [e_2, a_\circ] \\ x & if \quad x \in [a_\circ, a'_\circ] \\ \dfrac{x + a'_\circ}{2} & if \quad x \in [a'_\circ, e_1] \end{cases}.$$

It is clear that for any $x \in S^1$ one has that $V_{S^1}(x) \in [x, x']$. This means that V_{S^1} is a continuous DSO and $Fix(V_{S^1}) = [a_\circ, a'_\circ]$. Moreover, we get that

$$V_{S^1}^{\circ n}(x) = \begin{cases} \dfrac{x - a_\circ}{2^n} + a_\circ & if \quad x \in [e_2, a_\circ] \\ x & if \quad x \in [a_\circ, a'_\circ] \\ \dfrac{x - a'_\circ}{2^n} + a'_\circ & if \quad x \in [a'_\circ, e_1] \end{cases}.$$

Therefore, $V_{S^1}^{\circ n}(x) \to a_\circ$ if $x \in [e_2, a_\circ]$ and $V_{S^1}^{\circ n}(x) \to a'_\circ$ if $x \in [a'_\circ, e_1]$.

This example shows that the trajectory of DSO may converge to any interior point of S^1.

Example 2.2. *We define an operator $W_{S^1} : S^1 \to S^1$ as follows:*

$$W_{S^1}(x) = \begin{cases} \dfrac{x' + a'_\circ}{2} & if \quad x \in [e_2, a_\circ] \\ x' & if \quad x \in [a_\circ, a'_\circ] \\ \dfrac{x' + a_\circ}{2} & if \quad x \in [a'_\circ, e_1] \end{cases}.$$

It is clear that for any $x \in S^1$ one has that $W_{S^1}(x) \in [x, x']$. This means that W_{S^1} is a continuous DSO. The set of periodic-2 points of W_{S^1} is $[a_\circ, a'_\circ]$. Moreover, we have that

$$W_{S^1}^{\circ(2n+1)}(x) = \begin{cases} \dfrac{x' - a'_\circ}{2^{2n+1}} + a'_\circ & if \quad x \in [e_2, a_\circ] \\ x' & if \quad x \in [a_\circ, a'_\circ] \\ \dfrac{x' - a_\circ}{2^{2n+1}} + a_\circ & if \quad x \in [a'_\circ, e_1] \end{cases},$$

$$W_{S^1}^{\circ(2n)}(x) = \begin{cases} \dfrac{x - a_\circ}{2^{2n}} + a_\circ & if \quad x \in [e_2, a_\circ] \\ x & if \quad x \in [a_\circ, a'_\circ] \\ \dfrac{x - a'_\circ}{2^{2n}} + a'_\circ & if \quad x \in [a'_\circ, e_1] \end{cases}.$$

This yields that the trajectory of W_{S^1} converges to periodic-2 points, i.e., $\omega(x^{(0)}) = \{a_\circ, a'_\circ\}$ whenever $x^{(0)} \in [e_2, a_\circ] \cup [a'_\circ, e_1]$.

We see that a trajectory of a DSO might converge to its periodic points.

Now, we shall provide an example for a DSO in which the center of the simplex S^1 is its isolated fixed point. To do so, we are going to study the dynamics of the following function $f_{a_\circ} : [a_1^\circ, a_2^\circ] \to [a_1^\circ, a_2^\circ]$

$$f_{a_\circ}(x) = (x - a_1^\circ)\left(x - \frac{1}{2}\right)(x - a_2^\circ) + x, \tag{2.1}$$

where $a_\circ = (a_1^\circ; a_2^\circ) \in S^1$ and $a_1^\circ \leq a_2^\circ$.

Proposition 2.3. *Let $f_{a_\circ} : [a_1^\circ, a_2^\circ] \to [a_1^\circ, a_2^\circ]$ be a function given by (2.1). Then the following statements hold true:*

(i) The function $f_{a_\circ}$ is increasing;

(ii) One has that $Fix(f_{a_\circ}) = \{a_1^\circ, \frac{1}{2}, a_2^\circ\}$;

(iii) If $x \in [a_1^\circ, \frac{1}{2}]$ then $f_{a_\circ}(x) \geq x$ and if $x \in [\frac{1}{2}, a_2^\circ]$ then $f_{a_\circ}(x) \leq x$;

(iv) One has that $\omega_{f_{a_\circ}}(x_\circ) = \{\frac{1}{2}\}$ for any $x_\circ \in (a_1^\circ, a_2^\circ)$.

The proof of the proposition is straightforward.

Example 2.4. *We define an operator $Z_{S^1} : S^1 \to S^1$ as follows:*

$$Z_{S^1}(x) = \begin{cases} \dfrac{x + a_\circ}{2} & if \quad x \in [e_2, a_\circ] \\ F_{a_\circ}(x) & if \quad x \in [a_\circ, a'_\circ] \\ \dfrac{x + a'_\circ}{2} & if \quad x \in [a'_\circ, e_1] \end{cases},$$

where $F_{a_\circ}(x) = (f_{a_\circ}(x_1), 1 - f_{a_\circ}(x_1))$, a function $f_{a_\circ}$ is defined by (2.1), and $x = (x_1, x_2) \in [a_\circ, a'_\circ] \subset S^1$. This operator is continuous. We want to show that it is a DSO. Thanks to Proposition 2.3 (iii), one has that $F_{a_\circ}(x) \in [x, x']$ for any $x \in [a_\circ, a'_\circ]$. This yields that $Z_{S^1}(x) \in [x, x']$ for any $x \in S^1$. Due to Definition 1.4, the operator Z_{S^1} is a DSO. Moreover, we have that $Fix\,(Z_{S^1}) = \{a_\circ, a'_\circ, c\}$, where $c = \left(\frac{1}{2}, \frac{1}{2}\right)$ is the center of the simplex S^1. One can easily see that

$$Z_{S^1}^{\circ n}(x) = \begin{cases} \dfrac{x - a_\circ}{2^n} + a_\circ & if \quad x \in [e_2, a_\circ] \\ F_{a_\circ}^{\circ n}(x) & if \quad x \in [a_\circ, a'_\circ] \\ \dfrac{x - a'_\circ}{2^n} + a'_\circ & if \quad x \in [a'_\circ, e_1] \end{cases},$$

where $F_{a_\circ}^{\circ n}(x) = \left(f_{a_\circ}^{\circ n}(x_1), 1 - f_{a_\circ}^{\circ n}(x_1)\right)$. Consequently, we have that

$$\omega\left(x^{(0)}\right) = \begin{cases} \{a_\circ\} & if \quad x^{(0)} \in [e_2, a_\circ] \\ \{c\} & if \quad x^{(0)} \in (a_\circ, a'_\circ) \\ \{a'_\circ\} & if \quad x^{(0)} \in [a'_\circ, e_1] \end{cases}.$$

In all examples, operators were defined on S^1. However, we can get similar pictures in the higher dimensional simplex.

Let S^{m-1} be an $(m-1)-$dimensional simplex, $e_1, \cdots, e_m$ be vertexes of the simplex S^{m-1}, and $c = \left(\frac{1}{m}, \cdots, \frac{1}{m}\right)$ be a center of the simplex S^{m-1}, where $e_i = \left(0, \cdots, 0, \underbrace{1}_{i}, 0, \cdots, 0\right)$.

Similarly, we define $[a,b]$ as a closed line segment connecting points $a, b \in S^{m-1}$, i.e., $[a,b] = \{(1-\lambda)a + \lambda b\}_{0\leq\lambda\leq 1}$.

Let us fix an interior point $a_\circ = (a_1^\circ, \cdots, a_m^\circ) \in intS^{m-1}$. Let $Per_{m!}(a_\circ)$ be a set of all permutation of $a_\circ$ and $Conv(Per_{m!}(a_\circ))$ be a convex hull of $Per_{m!}(a_\circ)$. Let $\partial(Conv(Per_{m!}(a_\circ)))$ be a boundary of the set $Conv(Per_{m!}(a_\circ))$. One can easily check that for any point $x \in S^{m-1} \setminus Conv(Per_{m!}(a_\circ))$ the set $[x,c] \cap \partial(Conv(Per_{m!}(a_\circ)))$ is a singleton. We denote it by x^*, i.e., $x^* = [x,c] \cap \partial(Conv(Per_{m!}(a_\circ)))$.

Example 2.5. *We define an operator $V_{S^{m-1}} : S^{m-1} \to S^{m-1}$ as follows:*

$$V_{S^{m-1}}(x) = \begin{cases} \dfrac{x+x^*}{2} & if \quad x \in S^{m-1} \setminus Conv(Per_{m!}(a_\circ)) \\ x & if \quad x \in Conv(Per_{m!}(a_\circ)) \end{cases}.$$

This operator is continuous. Let us show that $V_{S^{m-1}}$ is a DSO. Since x, x^*, and $\frac{x+x^*}{2}$ belong to $Conv(Per_{m!}(x))$, one has that $V_{S^{m-1}}(Conv(Per_{m!}(x))) \subset Conv(Per_{m!}(x))$ for any $x \in S^{m-1}$. Due to Definition 1.4, the operator $V_{S^{m-1}}$ is a DSO. Moreover, we have that $Fix(V_{S^{m-1}}) = Conv(Per_{m!}(a_\circ))$.

One can easily check that for any $x \in S^{m-1} \setminus Conv(Per_{m!}(a_\circ))$ one has that $V_{S^{m-1}}([x,c]) \subset [x,c]$ and $(V_{S^{m-1}}(x))^* = x^*$, where c is the center of the simplex. Therefore, we get that

$$V_{S^{m-1}}^{\circ n}(x) = \begin{cases} \dfrac{x-x^*}{2^n} + x^* & if \quad x \in S^{m-1} \setminus Conv(Per_{m!}(a_\circ)) \\ x & if \quad x \in Conv(Per_{m!}(a_\circ)) \end{cases}.$$

Consequently, $V_{S^{m-1}}^{\circ n}(x) \to x^*$ for any $x \in S^{m-1} \setminus Conv(Per_{m!}(a_\circ))$.

This examples shows that any interior point of the simplex S^{m-1} might be a limiting point of a trajectory of some DSO defined on S^{m-1}.

Let us give an example for a DSO defined on the simplex S^{m-1} in which its omega limiting set is not a singleton.

Example 2.6. *We define an operator $W_{S^{m-1}} : S^{m-1} \to S^{m-1}$ as $W_{S^{m-1}}(x) = V_{S^{m-1}}(P_\circ x)$ for any $x \in S^{m-1}$, where $P_\circ$ is some permutation matrix and $V_{S^{m-1}}$ is a DSO defined in Example 2.5.*

One can easily see that $P_\circ \circ V_{S^{m-1}} = V_{S^{m-1}} \circ P_\circ$. In fact, we have for any $x \in S^{m-1} \setminus Conv(Per_{m!}(a_\circ))$ that

$$P_\circ(V_{S^{m-1}}(x)) = P_\circ\left(\frac{x+x^*}{2}\right) = \frac{P_\circ x + P_\circ x^*}{2} = \frac{P_\circ x + (P_\circ x)^*}{2} = V_{S^{m-1}}(P_\circ(x)).$$

Therefore, we get that $W_{S^{m-1}}^{\circ n}(x) = P_\circ^{\circ n}(V_{S^{m-1}}^{\circ n}(x))$ for any $x \in S^{m-1}$.

It follows from Example 2.5 that

$$Fix(V_{S^{m-1}}) = Conv(Per_{m!}(a_\circ)), \quad \omega_{V_{S^{m-1}}}(x) = \{x^*\}$$

for any $x \in S^{m-1} \setminus Conv(Per_{m!}(a_\circ))$. Consequently, we have that

$$Per_k(W_{S^{m-1}}) = Conv(Per_{m!}(a_\circ)), \quad \omega_{W_{S^{m-1}}}(x) = \{P_\circ^{\circ i} x^*\}_{i=1}^k$$

for any $x \in S^{m-1} \setminus Conv(Per_{m!}(a_\circ))$, where $P_\circ^{\circ k} = \mathbb{I}$ and $\mathbb{I}$ is an identity matrix.

Example 2.7. *The following quadratic operator $V : S^2 \to S^2$ was studied in the paper [25]*

$$V : \begin{cases} (V(x))_1 = x_1x_2 + x_1x_3 + x_3^2 \\ (V(x))_2 = x_2^2 + x_1x_2 + x_2x_3 \\ (V(x))_3 = x_3x_2 + x_1^2 + x_1x_3. \end{cases}$$

Let $I(x) = (x_1, x_2, x_3)$ $\Pi(x) = (x_3, x_2, x_1)$ be the identity and permutation operators. It is clear that, one has for any $x \in S^2$ that

$$V(x) = x_2I(x) + (x_1 + x_3)\Pi(x). \tag{2.2}$$

This means that $V(x) \in Conv(Per_{3!}(x))$ for any $x \in S^2$, i.e., V is a double stochastic operator. Due to (2.2), if $x_1 = x_3$ then $I(x) = \Pi(x)$ and $V(x) = x_2I(x) + (x_1 + x_3)I(x) = I(x) = x$, if $x_2 = 0$ then $V(x) = (x_1 + x_3)\Pi(x) = \Pi(x)$. Therefore, $\Gamma_{13} = \{x \in S^2 : x_1 = x_3\}$ is a fixed point set and $\Gamma_2 = \{x \in S^2 : x_2 = 0\}$ is a periodic-2 point set.

Theorem 4 of the paper [25] says that if $x^{(0)} \in S^2 \setminus \Gamma_2$ then the trajectory $\{V^{\circ n}(x^{(0)})\}_{n\in\mathbb{N}}$ converges to the center $(\frac{1}{3}, \frac{1}{3}, \frac{1}{3})$ of the simplex S^2. However, this is obviously wrong because of $\left(x^{(n)}\right)_2 = \left(V^{\circ n}(x^{(0)})\right)_2 = x_2^{(0)}$ for any $n \in \mathbb{N}$.

Since $x_2^{(n)} = x_2^{(0)}$ and $\left|x_1^{(n+1)} - x_3^{(n+1)}\right| = \left|2x_2^{(0)} - 1\right| \left|x_1^{(n)} - x_3^{(n)}\right|$ for any $n \in \mathbb{N}$, we have that

$$\omega_V(x^{(0)}) = \left\{ \left(\frac{1 - x_2^{(0)}}{2}, x_2^{(0)}, \frac{1 - x_2^{(0)}}{2} \right) \right\}$$

for any $x^{(0)} \in S^2 \setminus \Gamma_2$.

These examples show that, in general, a trajectory of DSO may converge to any interior point of the simplex. In the next section, we provide a criterion for a regularity of a double stochastic operator in terms of absences of its periodic points. This answers Problem 2.35 mentioned in the survey paper [8]. Some concrete examples to Problem 2.35 are also given in the papers [23, 24].

3. Regularity of Double Stochastic Operators

Let (X, d) be a compact convex manifold and $V : X \to X$ be a continuous operator. Let $\{x^{(n)}\}_{n=0}^{\infty}$, where $x^{(n)} = V(x^{(n-1)}) = V^{\circ n}(x^{(0)})$, $x^{(0)} \in X$, be a trajectory of V starting from an initial point $x^{(0)}$. An operator V is called *regular* if its trajectory $\{x^{(n)}\}_{n=0}^{\infty}$ converges for all $x^{(0)} \in X$ (a limiting point might be depended on an initial point). It is clear that if V is regular then V does not have any order periodic points except fixed points. One of the fascinating results in 1D dynamical system is that *if $X = [a, b]$ then V is regular if and only if it does not have any order periodic points except fixed points* (see [4]). It is natural to seek an analogy of this incredible result in the higher dimensional case. However, in general, this result does not hold true in the higher dimensional case. In the paper [20], it was shown that, in the class of Volterra-QSO, absences of periodic points do not imply a regularity of Volterra-QSO. It turns out that, in the class of DSO, the regularity of DSO can be described in terms of absences of periodic points of DSO.

Theorem 3.1. *Let $V : S^{m-1} \to S^{m-1}$ be a DSO. Then V is regular if and only if it does not have any order periodic points except fixed points.*

Proof. Let $V : S^{m-1} \to S^{m-1}$ be a DSO. It is clear that if V is regular then it does not have any order periodic points except fixed points. Let us prove the *"if"* part of the theorem.

The "if" part. Suppose that V does not have any order periodic points except fixed points. Let $\{x^{(n)}\}_{n=0}^{\infty}$ be a trajectory of V starting from any initial point $x^{(0)} \in S^{m-1}$. Since $V : S^{m-1} \to S^{m-1}$ is a DSO, we have that

$$\sum_{i=1}^{k} x_{[i]}^{(n+1)} \leq \sum_{i=1}^{k} x_{[i]}^{(n)}, \quad \forall\, k = \overline{1, m-1}, \tag{3.1}$$

$$\sum_{i=1}^{m} x_{[i]}^{(n+1)} = \sum_{i=1}^{m} x_{[i]}^{(n)} = 1. \tag{3.2}$$

It follows from (3.1) that $\left\{ \sum\limits_{i=1}^{k} x_{[i]}^{(n)} \right\}_{n=0}^{\infty}$ is a bounded decreasing sequence (therefore it is convergent) for every $k = \overline{1, m-1}$.

Consequently, it follows from $x_{[k]}^{(n)} = \sum\limits_{i=1}^{k} x_{[i]}^{(n)} - \sum\limits_{i=1}^{k-1} x_{[i]}^{(n)}$ and (3.2) that a sequence $\{x_{[k]}^{(n)}\}_{n=0}^{\infty}$ is convergent for any $k = \overline{1, m}$. This means that the sequence $x_{\downarrow}^{(n)} = \left(x_{[1]}^{(n)}, x_{[2]}^{(n)}, \cdots, x_{[m]}^{(n)}\right)$ converges to some point $x^* = (x_1^*, \cdots, x_m^*)$ whenever $n \to \infty$.

Let $\omega\left(x^{(0)}\right)$ be an omega limiting set of the trajectory $\{x^{(n)}\}_{n=0}^{\infty}$ and $x^{\circ} \in \omega\left(x^{(0)}\right)$ be any point. Then there is a sub-sequence $\{x^{(n_k)}\}_{k=0}^{\infty}$ of the sequence $\{x^{(n)}\}_{n=0}^{\infty}$ such that $x^{(n_k)} \to x^{\circ}$ whenever $k \to \infty$. Since a down arrow mapping $\downarrow: \mathbb{R}^m \to \mathbb{R}^m_{\downarrow}$, where $\downarrow(x) = x_{\downarrow}$, is continuous, we get that $x_{\downarrow}^{(n_k)} \to x_{\downarrow}^{\circ}$ whenever $k \to \infty$. On the other hand, we know that $x_{\downarrow}^{(n)} \to x^*$ whenever $n \to \infty$. Therefore, we have that $x_{\downarrow}^{\circ} = x^*$. This means that there is a permutation matrix P such that $x^{\circ} = Px^*$. As a conclusion, we may say that any omega limiting point is some permutation of x^*. It means that $\omega\left(x^{(0)}\right)$ is a finite set, i.e., $\omega\left(x^{(0)}\right) = \{P_i x^*\}_{i=1}^{r}$ for some permutation matrices P_i, where $i = \overline{1, r}$.

Since the simplex S^{m-1} is compact and V is continuous, we obtain that $V\left(\omega\left(x^{(0)}\right)\right) = \omega\left(x^{(0)}\right)$ and every omega limiting point is a periodic point of (minimum) period $|\omega\left(x^{(0)}\right)|$. We know that V does not have any order periodic points except fixed points, therefore, we get that $|\omega\left(x^{(0)}\right)| = 1$ for any $x^{(0)} \in S^{m-1}$. This means that V is regular. □

If a DSO is regular then a limiting point of its trajectory is a fixed point. The structure of the fixed point set can be arbitrary. The fixed point set can be finite as well as infinite (see examples section). It is clear that the center $c = (\frac{1}{m}, \cdots, \frac{1}{m})$ of the simplex S^{m-1} is always the fixed point of any DSO. In fact, one has that $c \prec V(c) \prec c$. This means that $V(c) = Pc$ for some permutation matrix P. On the other hand, $Pc = c$ for any permutation matrix P. Therefore, $V(c) = c$. Suppose that the center c of the simplex is the isolated fixed point, i.e., there is a neighborhood U_c of the point c such that U_c does not contain any periodic and fixed points of DSO except c. We then have the following result.

Proposition 3.2. *Let $V : S^{m-1} \to S^{m-1}$ be a DSO. If the center of the simplex is the isolated fixed point then it is locally attracting.*

Proof. Let $V : S^{m-1} \to S^{m-1}$ be a DSO and the center $c = (\frac{1}{m}, \cdots, \frac{1}{m})$ of the simplex be its isolated fixed point. This means that, without loss of any generality, there is a convex symmetric neighborhood U_c of c (a set is symmetric if it is invariant under any permutation matrix) such that U_c does not contain any periodic and fixed points of DSO except c. Due to Definition 1.4, we have that $V(U_c) \subset U_c$. Moreover, since the set U_c does not contain any periodic points of DSO, due to Theorem 3.1, V is regular in U_c. We know that the only fixed point in the set U_c is the center of the simplex, therefore, any trajectory of V starting from any point in U_c converges to the center of the simplex. It means that the center of the simplex is locally attracting. □

Therefore, the dynamics of DSO are diverse. It is worth pointing out that a convergence criterion and several properties of Schur decreasing sequences were studied in [10] while a trajectory of DSO is but an example for Schur decreasing sequences.

Acknowledgments
This work was partially supported by the MOHE grant FRGS14-141-0382. The Author is grateful to the Junior Associate scheme of the Abdus Salam International Centre for Theoretical Physics, Trieste, Italy.

References

[1] Ando T 1989 *Linear Algebra and Its Application* **118** 163–248
[2] Ando T 1994 *Linear Algebra and its Application* **199** 17–67
[3] Bernstein S 1942 *Annals of Mathematical Statistics* **13** 53–61
[4] Coppel W A 1955 *Proc. Cambridge Philos. Soc.* **51** 41–43
[5] Ganikhodjaev N, Saburov M and Jamilov U 2013 *App. Math. Info. Sci.* **7**(5) 1721–1729
[6] Ganikhodjaev N, Saburov M and Nawi A M 2014 *The Scientific World Journal* **2014** 1–11
[7] Ganikhodzhaev R 1992 *Russ. Math. Surveys* **48** 244–246
[8] Ganikhodzhaev R, Mukhamedov F and Rozikov U 2011 *Inf. Dim. Anal. Quan. Prob. Rel. Top.* **14**(2) 279–335
[9] Ganikhodzhaev R, Mukhamedov F and Saburov M 2012 *Linear Algebra and its Application* **436** 1344–1366
[10] Ganikhodjaev R, Saburov M and Saburov Kh 2013 *AIP Conference Proceedings* **1557** 108–111
[11] Kesten H 1970 *Adv. App. Prob.* **2** 1–82
[12] Lyubich Yu I 1992 *Mathematical structures in population genetics* (Berlin:Springer–Verlag)
[13] Marshall A, Olkin I and Arnold B 2011 *Inequalities: Theory of majorization and its applications* (Springer)
[14] Mukhamedov F and Ganikhodjaev N 2015 *Lect. Notes Math.* **2133** (Berlin: Springer)
[15] Mukhamedov F, Akin H and Temir S 2005 *J. Math. Anal. Appl.* **310** 533–556
[16] Mukhamedov F and Embong A F 2015 *J. Inequal. Appl.* **2015** 226
[17] Mukhamedov F and Saburov M 2010 *App. Math. Info. Sci.* **4** 47–62
[18] Mukhamedov F and Saburov M 2014 *Bull. Malay. Math. Sci. Soc* **37** 59–64
[19] Mukhamedov F, Saburov M and Qaralleh I 2013 *Abst. Appl. Anal.* **2013** 942038
[20] Saburov M 2013 *World Applied Science Journal* **21** 94–97
[21] Saburov M 2015 *Math Notes* **97**(5-6) 759–763
[22] Saburov M 2015 *Ann. Funct. Anal.* **6**(4) 247–254
[23] Saburov M, Saburov Kh 2014 *ScienceAsia* **40** (4) 306–312
[24] Saburov M and Saburov Kh 2014 *Inter. Jour.Cont. Auto. Sys.* 12(6) 1276–1282
[25] Shahidi F, Ganikhodzhaev R and Abdulghafor R 2013 *Middle-East Jour.f Sci. Res.* **13** 59–63
[26] Ulam S 1960 *A collection of mathematical problems* (New-York, London)

Journal of Physics: Conference Series **697** (2016) 012015 doi:10.1088/1742-6596/697/1/012015

Quadratic Stochastic Sarymsakov Operators

Mansoor Saburov

Department of Computational & Theoretical Sciences, Faculty of Science, International Islamic University Malaysia, P.O. Box, 25200, Kuantan, Pahang, MALAYSIA

E-mail: msaburov@iium.edu.my, msaburov@gmail.com

Abstract. In this paper, we introduce a new class of quadratic stochastic operators so-called Sarymsakov operators. We study the dynamics of such kind of operators. We establish regularity of each operator from this class. Note that there are some quadratic stochastic Sarymsakov operators which are not contraction mappings.

1. Introduction

Let $\|\mathbf{x}\|_1 = \sum\limits_{k=1}^{m} |x_k|$ be a norm of a vector $\mathbf{x} = (x_1, \cdots, x_m) \in \mathbb{R}^m$. We say that $\mathbf{x} \geq 0$ (resp. $\mathbf{x} > 0$) if $x_k \geq 0$ (resp. $x_k > 0$) for all $k = \overline{1,m}$. Let $\mathbb{S}^{m-1} = \{\mathbf{x} \in \mathbb{R}^m : \|\mathbf{x}\|_1 = 1, \ \mathbf{x} \geq 0\}$ be the $(m-1)-$dimensional simplex. An element of the simplex $\mathbb{S}^{m-1}$ is called *a stochastic vector*. Recall that a square matrix $\mathbb{P} = (p_{ij})_{i,j=1}^m$ is called *stochastic* if every row is a stochastic vector. A square stochastic matrix $\mathbb{P} = (p_{ij})_{i,j=1}^m$ is called *positive* if $p_{ij} > 0, \ \forall \ i,j = \overline{1,m}$. The classical Perron-Frobenius theorem states that any positive square stochastic matrix has a unique fixed point in the simplex. However, in general, the similar result for higher dimensional hyper-matrices (tensors) does not hold true.

A cubic matrix $\mathcal{P} = (p_{ijk})_{i,j,k=1}^m$ is called *stochastic* if $\sum\limits_{k=1}^{m} p_{ijk} = 1, \ p_{ijk} \geq 0, \ \forall i,j,k = \overline{1,m}$. Every cubic stochastic matrix is associated with a quadratic stochastic operator $\mathcal{Q} : \mathbb{S}^{m-1} \to \mathbb{S}^{m-1}$ as follows

$$(\mathcal{Q}(\mathbf{x}))_k = \sum_{i,j=1}^{m} x_i x_j p_{ijk}, \quad \forall \ k = \overline{1,m}. \tag{1.1}$$

By being the simplest nonlinear mapping, a quadratic stochastic operator has an incredible application in population genetics [1, 2, 3, 10, 11], control systems [21, 22]. In population genetics, the quadratic stochastic operator describes a distribution of the next generation of the system if the current distribution is given [11, 31]. In this sense, the quadratic stochastic operator is a primary source for investigations of evolution of population genetics. The detailed exposure of the theory of quadratic stochastic operators is presented in [5, 6],[13]-[20].

A cubic stochastic matrix $\mathcal{P} = (p_{ijk})_{i,j,k=1}^m$ is said to be *positive* (written $\mathcal{P} > 0$) if $p_{ijk} > 0, \ \forall i,j,k = \overline{1,m}$. A quadratic stochastic operator associated with a positive cubic stochastic matrix is called *positive*. Let $\mathbf{Fix}(\mathcal{Q}) = \{\mathbf{x} \in \mathbb{S}^{m-1} : \mathcal{Q}(\mathbf{x}) = \mathbf{x}\}$ be a fixed point set. Due to Brouwer's theorem, $\mathbf{Fix}(\mathcal{Q}) \neq \emptyset$. Meanwhile, a fixed point of the quadratic operator is an equilibrium for the system. In contrast to the linear case, the fixed point set of the quadratic operator is sophisticated. In general, if $\mathcal{P} > 0$ then it is not necessary to be true that $|\mathbf{Fix}(\mathcal{Q})| = 1$. Recall that the quadratic stochastic operator $\mathcal{Q} : \mathbb{S}^{m-1} \to \mathbb{S}^{m-1}$ given by (1.1) is said to be $\alpha-$*contraction* (where $0 < \alpha < 1$) if one has that

$$\|\mathcal{Q}(x) - \mathcal{Q}(y)\|_1 \leq \alpha\|x - y\|_1, \quad \forall\, x, y \in \mathbb{S}^{m-1}.$$

All $\alpha-$contraction QSO were described in [11]. Due to Banach's contraction principle, the trajectory of an $\alpha-$contraction quadratic stochastic operator starting from any point of the simplex converges to a unique fixed point.

In this paper, we introduce a new class of quadratic stochastic operators so-called Sarymsakov operators. We study the dynamics of such kind of operators. We establish regularity of each operator from this class. We also provide an example for the non-contraction (expansive) quadratic stochastic operator which is regular, i.e., its trajectory starting from any point of the simplex converges to a unique fixed point.

2. The Sarymsakov Class of Stochastic Matrices

Huge efforts have been made to construct various necessary and/or sufficient conditions for the convergence of non-homogeneous Markov chains. One of the major areas of study in non-homogeneous Markov chains is that of finding conditions under which a chain is weakly ergodic. A basic technique for doing this is to establish that all finite products are regular and then require some condition on the size of the entries in the transition matrices. Thus, in looking for sets of stochastic matrices which can be used in forming weakly ergodic chains, we need to find subsets of regular stochastic matrices which form semigroups. So far, the largest semigroup of regular stochastic matrices is a Sarymsakov class of stochastic matrices [8]. The Sarymsakov class of stochastic matrices, which was introduced by Sarymsakov in [25, 26] and redefined by Seneta [29] (the terminology "*the Sarymsakov class*" was coined by Seneta), is a semigroup under matrix multiplication and contains the set of scrambling matrices as a subclass. It has been shown that any product of matrices from the Sarymsakov class becomes scrambling provided the length of the product is no shorter than the order of the matrix minus one. Ergodic and limiting theorems for the Sarymsakov class of stochastic matrices were studied in the papers [23, 24, 25, 26, 27].

Let $\mathbb{P}$ be a stochastic matrix. For all $\mathcal{A} \subset \mathcal{I} = \{1, \cdots, m\}$, let $\mathcal{F}_{\mathbb{P}}(\mathcal{A}) = \{j : p_{ij} > 0 \text{ for some } i \in \mathcal{A}\}$ be a set of one-stage consequent indices of $\mathcal{A}$. We say that $\mathbb{P}$ belongs the Sarymsakov class $\mathfrak{N}$ of stochastic matrices if for any two disjoint nonempty subsets $\mathcal{A}, \bar{\mathcal{A}} \subset \mathcal{I}$, either $\mathcal{F}_{\mathbb{P}}(\mathcal{A}) \cap \mathcal{F}_{\mathbb{P}}(\bar{\mathcal{A}}) \neq \emptyset$, or $\mathcal{F}_{\mathbb{P}}(\mathcal{A}) \cap \mathcal{F}_{\mathbb{P}}(\bar{\mathcal{A}}) = \emptyset$ and $|\mathcal{F}_{\mathbb{P}}(\mathcal{A}) \cup \mathcal{F}_{\mathbb{P}}(\bar{\mathcal{A}})| > |\mathcal{A} \cup \bar{\mathcal{A}}|$ where $|\mathcal{A}|$ denotes the cardinality of $\mathcal{A}$. The following lemma is an alternative characterization of the Sarymsakov class $\mathfrak{N}$ of stochastic matrices (see [8]).

Lemma 2.1 ([8]). *A stochastic matrix $\mathbb{P} \in \mathfrak{N}$ if and only if for any nonempty subset $\mathcal{C} \subset \mathcal{I}$ satisfying $|\mathcal{F}_{\mathbb{P}}(\mathcal{C})| \leq |\mathcal{C}|$ the following holds true: for any proper subset $\mathcal{B} \subset \mathcal{C}$ one has that $\mathcal{F}_{\mathbb{P}}(\mathcal{B}) \cap \mathcal{F}_{\mathbb{P}}(\mathcal{C} \setminus \mathcal{B}) \neq \emptyset$.*

By means of this lemma, we may obtain the following result.

Proposition 2.2. *The Sarymsakov class $\mathfrak{N}$ of stochastic matrices is a convex set.*

Proof. Let $\mathbb{P}_1, \mathbb{P}_2$ be Sarymsakov matrices and $\mathbb{P} = \lambda_1\mathbb{P}_1 + \lambda_2\mathbb{P}_2$ for $\lambda_1 + \lambda_2 = 1, 0 < \lambda_1, \lambda_2 < 1$. It is clear that $\mathcal{F}_{\mathbb{P}}(\mathcal{C}) = \mathcal{F}_{\mathbb{P}_1}(\mathcal{C}) \cup \mathcal{F}_{\mathbb{P}_2}(\mathcal{C})$ for any $\mathcal{C} \subset \mathcal{I}$. Let $|\mathcal{F}_{\mathbb{P}}(\mathcal{C})| \leq |\mathcal{C}|$ for some nonempty subset $\mathcal{C} \subset \mathcal{I}$. We then have that $|\mathcal{F}_{\mathbb{P}_1}(\mathcal{C})| \leq |\mathcal{C}|$ and $|\mathcal{F}_{\mathbb{P}_2}(\mathcal{C})| \leq |\mathcal{C}|$. Since $\mathbb{P}_1, \mathbb{P}_2$ are Sarymsakov matrices, one has that $\mathcal{F}_{\mathbb{P}_1}(\mathcal{B}) \cap \mathcal{F}_{\mathbb{P}_1}(\mathcal{C} \setminus \mathcal{B}) \neq \emptyset$ and $\mathcal{F}_{\mathbb{P}_2}(\mathcal{B}) \cap \mathcal{F}_{\mathbb{P}_2}(\mathcal{C} \setminus \mathcal{B}) \neq \emptyset$ for any proper subset $\mathcal{B} \subset \mathcal{C}$. Therefore, $\mathcal{F}_{\mathbb{P}}(\mathcal{B}) \cap \mathcal{F}_{\mathbb{P}}(\mathcal{C} \setminus \mathcal{B}) \supset (\mathcal{F}_{\mathbb{P}_1}(\mathcal{B}) \cap \mathcal{F}_{\mathbb{P}_1}(\mathcal{C} \setminus \mathcal{B})) \cup (\mathcal{F}_{\mathbb{P}_2}(\mathcal{B}) \cap \mathcal{F}_{\mathbb{P}_2}(\mathcal{C} \setminus \mathcal{B})) \neq \emptyset$. □

3. Quadratic Stochastic Sarymsakov Operators

Let us first recall some notions and notations in the theory of majorization (see [12]).

Definition 3.1 ([9, 12]). *Let* $\mathbf{x}, \mathbf{y}$ *and* $\mathbf{q} > 0$ *be stochastic vectors. We say that* $\mathbf{x}$ *is* $r-$*majorized by* $\mathbf{y}$ *with respect to (w.r.t.)* $\mathbf{q}$ *(written* $\mathbf{x} \prec^r_{\mathbf{q}} \mathbf{y}$*) if one has that*

$$\sum_{i=1}^{m} |x_i - tq_i| \leq \sum_{i=1}^{m} |y_i - tq_i|, \quad \forall\, t \in \mathbb{R}. \tag{3.1}$$

The r in $r-$majorization stands for *relative* or *ratio*.

Note that if $\mathbf{q} = (\frac{1}{m}, \cdots, \frac{1}{m})$ then the $r-$majorization w.r.t. $\mathbf{q}$ is nothing but usual majorization [9, 12]. In this case, we shall use usual notation $\prec$ for the $r-$majorization.

We denote the set of all $m \times m$ stochastic (resp. doubly stochastic) matrices by SM_m (resp. DSM_m). Let us introduce the following set of stochastic matrices for a positive stochastic vector $\mathbf{q} > 0$

$$\mathsf{SM}_m[\mathbf{q}] = \{\mathbb{P} \in \mathsf{SM}_m : \mathbf{q}\mathbb{P} = \mathbf{q}\}. \tag{3.2}$$

The set $\mathsf{SM}_m[\mathbf{q}]$ of all stochastic matrices having a common fixed distribution (point) $\mathbf{q} > 0$ is a convex compact set. Extreme points of the set $\mathsf{SM}_m[\mathbf{q}]$ and some related problems were discussed in [7]. It is worth of mentioning that if $\mathbf{q} = (\frac{1}{m}, \cdots, \frac{1}{m})$ then $\mathsf{SM}_m[\mathbf{q}]$ is a set of all doubly stochastic matrices, i.e., $\mathsf{SM}_m[\mathbf{q}] = \mathsf{DSM}_m$.

The following result was proven in [9, 12].

Theorem 3.2 ([9, 12]). *The following statements are equivalent*

(i) *One has that* $\mathbf{x} \prec^r_{\mathbf{q}} \mathbf{y}$;

(ii) *There is a stochastic matrix* $\mathbb{P} \in \mathsf{SM}_m[\mathbf{q}]$ *such that* $\mathbf{x} = \mathbf{y}\mathbb{P}$;

(iii) *One has that*

$$\sum_{i=1}^{m} q_i \varphi\left(\frac{x_i}{q_i}\right) \leq \sum_{i=1}^{m} q_i \varphi\left(\frac{y_i}{q_i}\right)$$

for all convex continuous functions φ.

Let us now recall weak and strong ergodicity of a sequence of stochastic matrices (see [30]).

Let $\{\mathbb{P}_n\}_{n=1}^{\infty} \subset \mathsf{SM}_m$ be a sequence of stochastic matrices. Let $\mathbb{P}^{[r,s]} \equiv \mathbb{P}_r\mathbb{P}_{r+1}\cdots\mathbb{P}_s$ be a finite forward product and $\mathbb{P}^{[r,\infty]} \equiv \mathbb{P}_r\mathbb{P}_{r+1}\cdots\mathbb{P}_n\mathbb{P}_{n+1}\cdots$ be a infinity forward product of stochastic matrices.

Definition 3.3 ([30]). *A sequence* $\{\mathbb{P}_n\}_{n=1}^{\infty}$ *of stochastic matrices is said to be*

(i) *weakly ergodic if* $\left(\mathbb{P}^{[r,s]}\right)_{ik} - \left(\mathbb{P}^{[r,s]}\right)_{jk} \to 0$ *as* $s \to 0$ *for every* i, j, k, r

(ii) *strongly ergodic if it is weakly ergodic and* $\left(\mathbb{P}^{[r,s]}\right)_{ij} \to p_j^{[r]}$ *as* $s \to 0$ *for every* i, j, r.

We need the following result proven by T.A. Sarymsakov in the papers [25, 26].

Theorem 3.4 ([25, 26]). *A non-homogeneous Markov chain with transition matrices from the Sarymsakov class* $\mathfrak{N}$ *of stochastic matrices is weakly ergodic.*

We define the following stochastic vectors and square stochastic matrices associated with the cubic stochastic matrix $\mathcal{P} = (p_{ijk})_{i,j,k=1}^m$

$$\begin{aligned}
\mathbf{p}_{ij\bullet} &= (p_{ij1}, p_{ij2}, \cdots, p_{ijm}), & \forall\, i,j = \overline{1,m}, \\
\mathbb{P}_i \equiv \mathbb{P}_{i\bullet\bullet} &= (p_{ijk})_{j,k=1}^m, & \forall\, i = \overline{1,m}, \\
\mathbb{P}_{\mathbf{x}} &= \sum_{i=1}^m x_i \mathbb{P}_i, & \forall\, \mathbf{x} \in \mathbb{S}^{m-1}.
\end{aligned}$$

It is easy to check that the QSO has the following vector and matrix forms

$$\mathcal{Q}(\mathbf{x}) = \sum_{i,j=1}^m x_i x_j \mathbf{p}_{ij\bullet} \qquad (Vector\ form) \tag{3.3}$$

$$\mathcal{Q}(\mathbf{x}) = \mathbf{x}\mathbb{P}_{\mathbf{x}} = \sum_{i=1}^m x_i\,(\mathbf{x}\mathbb{P}_i) \qquad (Matrix\ form) \tag{3.4}$$

We write the following notation for the matrix form (3.4) of QSO

$$\mathcal{Q} = \mathbb{P}_1 \oplus \mathbb{P}_2 \oplus \cdots \oplus \mathbb{P}_m \tag{3.5}$$

Remark 3.5. *It is worth of mentioning that $\mathbb{P}_{\mathbf{e}_i} = \mathbb{P}_i$ for any vertex $\mathbf{e}_i = (\delta_{i1}, \cdots, \delta_{im})$ of the simplex $\mathbb{S}^{m-1}$, where δ_{ij} is a Kronecher symbol.*

Without loss of generality, we may assume that $p_{ijk} = p_{jik}$ for any $i,j,k = \overline{1,m}$. This means that

$$\mathbf{p}_{ij\bullet} = \mathbf{p}_{ji\bullet}, \quad \forall\, i,j = \overline{1,m}.$$

To put this another way, in the matrix form (3.4) of the QSO, the jth row of the square stochastic matrix $\mathbb{P}_i$ is the i^{th}row of the square stochastic matrix $\mathbb{P}_j$. Therefore, this allows us to have the following relation

$$\mathbb{P}_{\mathbf{x}} = \sum_{i=1}^m x_i \mathbb{P}_i = \begin{pmatrix} \mathbf{x}\mathbb{P}_1 \\ \mathbf{x}\mathbb{P}_2 \\ \vdots \\ \mathbf{x}\mathbb{P}_m \end{pmatrix} \tag{3.6}$$

We then immediately derive the following results.

Proposition 3.6. *The following statements hold true*

(i) One has that $\mathbb{P}_{\lambda\mathbf{x}+(1-\lambda)\mathbf{y}} = \lambda\mathbb{P}_{\mathbf{x}} + (1-\lambda)\mathbb{P}_{\mathbf{y}}$ for any $\mathbf{x}, \mathbf{y} \in \mathbb{S}^{m-1}$;

(ii) One has that $\mathbf{x}\mathbb{P}_{\mathbf{y}} = \mathbf{y}\mathbb{P}_{\mathbf{x}}$ for any $\mathbf{x}, \mathbf{y} \in \mathbb{S}^{m-1}$;

(iii) One has that $\mathbb{P}_{\mathbf{x}}\mathbb{P}_{\mathbf{y}} = \mathbb{P}_{\mathbf{y}}\mathbb{P}_{\mathbf{x}}$ for any $\mathbf{x}, \mathbf{y} \in \mathbb{S}^{m-1}$ if and only if $\mathbb{P}_i\mathbb{P}_j = \mathbb{P}_j\mathbb{P}_i$, $\forall\, i,j = \overline{1,m}$;

(iv) One has that $\mathbb{P}_{\mathbf{x}}\mathbb{P}_{\mathbf{y}} = \mathbb{P}_{\mathbf{y}}\mathbb{P}_{\mathbf{x}}$ and $\mathbb{P}_{\mathbf{x}}^T\mathbb{P}_{\mathbf{y}} = \mathbb{P}_{\mathbf{y}}\mathbb{P}_{\mathbf{x}}^T$ for any $\mathbf{x} \neq \mathbf{y}$, where $\mathbf{x}, \mathbf{y} \in \mathbb{S}^{m-1}$, if and only if a quadratic operator $\mathcal{Q}_{\mathcal{E}} : \mathbb{R}^m \to \mathbb{R}^m$, $\mathcal{Q}_{\mathcal{E}}(\mathbf{x}) = ((\mathbb{P}_1\mathbf{x}, \mathbf{x}), \cdots, (\mathbb{P}_m\mathbf{x}, \mathbf{x}))$ is elliptic[1].

We are ready to introduce a new class of QSOs which would be the main topic of this paper.

[1] For the definition of an elliptic quadratic operator see [4].

Definition 3.7 ($r-$majorizing QSO). *The QSO $\mathcal{Q}:\mathbb{S}^{m-1}\to\mathbb{S}^{m-1}$ given by (3.4) is said to be $r-$majorizing w.r.t. a stochastic vector $\mathbf{q}>0$ if square stochastic matrices $\mathbb{P}_1,\cdots,\mathbb{P}_m$ have a common fixed point $\mathbf{q}>0$, i.e., $\mathbb{P}_1,\cdots,\mathbb{P}_m\in\mathsf{SM}_m[\mathbf{q}]$.*

Definition 3.8 (Sarymsakov QSO). *The QSO $\mathcal{Q}:\mathbb{S}^{m-1}\to\mathbb{S}^{m-1}$ given by (3.4) is said to be a Sarymsakov operator if square stochastic matrices $\mathbb{P}_1,\cdots,\mathbb{P}_m$ belong to the Sarymsakov class $\mathfrak{N}$ of stochastic matrices.*

We provide an example for the $r-$majorizing Sarymsakov QSO which is not contraction.

Example 3.9. *Let $\mathbf{a}=(0.1,0.3,0.6)$, $\mathbf{b}=(0.7,0.1,0.2)$, $\mathbf{c}=(0.2,0.6,0.2)$. Let us consider the following QSO $\mathcal{Q}_{\mathbf{abc}}:\mathbb{S}^2\to\mathbb{S}^2$*

$$\mathcal{Q}_{\mathbf{abc}}(\mathbf{x})=\mathbf{a}x_1^2+\mathbf{b}x_2^2+\mathbf{c}x_3^2+2\mathbf{c}x_1x_2+2\mathbf{b}x_1x_3+2\mathbf{a}x_2x_3.$$

One can see that

$$\mathbb{P}_1=\begin{pmatrix}\mathbf{a}\\ \mathbf{c}\\ \mathbf{b}\end{pmatrix},\qquad \mathbb{P}_2=\begin{pmatrix}\mathbf{c}\\ \mathbf{b}\\ \mathbf{a}\end{pmatrix},\qquad \mathbb{P}_3=\begin{pmatrix}\mathbf{b}\\ \mathbf{a}\\ \mathbf{c}\end{pmatrix}$$

are doubly stochastic matrices, i.e., $\mathbb{P}_1,\mathbb{P}_2,\mathbb{P}_3\in\mathsf{DSM}_3$. Therefore, the QSO $\mathcal{Q}_{\mathbf{abc}}:\mathbb{S}^2\to\mathbb{S}^2$ is the $r-$majorizing Sarymsakov QSO. However, it is not a contraction mapping. Indeed, if $\mathbf{x}_0=(0.85,0.1,0.05)$ and $\mathbf{y}_0=(0.9,0.1,0)$ then we have that $\|\mathcal{Q}_{\mathbf{abc}}(\mathbf{x}_0)-\mathcal{Q}_{\mathbf{abc}}(\mathbf{y}_0)\|_1=0.1005>0.1=\|\mathbf{x}_0-\mathbf{y}_0\|_1$.

Example 3.10. *Let $\mathbf{p}\in\mathbb{S}^2$ and $\mathbf{p}=(p_1,p_2,p_3)>0$ such that $p_3\le p_2\le p_1$ and $0<\varepsilon\le\frac{1}{2}$. We define the following stochastic vectors $\mathbf{r_3}=(\varepsilon p_1,\varepsilon p_2,1-\varepsilon+\varepsilon p_3)$ and*

$$\mathbf{r}=\frac{1}{1-p_3}\mathbf{p}-\frac{p_3}{1-p_3}\mathbf{r}_3,\qquad \mathbf{r}_1=\frac{p_3}{p_1}\mathbf{r}_3+(1-\frac{p_3}{p_1})\mathbf{r},\qquad \mathbf{r}_2=\frac{p_3}{p_2}\mathbf{r}_3+(1-\frac{p_3}{p_2})\mathbf{r}.$$

Let us consider the following QSO $\mathcal{Q}_{\mathbf{r_1},\mathbf{r_2},\mathbf{r_3},\mathbf{r}}:\mathbb{S}^2\to\mathbb{S}^2$

$$\mathcal{Q}_{\mathbf{r_1},\mathbf{r_2},\mathbf{r_3},\mathbf{r}}(\mathbf{x})=x_1^2\mathbf{r}_1+x_2^2\mathbf{r}_2+x_3^2\mathbf{r}_3+2(x_1x_2+x_1x_3+x_2x_3)\mathbf{r}. \tag{3.7}$$

One can see that the stochastic matrices

$$\mathbb{P}_1=\begin{pmatrix}\mathbf{r}_1\\ \mathbf{r}\\ \mathbf{r}\end{pmatrix},\qquad \mathbb{P}_2=\begin{pmatrix}\mathbf{r}\\ \mathbf{r}_2\\ \mathbf{r}\end{pmatrix},\qquad \mathbb{P}_3=\begin{pmatrix}\mathbf{r}\\ \mathbf{r}\\ \mathbf{r}_3\end{pmatrix}$$

have a common fixed point $\mathbf{p}$, i.e., $\mathbb{P}_1,\mathbb{P}_2,\mathbb{P}_3\in\mathsf{SM}_3[\mathbf{p}]$. Therefore, $\mathcal{Q}_{\mathbf{r_1},\mathbf{r_2},\mathbf{r_3},\mathbf{r}}$ is the $r-$majorizing Sarymsakov QSO. However, it is not a contraction mapping. Indeed, the Lipschitz constant is $Lip(\mathcal{Q}_{\mathbf{r_1},\mathbf{r_2},\mathbf{r_3},\mathbf{r}})=2(1-\varepsilon)\ge 1$ (see Theorem 8.2.1, Corollary 8.2.4, pages 301–302, [11]).

The following theorem is the main result of this paper.

Theorem 3.11. *Let $\mathcal{Q}:\mathbb{S}^{m-1}\to\mathbb{S}^{m-1}$ be an $r-$majorizing Sarymsakov QSO with respect to a stochastic vector $\mathbf{q}>0$. Then the following statements hold true.*

(i) One has that $\mathcal{Q}(\mathrm{int}\mathbb{S}^{m-1})\subset\mathrm{int}\mathbb{S}^{m-1}$;

(ii) The functional $\varphi_{\mathbf{q}}(\mathbf{x})=\|\mathbf{x}-\mathbf{q}\|_1$ is a Lyapunov functional;

(iii) The $r-$majorizing Sarymsakov QSO is regular, i.e., its trajectory starting from any initial point of the simplex converges to a unique fixed point $\mathbf{q}$.

Proof. Let $\mathcal{Q} : \mathbb{S}^{m-1} \to \mathbb{S}^{m-1}$ be an $r-$majorizing Sarymsakov QSO with respect to a stochastic vector $\mathbf{q} > 0$.

(i) We then have that

$$\mathbb{P}_{\mathbf{q}} = q_1\mathbb{P}_1 + q_2\mathbb{P}_2 + \cdots + q_m\mathbb{P}_m = \begin{pmatrix} \mathbf{q}\mathbb{P}_1 \\ \mathbf{q}\mathbb{P}_2 \\ \vdots \\ \mathbf{q}\mathbb{P}_m \end{pmatrix} = \begin{pmatrix} \mathbf{q} \\ \mathbf{q} \\ \vdots \\ \mathbf{q} \end{pmatrix} > 0.$$

Since $\mathbf{q} > 0$, we get that $\mathbb{P}_1 + \mathbb{P}_2 + \cdots + \mathbb{P}_m > 0$. Moreover, we obtain that $\mathbb{P}_{\mathbf{x}} > 0$ for any $\mathbf{x} > 0$. This means that, for any $\mathbf{x} \in \mathbb{S}^{m-1}$ with $\mathbf{x} > 0$, the stochastic matrix $\mathbb{P}_{\mathbf{x}}$ is positive. Consequently, we have that $\mathcal{Q}(\mathbf{x}) = \mathbf{x}\mathbb{P}_{\mathbf{x}} > 0$ for any $\mathbf{x} > 0$. This means that $\mathcal{Q}(\mathrm{int}\mathbb{S}^{m-1}) \subset \mathrm{int}\mathbb{S}^{m-1}$.

(ii) It is clear that

$$\mathbf{q}\mathbb{P}_{\mathbf{x}} = \mathbf{x}\mathbb{P}_{\mathbf{q}} = x \begin{pmatrix} \mathbf{q} \\ \mathbf{q} \\ \vdots \\ \mathbf{q} \end{pmatrix} = \mathbf{q}.$$

Therefore, the stochastic matrix $\mathbb{P}_{\mathbf{x}}$ has a fixed point $\mathbf{q}$, i.e., $\mathbb{P}_{\mathbf{x}} \in \mathsf{SM}_m[\mathbf{q}]$ for any $\mathbf{x} \in \mathbb{S}^{m-1}$.

Since $\mathbf{x}^{(n+1)} = \mathbf{x}^{(n)}\mathbb{P}_{\mathbf{x}^{(n)}}$, due to Theorem 3.2, ($i$) and ($ii$), we get that

$$|x_1^{(n+1)} - tq_1| + \cdots + |x_m^{(n+1)} - tq_m| \leq |x_1^{(n)} - tq_1| + \cdots + |x_m^{(n)} - tq_m|$$

for any $t \in \mathbb{R}$. If we let $t = 1$ then we obtain that $\varphi_{\mathbf{q}}(\mathbf{x}^{(n+1)}) \leq \varphi_{\mathbf{q}}(\mathbf{x}^{(n)})$ for any $n \in \mathbb{N}$. This means that $\varphi_{\mathbf{q}}(\mathbf{x})$ is the Lyapunov functional.

(iii) Since the set of Sarymsakov matrices is convex (see Proposition 2.2), the stochastic matrix

$$\mathbb{P}_{\mathbf{x}} = x_1\mathbb{P}_1 + \cdots + x_m\mathbb{P}_m$$

is a Sarymsakov matrix for any $\mathbf{x} \in \mathbb{S}^{m-1}$. We know that

$$\mathbf{x}^{(n+1)} = \mathbf{x}^{(n)}\mathbb{P}_{\mathbf{x}^{(n)}} = \mathbf{x}^{(0)}\mathbb{P}_{\mathbf{x}^{(0)}}\mathbb{P}_{\mathbf{x}^{(1)}}\cdots\mathbb{P}_{\mathbf{x}^{(n)}} = \mathbf{x}^{(0)}\mathbb{P}^{[0,n]}.$$

We can see that the trajectory of QSO is associated with the forward product of stochastic matrices. Due to Theorem 3.4, the Sarymsakov matrices $\mathbb{P}_{\mathbf{x}^{(n)}}$ are weakly ergodic. Since the Sarymsakov matrices $\mathbb{P}_{\mathbf{x}^{(n)}}$ share a common fixed point $\mathbf{q}$, we have that

$$\mathbf{q}\mathbb{P}^{[0,n]} = \mathbf{q}\mathbb{P}_{\mathbf{x}^{(0)}}\mathbb{P}_{\mathbf{x}^{(1)}}\cdots\mathbb{P}_{\mathbf{x}^{(n)}} = \mathbf{q}, \quad \forall\, n \in \mathbb{N}.$$

This means the Sarymsakov matrices $\mathbb{P}_{\mathbf{x}^{(n)}}$ are asymptotically stationary (see [28]). Consequently, the Sarymsakov matrices $\mathbb{P}_{\mathbf{x}^{(n)}}$ are strongly ergodic (see [28]). This means that the sequence $\{\mathbb{P}^{[0,n]}\}$, where $\mathbb{P}^{[0,n]} = \mathbb{P}_{\mathbf{x}^{(0)}}\mathbb{P}_{\mathbf{x}^{(1)}}\cdots\mathbb{P}_{\mathbf{x}^{(n)}}$ converges to a stable matrix $\mathbb{P}^*$ with identical rows $\mathbf{q}$. Therefore, the trajectory $\mathbf{x}^{(n+1)} = \mathbf{x}^{(0)}\mathbb{P}^{[0,n]}$ starting from any $\mathbf{x}^{(0)} \in \mathbb{S}^{m-1}$ converges to $\mathbf{q}$. This completes the proof. □

Due to Examples 3.9 and 3.10, there exist some $r-$majorizing Sarymsakov QSOs which are not contraction. This shows that there are some non-contraction QSOs which are still regular.

Acknowledgments
This work was partially supported by the MOHE grant FRGS14-141-0382. The Author is grateful to the Junior Associate scheme of the Abdus Salam International Centre for Theoretical Physics, Trieste, Italy.

References

[1] Bernstein S 1942 *Annals Math. Statis.* **13** 53–61
[2] Ganikhodjaev N, Saburov M and Jamilov U 2013 *App. Math. Info. Sci.* **7**(5) 1721–1729
[3] Ganikhodjaev N, Saburov M and Nawi A M 2014 *The Scientific World Jour.* **2014** 1–11
[4] Ganikhodzhaev R 1993 *Studying of quadratic stochastic operators.* Doctor of science degree thesis (Russian)
[5] Ganikhodzhaev R, Mukhamedov F and Rozikov U 2011 *Inf. Dim. Anal. Quan. Prob. Rel. Top.* **14**(2) 279–335
[6] Ganikhodjaev R, Saburov M and Saburov Kh 2013 *AIP Conf. Proc.* **1557** 108–111
[7] Hartfiel D J 1974 *Pacific J. Math.* **52**(2) 405–418
[8] Hartfiel D J and Seneta E 1990 *Lin. Alg. Appl.* **141** 47–51
[9] Joe H 1990 *J. Math. Anal. App.* **148** 287–305
[10] Kesten H 1970 *Adv. App. Prob.* **2** 1–82
[11] Lyubich Yu I 1992 *Mathematical structures in population genetics* (Berlin:Springer–Verlag)
[12] Marshall A, Olkin I and Arnold B 2011 *Inequalities: Theory of majorization and its applications* (Springer)
[13] Mukhamedov F and Ganikhodjaev N 2015 *Lect. Notes Math.* **2133** (Berlin: Springer)
[14] Mukhamedov F, Akin H and Temir S 2005 *J. Math. Anal. Appl.* **310** 533–556
[15] Mukhamedov F and Saburov M 2010 *App. Math. Info. Sci.* **4** 47–62
[16] Mukhamedov F and Saburov M 2014 *Bull. Malay. Math. Sci. Soc* **37**59–64
[17] Mukhamedov F, Saburov M and Qaralleh I 2013 *Abst. Appl. Anal.* **2013** 942038
[18] Saburov M 2013 *World Appl. Sci. Jour.* **21** 94–97
[19] Saburov M 2015 *Math Notes* **97**(5-6) 759–763
[20] Saburov M 2015 *Ann. Funct. Anal.* **6**(4) 247–254
[21] Saburov M and Saburov Kh 2014 *ScienceAsia* **40**(4) 306–312
[22] Saburov M and Saburov Kh 2014 *Inter. J. Cont. Auto. Sys.* **12**(6) 1276–1282
[23] Sarymsakov T A 1953 *Dokl. Akad. Nauk SSSR,* **90** (1) 25–28 (in Russian)
[24] Sarymsakov T A 1956 *Dokl. Akad. Nauk UzSSR* **8** 3–7 (in Russian)
[25] Sarymsakov T A 1958 *Dokl. Akad. Nauk SSSR,* **120**(3) 465–468 (in Russian)
[26] Sarymsakov T A 1961 *Theor. Prob. Appl* **6**(2) 178–185
[27] Sarymsakov T A and Mustafin Kh A 1957 *Tashkent, Tr. Sred Gosuniver 74, Fiz.-Mat. Nauki* **15** 1–38
[28] Seneta E 1973 *Studia Math.* **46**(3) 241–247
[29] Seneta E 1979 *Adv. Prob.* **11** 576–590
[30] Seneta E 1981 *Nonnegative Matrices and Markov Chains* (New York:Springer-Verlag)
[31] Ulam S 1960 *A collection of mathematical problems* (New-York, London)

Quadratic Stochastic Operators with Countable State Space

Nasir Ganikhodjaev

Department of Computational & Theoretical Sciences, Faculty of Science, International Islamic University Malaysia, P.O. Box 25200, Kuantan,Pahang, Malaysia

E-mail: gnasir@iium.edu.my

Abstract. In this paper, we provide the classes of Poisson and Geometric quadratic stochastic operators with countable state space, study the dynamics of these operators and discuss their application to economics.

1. Introduction

Let $(X, \mathcal{F})$ be a measurable space, $S(X, \mathcal{F})$ be the set of all probability measures on $(X, \mathcal{F})$, and $\{P(x, y, A) : x, y \in X, A \in \mathcal{F}\}$ be a family of functions on $X \times X \times \mathcal{F}$ that satisfy the following conditions:

(i) $P(x, y, \cdot) \in S(X, \mathcal{F})$, for any fixed $x, y \in X$;

(ii) $P(x, y, A)$ regarded as a function of two variables x and y with fixed $A \in \mathcal{F}$ is measurable function on $(X \times X, \mathcal{F} \otimes \mathcal{F})$;

(iii) $P(x, y, A) = P(y, x, A)$ for any $x, y \in X, A \in \mathcal{F}$.

We consider a nonlinear transformation (quadratic stochastic operator) $V : S(X, \mathcal{F}) \to S(X, \mathcal{F})$ defined by

$$(V\lambda)(A) = \int_X \int_X P(x, y, A) d\lambda(x) d\lambda(x), \tag{1}$$

where $\lambda \in S(X, \mathcal{F})$ is an arbitrary initial probability measure and $A \in \mathcal{F}$ is an arbitrary measurable set.

Note that the third condition $P(x, y, A) = P(y, x, A)$ is not overloaded, since otherwise one can determine a new function

$$Q(x, y, A) = \frac{P(x, y, A) + P(y, x, A)}{2}$$

with preserving quadratic stochastic operator (qso) V, i.e.

$$(V\lambda)(A) = \int_X \int_X P(x, y, A) d\lambda(x) d\lambda(x) = \int_X \int_X Q(x, y, A) d\lambda(x) d\lambda(x).$$

Definition 1.1 *An element $\mu \in S(X, \mathcal{F})$ is called a fixed point of a qso V if $V\mu = \mu$.*

Let $\{V^n\lambda : n = 0, 1, 2, \cdots\}$ be the trajectory of the initial point $\lambda \in S(X, \mathcal{F})$, where $V^{n+1}\lambda = V(V^n\lambda)$ for all $n = 0, 1, 2, \cdots$.
In measure theory, there are various notions of the convergence of measures: weak convergence, strong convergence, total variation convergence. Below we consider strong convergence.

Definition 1.2 *For $(X, \mathcal{F})$ a measurable space, a sequence $\{\mu_n\}$ is said to converge strongly to a limit μ if*

$$\lim_{n\to\infty} \mu_n(A) = \mu(A)$$

for every set $A \in \mathcal{F}$.

Definition 1.3 *A qso V is called regular if for any initial point $\mu \in S(X, \mathcal{F})$ the strong limit*

$$\lim_{n\to\infty} V^n(\mu)$$

exists.

If a state space $X = \{1, 2, \cdots, m\}$ is finite and the corresponding σ-algebra is the power set $\mathcal{P}(X)$, i.e. the set of all subsets of X, then the set of all probability measures on $(X, \mathcal{P}(X))$ coincides with

$$S^{m-1} = \{\mathbf{x} = (x_1, x_2, \cdots, x_m) \in R^m : x_i \geq 0 \text{ for any } i, \text{ and } \sum_{i=1}^{m} x_i = 1\} \tag{2}$$

and corresponding qso V has the following form

$$(V\mathbf{x})_k = \sum_{i,j=1}^{m} P_{ij,k} x_i x_j, \tag{3}$$

where $P_{ij,k} \equiv P(i, j, \{k\})$.
Such operators can be reinterpreted in terms of evolutionary operator of free population, evolutionary games and a gene conversion and in those forms it has a fair history. Note that the theory of qso on finite state space is well developed. The detailed exposure of the theory of quadratic stochastic operators is presented in [1]-[2], [4]-[5], [11-30]. In [6]-[10], [20] it was studied qso defined on countable and continual state spaces.

In statistical mechanics the ergodic hypothesis proposes a connection between dynamics and statistics. In the classical theory the assumption was made that the average time spent in any region of phase space is proportional to the volume of the region in terms of the invariant measure, more generally, that time averages may be replaced by space averages. For nonlinear dynamical systems Ulam [29] suggested as analogue of measure-theoretic ergodicity, following ergodic hypothesis:

Definition 1.4 *A nonlinear operator V defined on $S(X, \mathcal{F})$ is called ergodic, if the limit*

$$\lim_{n\to\infty} \frac{1}{n} \sum_{k=0}^{n-1} V^k \lambda$$

exists for any $\lambda \in S(X, \mathcal{F})$.

On the basis of numerical calculations for quadratic stochastic operators defined on $S(X, \mathcal{F})$ with finite X, Ulam conjectured [28] that the ergodic theorem holds for any such qso V. In 1977 Zakharevich [29] proved that this conjecture is false in general. He considered following operator on S^2

$$x_1' = x_1^2 + 2x_1x_2,$$
$$x_2' = x_2^2 + 2x_2x_3$$
$$x_3' = x_3^2 + 2x_1x_3$$

and proved that it is non-ergodic transformation. Later in [13] it was established sufficient condition to be non-ergodic transformation for qso defined on S^2 and in [5] described non-ergodic qso defined on S^3 and S^4. In next sections we will show that Ulam's conjecture is true for some classes of qso defined on infinite state space.

2. Countable State Space

In this section we consider non-linear transformations defined on countable state space and investigate their trajectory behavior. Let $X = \{0, 1, \cdots\}$ be a countable sample space and corresponding σ-algebra $\mathcal{F}$ be the power set $\mathcal{P}(X)$. Remind that to define a probability measure μ on countable sample space X it is enough to define the measure $\mu(\{k\})$ of each singleton $\{k\}, k = 0, 1, \cdots$. Below we will write $\mu(k)$ instead of $\mu(\{k\})$. For countable state space X a qso (1) has the following form:

$$V\mu(k) = \sum_{i=0}^{\infty}\sum_{j=0}^{\infty} P_{ij,k}\mu(i)\mu(j) \tag{4}$$

where $\mu \in S(X, \mathcal{F})$, $P_{ij,k} \equiv P(j, i, k)$, and $k \in X$.

2.1. Poisson QSO

Remind that a Poisson distribution P_λ with a positive real parameter λ is defined on X by the equation

$$P_\lambda(k) = e^{-\lambda}\frac{\lambda^k}{k!}, \quad k \in X$$

Definition 2.1 *A quadratic stochastic operator V (4) is called a* Poisson qso *if for any $i, j \in X$, the probability measure $P(i, j, \cdot)$ is the Poisson distribution $P_{\lambda(i,j)}$ with positive real parameter $\lambda(i, j)$.*
We select the following class of a Poisson qso.

Definition 2.2 *A qso V is called m-Poisson, where m is a positive integer, if $\{\lambda(i, j) : i, j \in X\} = \{\lambda_0, \lambda_1, \cdots, \lambda_{m-1}\}$.*

One can define m-Poisson qso as follows. Assume that for any $i, j \in X$, with $i + j \equiv s(mod\ m)$ we have $\lambda(i, j) = \lambda_s$, and

$$P_{ij,k} = e^{-\lambda_s}\frac{\lambda_s^k}{k!} \tag{5}$$

for any $k \in X$ with $k \equiv s(mod\ m)$, where $s = 0, 1, \cdots, m-1$. Then corresponding qso V is a m-Poisson qso.
For any initial measure $\mu \in S(X, \mathcal{F})$ let

$$A_s(\mu) = \sum_{n=0}^{\infty} \mu(mn + s),$$

where $s = 0, 1, \cdots, m-1$. It is easy to compute $A_s(P_\lambda)$ for the Poisson distribution P_λ. Let

$$B_q(P_\lambda) = \sum_{n=0}^{\infty} \frac{\lambda^{nm+q}}{(nm+q)!}$$

with $q = 0, 1, \cdots, m-1$, where $\sum_{q=0}^{m-1} B_q(P_\lambda) = e^\lambda$ and $A_s(P_\lambda) = e^{-\lambda} B_s(P_\lambda)$. Let us consider the equation $\nu^m = 1$ with roots $\nu_q = \cos\frac{2\pi q}{m} + i\sin\frac{2\pi q}{m}$, where $q = 0, 1, \cdots, m-1$. Then

$$e^{\nu_q \lambda} = e^{\cos\frac{2\pi q}{m}} \left(\cos\left(\lambda \sin\frac{2\pi q}{m}\right) + i\sin\left(\lambda\sin\frac{2\pi q}{m}\right)\right)$$

and

$$e^{\nu_q \lambda} = \sum_{s=0}^{m-1} \nu_q^s B_s(P_\lambda),$$

where $q = 0, 1, \cdots, m-1$.
For any initial measure $\mu \in S(X, \mathcal{F})$ we have

$$\begin{aligned} V\mu(k) &= \sum_{i=0}^{\infty}\sum_{j=0}^{\infty} P_{ij,k}\mu(i)\mu(j) \\ &= \sum_{s=0}^{m-1} e^{-\lambda_s}\frac{\lambda_s^k}{k!}\left[\sum_{p+q\equiv s(mod\ m)}^{m-1} A_p(\mu)A_q(\mu)\right] \end{aligned}$$

and

$$\begin{aligned} V^2\mu(k) &= \sum_{i=0}^{\infty}\sum_{j=0}^{\infty} P_{ij,k}V\mu(i)V\mu(j) \\ &= \sum_{s=0}^{m-1} e^{-\lambda_s}\frac{\lambda_s^k}{k!}\left[\sum_{p+q\equiv s(mod\ m)}^{m-1} A_p(V\mu)A_q(V\mu)\right] \end{aligned}$$

By simple calculations we have

$$A_t(V\mu) = \sum_{s=0}^{m-1} A_t(P_{\lambda_s})\left[\sum_{p+q\equiv s(mod\ m)}^{m-1} A_p(\mu)A_q(\mu)\right] \tag{6}$$

Thus by induction for the sequence $V^n(\mu)$ we produce the following recurrent equation

$$V^{n+1}\mu(k) = \sum_{s=0}^{m-1} e^{-\lambda_s}\frac{\lambda_s^k}{k!}\left[\sum_{p+q\equiv s(mod\ m)}^{m-1} A_p(V^n\mu)A_q(V^n\mu)\right] \tag{7}$$

where $n = 0, 1, \cdots,$ and for parameters $A_t(V^n\mu), t = 0, 1, \cdots, m-1$ we have the following recurrent equations

$$A_t(V^{n+1}\mu) = \sum_{s=0}^{m-1} A_t(P_{\lambda_s})\left[\sum_{p+q\equiv s(mod\ m)}^{m-1} A_p(V^n\mu)A_q(V^n\mu)\right] \tag{8}$$

Let us put $x_p = A_p(V^n\mu)$, where $p = 0, 1, \cdots, m-1$. It is evident that $(x_0, x_1, \cdots, x_{m-1}) \in S^{m-1}$, such that the equations (8) one can rewrite as

$$x'_t = \sum_{s=0}^{m-1} A_t(P_{\lambda_s}) \left[\sum_{p+q\equiv s(mod\ m)}^{m-1} x_p x_q \right], \tag{9}$$

where $t = 0, 1, \cdots, m-1$. Then these system of equations (9) is a qso on finite dimensional simplex S^{m-1} which is denoted by W_1. It is evident that if W_1 is a regular transformation, then from (7) it follows that the limit $\lim_{n\to\infty} V^{n+1}\mu(k)$ exists for any $k \in X$. Thus the problem of investigating limit behaviour of the trajectory qso (4) is reduced to similar problem for qso defined on finite dimensional simplex. In next subsection we consider Geometric qso.

2.2. Geometric qso

A Geometric distribution G_α with a real parameter α, $0 < \alpha < 1$, is defined on countable set X by the equation

$$G_\alpha(k) = \alpha(1-\alpha)^k, \quad k \in X.$$

Definition 2.3 *A quadratic stochastic operator V (4) is called a* Geometric qso *if for any $i, j \in X$, the probability measure $P(i, j, \cdot)$ is the Geometric distribution with parameter $\alpha(i, j)$. We select the following class of a Geometric qso.*

Definition 2.4 *A qso V is called m-Geometric, where m is a positive integer, if $\{\alpha(i, j) : i, j \in X\} = \{\alpha_0, \alpha_1, \cdots, \alpha_{m-1}\}$.*

One can define m-Geometric qso as follows. Assume that for any $i, j \in X$, with $i+j \equiv s(mod\ m)$ we have $\alpha(i, j) = \alpha_s$, and

$$P_{ij,k} = \alpha_s(1-\alpha_s)^k$$

for any $k \in X$, where $s = 0, 1, \cdots, m-1$. Then the corresponding qso V is a m-Geometric one. As above one can produce the following recurrent equations

$$V^{n+1}\mu(k) = \sum_{s=0}^{m-1} \alpha_s(1-\alpha_s)^k \left[\sum_{p+q\equiv s(mod\ m)}^{m-1} A_p(V^n\mu)A_q(V^n\mu) \right] \tag{10}$$

where $n = 0, 1, \cdots$, and for parameters $A_t(V^n\mu), t = 0, 1, \cdots, m-1$ we have the following recurrent equations

$$A_t(V^{n+1}\mu) = \sum_{s=0}^{m-1} A_t(G_{\alpha_s}) \left[\sum_{p+q\equiv s(mod\ m)}^{m-1} A_p(V^n\mu)A_q(V^n\mu) \right], \tag{11}$$

where

$$A_t(G_{\alpha_s}) = \alpha_s \frac{(1-\alpha_s)^t}{1-(1-\alpha_s)^m}.$$

Now we assume $x_p = A_p(V^n\mu)$, where $p = 0, 1, \cdots, m-1$. Then $(x_0, x_1, \cdots, x_{m-1}) \in S^{m-1}$, and the equations (8) can be rewritten as

$$x'_t = \sum_{s=0}^{m-1} A_t(G_{\alpha_s}) \left[\sum_{p+q\equiv s(mod\ m)}^{m-1} x_p x_q \right], \tag{12}$$

where $t = 0, 1, \cdots, m-1$. These system of equations (11) is a qso on finite dimensional simplex S^{m-1} which is denoted by W_2. It is evident that if W_2 is regular, then from (10) one finds that the limit $\lim_{n\to\infty} V^{n+1}\mu(k)$ exists for any $k \in X$.

In next Section 3 we show the regularity of W_1 and W_2.

3. Regularity of quadratic stochastic operators W_1 and W_2

First we recall a construction of qso [30]. Let G be a finite commutative group and $S(G)$ be a set of all probability measures on G. It is evident, that if $|G| = m$, then $S(G)$ coincides with S^{m-1}.

Let further $H \subset G$ be a subgroup of G and $\{g + H : g \in G\}$ be the cosets of H in G. Assume $\mu \in S(G)$ is a fixed positive measure, that is $\mu(g) > 0$ for any $g \in G$. Then we define the coefficients $p_{fg,h}$, where $f, g, h \in G$ in the following way:

$$p_{fg,h} = \begin{cases} \frac{\mu(h)}{\mu(f+g+H)} & \text{if } h \in f + g + H; \\ 0 & \text{otherwise,} \end{cases}$$

where $p_{fg,h} = p_{gf,h}$, $p_{fg,h} > 0$ for any f, g, h and $\sum_{h\in G} p_{fg,h} = 1$. It is evident that if $H = \{e\}$, where e is the neutral element of group G, then

$$p_{fg,h} = \begin{cases} 1 & \text{if } h = f + g; \\ 0 & \text{otherwise} \end{cases}$$

and corresponding qso V is defined as

$$(Vx)_h = \sum_{f,g\in G, f+g=h} x_f x_g. \tag{13}$$

In [31] the authors studied qso (13) and proved the following statement.

Theorem 3.1 *Almost all orbits of the qso V tend to the center of the simplex.*

Below we apply this construction to investigate the operators W_1 and W_2, respectively.

Let $G = Z_m$ and Q be a qso defined on $S(Z_m)$ by the trivial subgroup. Namely, the qso Q is defined as follows:

$$(Qx)_k = \sum_{i,j\in Z_m : i+j\equiv k(mod m)} x_i x_j, \tag{14}$$

It is evident that any heredity coefficient $p_{ij,k}$ of the operator Q is equal to either 0 or 1.

Let $\Pi = ||p_{ij}||_{i,j=0}^{m-1}$ be a right stochastic matrix with each row summing to 1, and $W = \Pi Q$ be a composition of the operators Q and Π, i.e.

$$(Wx)_k = \sum_{i=0}^{m-1} p_{ki}(Qx)_i, \tag{15}$$

where Qx and Wx are column vectors. One can see that W is also a qso such that, for any fixed k its heredity coefficient satisfies $P_{ij,k} \in \{p_{k0}, p_{k1}, \cdots, p_{k,m-1}\}$. Note that the quadratic stochastic operators (9) and (12) have the form (15) with right stochastic matrix $\Pi = ||p_{ij}||_{i,j=0}^{m-1}$, where $p_{ki} = A_k(G_{\lambda_i})$ and $p_{ki} = A_k(G_{\alpha_i})$, respectively, for all $i, j = 0, 1, \cdots, m-1$.

In [18] and [19] the authors proved that if all heredity coefficients of qso are positive, and $\frac{p_{i_1k,j}}{p_{i_1k,j}} \leq \mu < 3$, then the qso has a single fixed point, and all trajectories converge to this fixed point.

It is evident that for any Geometric distribution G_α we have $A_0(G_\alpha) > A_1(G_\alpha) > A_2(G_\alpha) > \cdots > A_{m-1}(G_\alpha)$ and $\frac{A_0(G_\alpha)}{A_{m-1}(G_\alpha)} < 3$ if $\alpha < 1 - \frac{1}{\sqrt[m]{3}}$. Thus we have the following statement.

Theorem 3.2 *Let V be an $m-$Geometric qso with parameters $\alpha_0, \alpha_1, \cdots, \alpha_{m-1}$. If $\alpha_0, \alpha_1, \cdots, \alpha_{m-1} < 1 - \frac{1}{\sqrt[m]{3}}$, then V is a regular qso.*

As shown above for any Poisson distribution P_λ one can compute explicitly $A_0(P_\lambda), A_1(P_\lambda), \cdots, A_{m-1}(P_\lambda)$ for any positive integer m. Then one can show that there exists λ^* such that $\frac{A_i(P_\lambda)}{A_j(P_\lambda)} < 3$ for any $\lambda > \lambda^*$. For example, if $m = 2$ we have $A_0(P_\lambda) = \frac{1+e^{-2\lambda}}{2}, A_1(P_\lambda) = \frac{1-e^{-2\lambda}}{2}$ and $\frac{A_0(P_\lambda)}{A_1(P_\lambda)} < 3$ if $\lambda > \frac{\ln 2}{2}$, i.e. $\lambda^* = \frac{\ln 2}{2}$. Hence, we have the following statement.

Theorem 3.3 *Let V be an $m-$Poisson qso with parameters $\lambda_0, \lambda_1, \cdots, \lambda_{m-1}$. If $\lambda_0, \lambda_1, \cdots, \lambda_{m-1} > \lambda^*$, then V is a regular qso.*

Remark 3.4 *In [19] Lyubich conjectured that any qso V with positive heredity coefficients has a unique or at least finitely many fixed points. By numerical analysis for small m ([6]-[9]) we have checked uniqueness of fixed point for the considered qso. Therefore, for any m, one can expect that $m-$Geometric and $m-$Geometric qso's have unique fixed points.*

4. Applications

In this section we consider an application to economics of qso with infinite state space, using a model introduced by Föllmer [3]. He considered a countably set A of economic agents (as in physical examples the number of agents considered is often quite large, and therefore, it was considered an infinite case to approximate such situations), each being in a state s specified by his preferences and resources. He then allowed an interaction between different agents in the following manner. First, the environment e of the economic agent a is a configuration on $A - a$ which specifies the states of the other agents. The collection of local (microeconomic) characteristics of the form $r_a(s|e)$ can then be given as the conditional probability that a is in state s given the environment e. Then any probability measure μ which possesses the local characteristics is given by r called a *global phase* of the economy.

A price is then defined as a vector $p = (p_1, \cdots, p_k)$. Then, based on some maximization scheme using the agent's preferences, a well-defined *individual excess demand* $\zeta(\omega(a), p)$ is determined. The individual's excess demand is to be thought of as the difference between his demand and what he already has. The price p is said to stabilize the global phase μ of the economy E if

$$\lim \frac{1}{|A_n|} \sum_{a \in A_n} \zeta(\omega(a), p) = 0, \tag{16}$$

whenever A_n is an increasing sequence of finite subsets of A which exhausts A. Equation (16) is interpreted as having the per capita excess demand going to 0. Föllmer [3] showed that for the Ising model, if there are two pure phases then there is no price which makes (16) true. In [4] the authors proposed a construction of quadratic stochastic operator V_μ using Gibbs state μ of some lattice model. Applying this construction for Föllmer model one can define qso on infinite state space, namely countable infinite set A of economic agents and investigate ergodicity of this qso. The question then consideredas follows: is there any correlation between ergodicity of the qso and stabilization the price (16)?

Theorem 4.1 *If for the Ising model there are two pure phases μ^* and μ^{**} then corresponding qso V_{μ^*} and $V_{\mu^{**}}$ are non ergodic transformation.*

The Sketch of the Proof Let $\{\mu_n\}$ be a sequence of conditional Gibbs measures that converges to pure phase μ. Using the construction suggested in [4] one can define the sequence of qso $\{V_{\mu_n}\}$ defined on finite state space and show that for $T < T_c$ (see [4]) qso V_{μ_n} is non ergodic transformation. By the standard calculus arguments, one can prove that there exists limit of the sequence qso $\{V_{\mu_n}\}$ and limit transformation is non ergodic qso. Thus if for the Ising model occurs the phase transition, then there is no price which makes (16) true and respectively the corresponding qso is non ergodic transformation.

4.1. Acknowledgments
The work has been supported by the MOHE grant FRGS14-116-0357 and IIUM Research Fund EDW B 14-191-1076.

5. References

[1] Akin E and Losert V 1984 *J. Math. Biology* **20** 231
[2] Bernstein S 1924 *Uchn. Zapiski. NI Kaf. Ukr. Otd. Mat.* **1** 83 (Russian)
[3] Föllmer H 1974 *Jour. Math. Economics***1** 51
[4] Ganikhodjaev N N and Rozikov U A 2006 *Regular Chaotic Dyn.* **11** 467
[5] Ganikhodjaev N, Ganikhodjaev R and Jamilov U 2015 *Ergodic Th. Dyn. Sys.* **35** 1443
[6] Ganikhodjaev N and Hamzah N Z A 2014 *The Scientific World Jour.* **2014** ID 832861
[7] Ganikhodjaev N and Hamzah N Z A *Third International Conf. Math. Appl. Engineering (ICMAE14)* (to appear)
[8] Ganikhodjaev N and Hamzah N Z A 2015 *AIP Conf. Proc.* **1643** 706
[9] Ganikhodjaev N and Hamzah N Z A *Simposium Kebangsaan Sains Matematik ke-22" (SKSM22)* (to appear)
[10] Ganikhodjaev N and Hamzah N Z A 2015 *AIP Conf. Proc.* **1660** 050025
[11] Ganikhodjaev N, Saburov M and Jamilov U 2013 *App Math Info Sci* **7** 1721
[12] Ganikhodjaev N, Saburov M and Nawi A M 2014 *The Scientific World Journal* **2014** 1
[13] Ganikhodjaev N and Zanin D 2004 *Russian Math. Surveys* **59** 571
[14] Ganikhodzhaev R 1989 *Dokl. Akad. Nauk Uz* **1** 3
[15] Ganikhodzhaev R 1993 *Acad. Sci. Sb.Math.* **76** 489
[16] Ganikhodzhaev R 1994 *Math. Notes* **56** 1125
[17] Ganikhodzhaev R, Mukhamedov F and Rozikov U 2011 *Infinite Dim. Anal. Quantum Probab. Related Topics* **14** 279
[18] Kesten H 1972 *Adv. App. Prob.* **2** 1
[19] Lyubich Yu 1992 *Mathematical structures in population genetics* (Berlin: Springer)
[20] Mukhamedov F, Akin H and Temir S. 2005 *J. Math. Anal. Appl.* **310** 533
[21] Mukhamedov F and Ganikhodjaev N 2015 *Lect. Notes Math.* **2133** (Berlin: Springer)
[22] Mukhamedov F and Saburov M 2014 *Bull. Malays. Math. Sci. Soc* **37** 59
[23] Nagylaki T 1983 *Proc. Natl. Acad. Sci. USA* **80** 5941
[24] Nagylaki T 1983 *Proc. Natl. Acad. Sci. USA* **80** 6278
[25] Saburov M 2007 *Dokl. Acad. N. Rep. Uz.* **6** 8
[26] Saburov M 2012 *Proc. Inter. Conf. Appl. Sci. & Tech.* 54
[27] Saburov M 2012 *Dokl. Acad. Nauk Rep. Uzb.* **3** 9
[28] Saburov M 2013 *World Applied Science Journal* **21** 94
[29] Ulam S 1960 *A collection of mathematical problems* (New-York: Interscience)
[30] Zakharevich M 1978 *Russian Math. Surv.* **33** 265
[31] Ganikhodjaev N, Wahiddin M R B and Zanin D 2007 arXiv:0708.0697

Journal of Physics: Conference Series **697** (2016) 012017 doi:10.1088/1742-6596/697/1/012017

On cubic stochastic operators and processes [1]

B. J. Mamurov[1], **U. A. Rozikov**[2]

[1]Bukhara State Medical Institute, 1, Navoi str., Bukhara, Uzbekistan
[2] Institute of Mathematics, National University of Uzbekistan, 29, Do'rmon Yo'li str., 100125, Tashkent, Uzbekistan

E-mail: `bmamurov.51@mail.ru`, `rozikovu@yandex.ru`

Abstract. In this paper analogically as quadratic stochastic operators and processes we define cubic stochastic operator (CSO) and cubic stochastic processes (CSP). These are defined on the set of all probability measures of a measurable space. The measurable space can be given on a finite or continual set. The finite case has been investigated before. So here we mainly work on the continual set. We give a construction of a CSO and show that dynamical systems generated by such a CSO can be studied by studying of the behavior of trajectories of a CSO given on a finite dimensional simplex. We define a CSP and drive differential equations for such CSPs with continuous time.

1. Introduction

Dynamical systems generated by quadratic operators have been proved to be a rich source of analysis for the study of dynamical properties and modeling in different domains, such as population dynamics [2, 5], physics [27], economy [3], mathematics [9],[7],[11],[12],[15]. The approach to population genetics, posed within that scheme the problem of an explicit description of evolutionary operators of free populations.

A *quadratic* stochastic operator (QSO) of a free population [15] is a (quadratic) mapping of the simplex

$$S^{m-1} = \{x = (x_1, ..., x_m) \in R^m : x_i \geq 0, \sum_{i=1}^{m} x_i = 1\}$$

into itself, of the form

$$V : x'_k = \sum_{i,j=1}^{m} p_{ij,k} x_i x_j, \quad (k = 1, ..., m) \tag{1.1}$$

where $p_{ij,k}$ are coefficients of heredity and

$$p_{ij,k} \geq 0, \quad \sum_{k=1}^{m} p_{ij,k} = 1, \quad (i, j, k = 1, ..., m) \tag{1.2}$$

Note that each element $x \in S^{m-1}$ is a probability distribution on $E = \{1, ..., m\}$.

[1] To the memory of our teacher T. A. Sarymsakov on the occasion of his 100th birthday.

Similarly, one can define a *cubic* stochastic operator (CSO) $W : S^{m-1} \to S^{m-1}$ as

$$W : x'_l = \sum_{i,j,k=1}^{m} P_{ijk,l} x_i x_j x_k, \quad (l = 1, ..., m) \tag{1.3}$$

where $P_{ijk,l}$ are coefficients such that

$$P_{ijk,l} \geq 0, \quad \sum_{l=1}^{m} P_{ijk,l} = 1, \quad (i, j, k, l = 1, ..., m) \tag{1.4}$$

The population evolves by starting from an arbitrary state (probability distribution on E) $x \in S^{m-1}$ then passing to the state Vx (or Wx) (in the next "generation"), then to the state V^2x (resp. W^2x), and so on. Thus, states of the population described by the following discrete-time dynamical system

$$x^0, \quad x' = V(x), \quad x'' = V^2(x), \quad x''' = V^3(x), \ldots \tag{1.5}$$

where $V^n(x) = \underbrace{V(V(...V(x))...)}_{n}$ denotes the n times iteration of V to x. The main problem for a given dynamical system (1.5) is to describe the limit points of $\{x^{(n)}\}_{n=0}^{\infty}$ for arbitrary given $x^{(0)}$.

In [10] (see also [19]) a review of the theory of QSOs is given. Note that each quadratic (resp. cubic) stochastic operator can be uniquely defined by a stochastic matrix $\mathbf{P} = \{p_{ij,k}\}_{i,j,k=1}^{m}$ (resp. $\mathbb{P} = \{P_{ijk,l}\}_{i,j,k,l=1}^{m}$). In [6] a constructive description of $\mathbf{P}$ (i.e. a QSO) is given. This construction depends on cardinality of E, which can be finite or continual. Some particular cases of this construction were defined in [7]. In [23] a similar construction for the CSOs on a finite set E is given. We note that for the continual set E one of the key problem is to determine the set of coefficients of heredity which is already infinite dimensional. In this paper we shall give a construction of CSO for a continual set E. In this construction the CSO depends on a probability measure μ given on a measurable space $(E, \mathcal{F})$. We will show that the dynamical systems generated by the CSO depending on μ can be reduced to a dynamical system generated by a Volterra CSO defined on a finite-dimensional simplex.

Next goal of the paper is to define cubic stochastic processes (CSP) and give differential equations for such processes. This investigations will be similar to works [8],[20, 21], [24]-[26], where the authors introduced a continuous-time dynamical system as quadratic stochastic processes (QSP). The reader is referred to very recent book [19] for the theory of QSPs.

2. Definitions and examples

Consider a measurable space $(E, \mathcal{F})$ and let $S(E, \mathcal{F})$ be the set of all probability measures on $(E, \mathcal{F})$.

Definition 2.1. *A mapping $W : S(E, \mathcal{F}) \to S(E, \mathcal{F})$ is called a cubic stochastic operator (CSO) if, for an arbitrary measure $\lambda \in S(E, \mathcal{F})$, the measure $\lambda' = W\lambda$ is defined by*

$$\lambda'(A) = \int_E \int_E \int_E P(x, y, z, A) d\lambda(x) d\lambda(y) d\lambda(z), \quad \forall A \in \mathcal{F},$$

where $P(x, y, z, A)$, satisfies the following conditions:

(i) $P(x, y, z, \cdot) \in S(E, \mathcal{F})$ for any fixed $x, y, z \in E$;

(ii) $P(x, y, z, A)$, regarded a function of three variables x, y, and z is measurable on $(E \times E \times E, \mathcal{F} \otimes \mathcal{F} \otimes \mathcal{F})$ for any fixed $A \in \mathcal{F}$.

When E is finite, a CSO on $S(E,\mathcal{F}) = S^{m-1}$ is defined as in (1.3) with $P_{ijk,l} = P(i,j,k,l)$.

Let $(E,\mathcal{F},\mathcal{M})$ be triple, where $\mathcal{F}$ is a σ -algebra of subsets of E and $\mathcal{M} = S(E,\mathcal{F})$, i.e. the set of all probability measures on $(E,\mathcal{F})$. For any three elements $x,y,z \in E$ and a given measure $m_{t_0} \in \mathcal{M}$ at some moment t_0 of the time we assume that we know the law of probability distribution $m_{t_1} \in \mathcal{M}$ of the system E at the moment $t_1 > t_0$ of time.

Denote by $P(t_1,x,y,z,t_2,A)$ the probability of obtaining an element from the set $A \in \mathcal{F}$ at the moment t_2, provided that the elements x,y and z of E interact starting at moment t_1, where $t_2 \geq t_1 + 1$.

Thus if at the moment t_1 we start with a probability measure $m_{t_1} \in \mathcal{M}$, then $m_{t_2} \in \mathcal{M}$ for any $t_2 \geq t_1 + 1$ is defined by

$$m_{t_2}(A) = \int_E \int_E \int_E P(t_1,x,y,z,t_2,A) m_{t_1}(dx) m_{t_1}(dy) m_{t_1}(dz). \tag{2.1}$$

Without loss of generality we assume that the process starts at the moment $t = 0$.

Definition 2.2. *A family* $\{P(t_1,x,y,z,t_2,A)\colon x,y,z \in E,\ A \in \mathcal{F},\ t_1,t_2 \in \mathbb{R}^+,\ t_2 - t_1 \geq 1\}$ *is called* cubic stochastic process *(CSP) if it satisfies the following conditions*

(I) $P(t,x,y,z,t+1,A) = P(0,x,y,z,1,A)$ *for any* $t \geq 1$*;*

(II) The value of $P(t_1,x,y,z,t_2,A)$ *is independent on any permutations of variables* x,y,z *for all* $x,y,z \in E$*, and* $A \in \mathcal{F}$*;*

(III) $P(t_1,x,y,z,t_2,A)$ *is a probability measure on* $(E,\mathcal{F})$ *for all* $x,y,z \in E$ *and* $t_1,t_2 \in \mathbb{R}^+$, $t_2 - t_1 \geq 1$*;*

(IV) $P(t_1,x,y,z,t_2,A)$ *as function of the three variables* x,y *and* z *is measurable with respect to* $(E \times E \times E,\quad \mathcal{F} \otimes \mathcal{F} \otimes \mathcal{F})$ *for all* $A \in \mathcal{F}$*;*

(V) For any $t_1 < t_2 < t_3$ *such that* $t_2 - t_1 \geq 1$ *and* $t_3 - t_2 \geq 1$ *the following holds*

$$P(t_1,x,y,z,t_3,A) = \int\limits_E \int\limits_E \int\limits_E P(t_1,x,y,z,t_2,du)\, P(t_2,u,\vartheta,q,t_3,A)\, m_{t_2}(d\vartheta) m_{t_2}(dq), \tag{2.2}$$

where m_{t_2} *on* $(E,\mathcal{F})$ *is defined by (2.1).*

Remark 2.3. *Now we point out the followings:*

1. *The consideration of CSO and CSP are motivated by their appearance in biology (for example in gene engineering, a triple crossings for different sorts of plants to obtain another sort, and free population with ternary production) [1], in physics (for example in spin systems with ternary interactions) [16]. We note that one may have more general form of stochastic operators (resp. processes) with order* $n \geq 1$*, and coefficients* $P_{i_1 i_2 \ldots i_n,j} \geq 0$ *(resp.* $P(t_1,x_1,x_2,\ldots,x_n,t_2,A)$*) in biology (gene engineering) and physics (*n*-nary interactions). In this paper, we restrict ourselves to the case* $n = 3$*, i.e. CSOs and CSPs. Because even for the considered setting associated dynamical systems are very complicated in comparison with* $n = 1$ *(i.e. Markov chains) and* $n = 2$ *(i.e. QSOs).*
2. *Note that if one defines a new process by*

$$Q(s,x,y,t,A) = \int_E P(s,x,y,z,t,A) m_s(dz)$$

then one can check that the defined process is QSP (see [20, 24]). We recall that the functions $Q(s,x,y,t,A)$ *denote the probability that under the interaction of the elements* x *and* y *at time* s *an event* A *comes into effect at time* t*. Since for physical, chemical and biological* *phenomena, a certain time is necessary for the realization of an interaction, it is taken the greatest such time to be equal to 1 (see the Boltzmann model [12] or the biological model [15]). Thus the probability $Q(s,x,y,t,A)$ is defined for $t-s\geq 1$. Hence, for CSP, the probabilities $P(s,x,y,z,t,A)$ are also defined for $t-s\geq 1$.*

3. *It should be noted that the CSPs are related to CSOs (see [12, 19]) in the same way as Markov processes are related to linear transformations.*
4. *The equation (2.2) in Definition 2.2 is an analogue of Chapman-Kolmogorov equation. In [8], [21],[24]-[26] such equations were extended to QSPs. We note that for QSPs there are two types of the Chapman-Kolmogorov equations: type A and type B. Similarly, one also can define (at least) two types of the Chapman-Kolmogorov equations for a CSP: The equation (2.2) corresponds to the type A of QSP, the following equation is an analogue of the type B for CSP:*

$$\begin{aligned}P(s,x,y,z,t,A) &= \int_E\int_E\int_E\int_E\int_E\int_E P(s,x,y_1,z_1,\tau,du)\,P(s,y,y_2,z_2,\tau,dv)\\ &\times P(s,z,y_3,z_3,\tau,dw)\,P(\tau,u,v,w,t,A)\,m_s(dy_1)m_s(dz_1)\\ &m_s(dy_2)m_s(dz_2)m_s(dy_3)m_s(dz_3).\end{aligned}$$

In this paper, for the sake of simplicity, we shall only consider CSPs which satisfy (2.2).

Let us provide some examples of CSPs.

Let $E=\{1,2,...,n\}$ and $x^{(0)}=\left\{x_1^{(0)},x_2^{(0)},...,x_n^{(0)}\right\}$ be an initial distribution on E.

Denote

$$P_{ijk,l}:=P\left(0,i,j,k,1,\{l\}\right),\quad P_{ijk,l}^{[s,t]}:=P\left(s,i,j,k,t,\{l\}\right).$$

By equation (2.1) at the moment $t=1$ the vector $x^{(1)}=\left(x_1^{(1)},x_2^{(1)},...,x_n^{(1)}\right)$ is defined as follows

$$x_l^{(1)}=\sum_{i,j,k=1}^{n}P_{ijk,l}x_i^{(0)}x_j^{(0)}x_k^{(0)},\quad l=\overline{1,n}.$$

In this case the condition (I) reduces to $P_{ijk,l}^{[t,t+1]}=P_{ijk,l}$.

In general, from (2.1),(2.2) one finds

$$x_l^{(t)}=\sum_{i,j,k}P_{ijk,l}^{[s,t]}x_i^{(s)},x_j^{(s)}x_k^{(s)} \tag{2.3}$$

$$P_{ijk,l}^{[s,t]}=\sum_{m,\gamma,\delta}P_{ijk,m}^{[s,\tau]}P_{m\gamma\delta,l}^{[\tau,t]}x_\gamma^{(\tau)}x_\delta^{(\tau)}, \tag{2.4}$$

where $\tau-s\geq 1$ and $t-\tau\geq 1$.

Example 2.4. *To define a CSP for $E=\{1,2\}$ we first define a CSO by the matrix $(P_{ijk,l})$, where*

$$P_{111,1}=1,\quad P_{112,1}=P_{121,1}=P_{211,1}=\frac{2}{3},\quad P_{122,1}=P_{212,1}=P_{221,1}=\frac{1}{3},\quad P_{222,1}=0;$$

$$P_{111,2}=0,\quad P_{112,2}=P_{121,2}=P_{211,2}=\frac{1}{3},\quad P_{122,2}=P_{212,2}=P_{221,2}=\frac{2}{3},\quad P_{222,2}=1.$$

It is easy to see that this CSO is the identity mapping, i.e. $x'_1 = x_1$, $x'_2 = x_2$. *Take an initial vector* $(x, 1-x)$ *and using this CSO by formulas (2.3),(2.4) one can define a CSP:*

$$\begin{aligned}
&P^{[s,t]}_{111,1} = \tfrac{1}{3^{t-s-1}}\left[1 + (3^{t-s-1}-1)x\right],\\
&P^{[s,t]}_{112,1} = P^{[s,t]}_{121,1} = P^{[s,t]}_{211,1} = \tfrac{1}{3^{t-s-1}}\left[\tfrac{2}{3} + (3^{t-s-1}-1)x\right],\\
&P^{[s,t]}_{122,1} = P^{[s,t]}_{212,1} = P^{[s,t]}_{221,1} = \tfrac{1}{3^{t-s-1}}\left[\tfrac{2}{3} + (3^{t-s-1}-1)(1-x)\right],\\
&P^{[s,t]}_{222,1} = \tfrac{3^{t-s-1}-1}{3^{t-s-1}}x,\\
&P^{[s,t]}_{ijk,2} = 1 - P^{[s,t]}_{ijk,1}, \quad \textit{for all} \quad i,j,k \in E = \{1,2\}.
\end{aligned}$$

Example 2.5. *Let* $E = \{1, 2, \ldots, n\}$. *Take a family of stochastic vectors:* $a(t) = (a_1(t), a_2(t), \ldots, a_n(t))$, *i.e.* $a_i(t) \geq 0$, $\sum_i a_i(t) = 1$ *for any* $t \geq 0$. *For each pari* s, t *we define a stochastic matrix* $Q^{[s,t]} = (q^{[s,t]}_{il})_{i,l \in E}$, *where* $q^{[s,t]}_{il} = a_l(t)$ *for all* $i \in E$, *i.e. it does not depend on* s. *It is easy to see that this matrix satisfies the Kolmogorov-Chapman equation:*

$$Q^{[s,t]} = Q^{[s,\tau]}Q^{[\tau,t]}, \quad \textit{for all} \quad 0 \leq s < \tau < t.$$

Now define functions

$$P(s, i, j, k, t, \{l\}) = q^{[s,t]}_{il} = a_l(t).$$

One can check that the defined family $\{P(s, i, j, k, t, \{l\})\}$ *is a CSP.*

Example 2.6. *(cf. Example 4.2.1 of [19]) Let* $(E, \mathcal{F})$ *be a measurable space and* m_0 *be an initial measure on this space. Consider the following functions*

$$P(s, x, y, z, t, A) = \frac{1}{3^{t-s-1}}\left(\frac{\delta_x(A) + \delta_y(A) + \delta_z(A)}{3} + (3^{t-s-1}-1)m_0(A)\right),$$

where $t - s \geq 1$, $x, y, z \in E$ *and* $A \in \mathcal{F}$,

$$\delta_x(A) = \begin{cases} 1, & \textit{if } x \in A \\ 0, & \textit{if } x \notin A. \end{cases}$$

It is easy to see that the defined family is CSP.

3. A construction of CSOs

Recall that a construction of CSO for finite E was given in [23]. Therefore, in this section we consider the continual case.

Let $G = (\Lambda, L)$ be a countable graph. For a finite set Φ denote by Ω the set of all functions $\sigma : \Lambda \to \Phi$. Let $S(\Omega, \Phi)$ be the set of all probability measures defined on $(\Omega, \mathcal{F})$, where $\mathcal{F}$ is the $\sigma-$algebra generated by the finite-dimensional cylindrical set. Let μ be a measure on $(\Omega, \mathcal{F})$ such that $\mu(B) > 0$ for any finite-dimensional cylindrical set $B \in \mathcal{F}$.

Let $M \subset \Lambda$ be a finite connected subgraph. Two elements $\sigma, \varphi \in \Omega$ are called equivalent if $\sigma(x) = \varphi(x)$ for any $x \in M$, i.e. $\sigma(M) = \varphi(M)$. Let $\xi = \{\Omega_i, i = 1, 2, ..., |\Phi|^{|M|}\}$, be the partition of Ω generated by this equivalent relation, where $|\cdot|$ denotes the cardinality of a set and Ω_i contains all equivalent elements[2].

[2] We note that ξ depends on M, therefore all quantities which we define using ξ also depend on M. But for simplicity of formulas we will omit M from our formulas.

Denote $\langle ijk, l\rangle = \delta_{il} + \delta_{jl} + \delta_{kl}$, where δ is the Kroneckers symbol, i.e.

$$\delta_{ij} = \begin{cases} 1, & \text{if } i = j \\ 0, & \text{if } i \neq j. \end{cases}$$

Consider

$$P_{\sigma_1\sigma_2\sigma_3,\sigma} = \frac{\langle ijk, l\rangle \mu(\Omega_l)}{\mu(\Omega_i) + \mu(\Omega_j) + \mu(\Omega_k)} \quad \text{if} \quad \sigma_1 \in \Omega_i,\ \sigma_2 \in \Omega_j,\ \sigma_3 \in \Omega_k,\ \sigma \in \Omega_l. \tag{3.1}$$

For arbitrary σ it is easy to see that $P_{\sigma_1\sigma_2\sigma_3,\sigma}$ is invariant with respect to any permutations of $\sigma_1, \sigma_2, \sigma_3$.

Then the coefficients $P(\sigma_1, \sigma_2, \sigma_3, A)$ $(\sigma_1, \sigma_2, \sigma_3 \in \Omega,\ \ A \in \mathcal{F})$ are defined as

$$P(\sigma_1, \sigma_2, \sigma_3, A) = Z(\sigma_1, \sigma_2, \sigma_3) \int_A P_{\sigma_1\sigma_2\sigma_3,\sigma} d\mu(\sigma) = Z(\sigma_1, \sigma_2, \sigma_3) \sum_{l=1}^{m} \int_{A\cap\Omega_l} P_{\sigma_1\sigma_2\sigma_3,\sigma} d\mu(\sigma),$$

where $m = |\Phi|^{|M|}$ and $Z(\sigma_1, \sigma_2, \sigma_3)$ is the normalizing factor, which is chosen by the condition that $P(\sigma_1, \sigma_2, \sigma_3, \Omega) = 1$.

It is easy to obtain the following

$$P(\sigma_1, \sigma_2, \sigma_3, A) = \begin{cases} \frac{\mu(\Omega_i)\mu(A\cap\Omega_i) + \mu(\Omega_j)\mu(A\cap\Omega_j) + \mu(\Omega_k)\mu(A\cap\Omega_k)}{\mu^2(\Omega_i) + \mu^2(\Omega_j) + \mu^2(\Omega_k)}, \\ \qquad \text{if } \sigma_1 \in \Omega_i, \sigma_2 \in \Omega_j, \sigma_3 \in \Omega_k, i \neq k, j \neq k, i \neq j \\ \frac{2\mu(\Omega_i)\mu(A\cap\Omega_i) + \mu(\Omega_j)\mu(A\cap\Omega_j)}{2\mu^2(\Omega_i) + \mu^2(\Omega_j)}, \quad \text{if } \sigma_1, \sigma_2 \in \Omega_i, \sigma_3 \in \Omega_j, i \neq j \\ \frac{\mu(A\cap\Omega_i)}{\mu(\Omega_i)} \quad \text{if } \sigma_1, \sigma_2, \sigma_3 \in \Omega_i. \end{cases} \tag{3.2}$$

The CSO W acting on the set $S(\Omega, \Phi)$ is determined by coefficients (3.2) is defined as follows: for an arbitrary measure $\lambda \in S(\Omega, \Phi)$, the measure $\lambda' = W\lambda$ is

$$\lambda'(A) = \int_\Omega \int_\Omega \int_\Omega P(\sigma_1, \sigma_2, \sigma_3, A) d\lambda(\sigma_1) d\lambda(\sigma_2) d\lambda(\sigma_3). \tag{3.3}$$

Using (3.2) from (3.3) we obtain

$$\lambda'(A) = \sum_{i=1}^{m} a_i(A)\lambda^3(\Omega_i) +$$

$$3\sum_{i=1}^{m} \sum_{\substack{j=1 \\ j\neq i}}^{m} b_{ij}(A)\lambda^2(\Omega_i)\lambda(\Omega_j) + 6 \sum_{1\leq i<j<k\leq m} c_{ijk}(A)\lambda(\Omega_i)\lambda(\Omega_j)\lambda(\Omega_k), \tag{3.4}$$

where

$$\begin{aligned} a_i(A) &= \frac{\mu(A\cap\Omega_i)}{\mu(\Omega_i)}, \\ b_{ij}(A) &= \frac{2\mu(\Omega_i)\mu(A\cap\Omega_i) + \mu(\Omega_j)\mu(A\cap\Omega_j)}{2\mu^2(\Omega_i) + \mu^2(\Omega_j)}, \\ c_{ijk}(A) &= \frac{\mu(\Omega_i)\mu(A\cap\Omega_i) + \mu(\Omega_j)\mu(A\cap\Omega_j) + \mu(\Omega_k)\mu(A\cap\Omega_k)}{\mu^2(\Omega_i) + \mu^2(\Omega_j) + \mu^2(\Omega_k)}. \end{aligned} \tag{3.5}$$

It is easy to see that

$$a_i(\Omega_l) = \begin{cases} 1, & \text{if } l = i, \\ 0, & \text{if } l \neq i \end{cases} \quad b_{ij}(\Omega_l) = \begin{cases} \frac{2\mu^2(\Omega_i)}{2\mu^2(\Omega_i)+\mu^2(\Omega_j)}, & l = i \\ \frac{\mu^2(\Omega_j)}{2\mu^2(\Omega_i)+\mu^2(\Omega_j)}, & l = j \\ 0, & l \neq i,\, l \neq j. \end{cases}$$
$$c_{ijk}(\Omega_l) = \begin{cases} \frac{\mu^2(\Omega_l)}{\mu^2(\Omega_i)+\mu^2(\Omega_j)+\mu^2(\Omega_k)}, & l \in \{i,j,k\} \\ 0, & l \notin \{i,j,k\}. \end{cases} \tag{3.6}$$

For a given measure $\lambda \in S(\Lambda, \Phi)$ the trajectory $\{\lambda^{(n)}\}, n = 1, 2, \ldots$ of the operator (3.3) is defined by $\lambda^{(n+1)}(A) = W(\lambda^{(n)})(A)$, where $n = 0, 1, 2, \ldots$ and $\lambda^{(0)} = \lambda$, $A \in \mathcal{F}$.

By (3.4) and (3.6) we have

$$\lambda'(\Omega_l) = \sum_{i=1}^{m} a_i(\Omega_l)\lambda^3(\Omega_i) +$$
$$3\sum_{i=1}^{m}\sum_{\substack{j=1\\ j\neq i}}^{m} b_{ij}(\Omega_l)\lambda^2(\Omega_i)\lambda(\Omega_j) + 6\sum_{1\leq i<j<k\leq m} c_{ijk}(\Omega_l)\lambda(\Omega_i)\lambda(\Omega_j)\lambda(\Omega_k). \tag{3.7}$$

A CSO (1.3) is called Volterra if the coefficients $P_{ijk,l}$ may be nonzero only when $l \in \{i,j,k\}$ and vanish in all the remaining cases (see [13], [14]).

It is easy to see that any Volterra CSO has the following form

$$W : \lambda'_l = \lambda_l \left(\lambda_l^2 + \lambda_l \sum_{\substack{i=1\\ i\neq l}}^{m} a_{i,l}\lambda_i + \sum_{\substack{i,j=1\\ i\neq l,\, j\neq l}}^{m} b_{ij,l}\lambda_i\lambda_j \right), \quad (l = 1, \ldots, m) \tag{3.8}$$

where $a_{i,l}$ and $b_{ij,l}$ are some coefficients depending on $P_{ijk,l}$.

Denoting $\lambda_i = \lambda(\Omega_i)$, and $a_{i,l} = 3b_{li}(\Omega_l)$, $b_{ij,l} = 6c_{ijl}(\Omega_l)$ the operator (3.7) can be written as (3.8).

Note that the $n-$th iteration $\lambda^{(n)} = W^{(n)}\lambda^{(0)}$ of the operator (3.3) (i.e. (3.4)) can be written as

$$\lambda^{(n+1)}(A) = \sum_{i=1}^{m} a_i(A)(\lambda_i^{(n)})^3 +$$
$$3\sum_{i=1}^{m}\sum_{\substack{j=1\\ j\neq i}}^{m} b_{ij}(A)(\lambda_i^{(n)})^2\lambda_j^{(n)} + 6\sum_{1\leq i<j<k\leq m} c_{ijk}(A)\lambda_i^{(n)}\lambda_j^{(n)}\lambda_k^{(n)}, \tag{3.9}$$

where $\lambda_j^{(n)}$, $j = 1, \ldots, m$ are coordinates of the trajectory of operator (3.8) for the given λ.

Thus in order to study the trajectory of operator (3.3) it is enough to know the behavior of trajectories of the operator (3.8), i.e. we proved the following

Theorem 3.1. *For any finite $M \subset \Lambda$ the dynamical system generated by the CSO (3.3) is reducible to a dynamical system generated by a Volterra CSO acting on $(m-1)$-dimensional simplex.*

From Theorem 3.1 we get

Corollary 3.2. *Assume for a given measure μ and $\lambda = (\lambda_1, ..., \lambda_m) \in S^{m-1}$ for the trajectory of the Volterra operator (3.8) we have*

$$\lim_{n\to\infty} \lambda^{(n)} = (\lambda_1^*, \lambda_2^*, ..., \lambda_m^*).$$

Then the corresponding trajectory $\{\lambda^{(n)}(A)\}$ of the operator (3.3) has the following limit

$$\lambda(A) = \lim_{n\to\infty} \lambda^{(n)}(A) = \sum_{i=1}^m a_i(A)(\lambda_i^*)^3 +$$

$$3\sum_{i=1}^m \sum_{\substack{j=1\\ j\neq i}}^m b_{ij}(A)(\lambda_i^*)^2\lambda_j^* + 6 \sum_{1\le i<j<k\le m} c_{ijk}(A)\lambda_i^*\lambda_j^*\lambda_k^*. \tag{3.10}$$

Remark 3.3. *As it was mentioned above the theory of Volterra QSOs is well studied (see for example [10]). But Volterra CSOs were not exhaustively studied, because such cubic operators are still complicated. There are just a few articles devoted to Volterra CSOs [13], [14], [22], [23]. Therefore formula (3.10) is already helpful by using the results for the Volterra CSOs studied in these papers.*

4. Integro-differential equations for CSP

The equations which we want to drive here were given in [17] for finite E. So we consider the continual case of E. Let E be a continual set. Consider a CSP on a measurable space $(E, \mathcal{F})$ with initial measure m_0. For $t > s+2$ from condition (V) of Definition 2.2 we get

$$P(s,x,y,z,t+\Delta,A) - P(s,x,y,z,t,A)$$

$$= \int_E\int_E\int_E P(s,x,y,z,t-1,du)\,\{P(t-1,u,\vartheta,q,t+\Delta,A)$$

$$-P(t-1,u,\vartheta,q,t,A)\}\, m_{t-1}(d\vartheta)m_{t-1}(dq).$$

Assume the following limit exists

$$C(t,u,\vartheta,q,A) = \lim_{\Delta\to 0} \frac{P(t-1,u,\vartheta,q,t+\Delta,A) - P(t-1,u,\vartheta,q,t,A)}{\Delta}.$$

Then taking limit $\Delta \to 0$ we obtain the *first integro-differential equation*:

$$\frac{\partial P(s,x,y,z,t,A)}{\partial t} = \int_E\int_E\int_E P(s,x,y,z,t-1,du)\; C(t,u,\vartheta,q,A) m_{t-1}(d\vartheta)m_{t-1}(dq). \tag{4.1}$$

Similarly, one gets the *second integro-differential equation*:

$$\frac{\partial P(s,x,y,z,t,A)}{\partial s} = -\int_E\int_E\int_E C(s+1,x,y,z,du)P(s+1,u,\vartheta,q,t,A)m_{s+1}(d\vartheta)m_{s+1}(dq). \tag{4.2}$$

Let $E = \mathbb{R}$ and $A_w = (-\infty, w]$, where $w \in \mathbb{R}$. Denote $F(s,x,y,z,t,w) = P(s,x,y,z,t,A_w)$. It is clear that $F(s,x,y,z,t,w)$ as the function of w is monotone, right-continuous and $F(s,x,y,z,t,-\infty) = 0$, $F(s,x,y,z,t,+\infty) = 1$.

The condition (V) of Definition 2.2 for the function $F(s,x,y,z,t,w)$ has the following form

$$F(s,x,y,z,t,w)=\int_E\int_E\int_E dF(s,x,y,z,\tau,u)F(\tau,u,\vartheta,q,t,w)m_\tau(d\vartheta)m_\tau(dq).$$

If the function $F(s,x,y,z,t,w)$ is absolutely continuous with respect to variable w then

$$F(s,x,y,z,t,w)=\int_{-\infty}^{w} f(s,x,y,z,t,u)du,$$

where $f(s,x,y,z,t,w)$ is a non-negative function and measurable with respect to variables x,y,z,w, moreover it satisfies the following conditions

$$\int_{-\infty}^{\infty} f(s,x,y,z,t,w)dw=1.$$

$$f(s,x,y,z,t,w)=\int_E\int_E\int_E f(s,x,y,z,\tau,u)f(\tau,u,\vartheta,q,t,w)m_\tau(d\vartheta)m_\tau(dq)du. \tag{4.3}$$

Form (4.1) and (4.2) for $F(s,x,y,z,t,w)$ we get

$$\frac{\partial F(s,x,y,z,t,w)}{\partial t}=\int_E\int_E\int_E \frac{\partial F(s,x,y,z,t-1,u)}{\partial u}C(t,u,\vartheta,q,w)dum_{t-1}(d\vartheta)m_{t-1}(dq),$$

$$\frac{\partial F(s,x,y,z,t,w)}{\partial s}=\int_E\int_E\int_E \frac{\partial C(s+1,x,y,z,u)}{\partial u}F(s+1,u,\vartheta,q,t,w)dum_{s+1}(d\vartheta)m_{s+1}(dq).$$

Assume the following limit exists:

$$a(t,u,\vartheta,q,w)=\lim_{\Delta\to 0}\frac{f(t-1,u,\vartheta,q,t+\Delta,w)-f(0,u,\vartheta,q,1,w)}{\Delta}.$$

Then we obtain the following integro-differential equations:

$$\frac{\partial f(s,x,y,z,t,w)}{\partial t}=\int_E\int_E\int_E a(t,u,\vartheta,q,w)f(s,x,y,z,t-1,u)dum_{t-1}(d\vartheta)m_{t-1}(dq) \tag{4.4}$$

$$\frac{\partial f(s,x,y,z,t,w)}{\partial s}=-\int_E\int_E\int_E a(s+1,x,y,z,u)f(s+1,u,\vartheta,q,t,w)dum_{s+1}(d\vartheta)m_{s+1}(dq). \tag{4.5}$$

5. Reduction of the integro-differential equations to differential equations

Under some conditions the integro-differential equations (4.4) and (4.5) can be reduced to differential equations. In this subsection we illustrate this for equation (4.5).

Let $t > s + 2$, then from (4.3) we get

$$\begin{array}{c} f(s,x,y,z,t,w) - f(s+\Delta,x,y,z,t,w) = \int\limits_E \int\limits_E \int\limits_E \{f(s,x,y,z,s+1+\Delta,u) - \\ -f(s+\Delta,x,y,z,s+1+\Delta,u)\} f(s+1+\Delta,u,\vartheta,q,t,w) m_{s+1+\Delta}(d\vartheta) m_{s+1+\Delta}(dq) du \end{array}. \quad (5.1)$$

We consider function f such that the decomposition of $f(s+1+\Delta,u,\vartheta,q,t,w)$ into Taylor's series in a neighborhood of the point (x,y,z) has the form:

$$f(s+1+\Delta,u,\vartheta,q,t,w) = f(s+1+\Delta,x,y,z,t,w) + \frac{\partial f(s+1+\Delta,x,y,z,t,w)}{\partial u}(u-x)$$

$$+\frac{\partial f(s+1+\Delta,x,y,z,t,w)}{\partial \vartheta}(\vartheta - y) + \frac{\partial f(s+1+\Delta,x,y,z,t,w)}{\partial q}(q-z)$$

$$+\frac{1}{2}\frac{\partial^2 f(s+1+\Delta,x,y,z,t,w)}{\partial u^2}(u-x)^2 + \frac{1}{2}\frac{\partial^2 f(s+1+\Delta,x,y,z,t,w)}{\partial \vartheta^2}(\vartheta-y)^2$$

$$+\frac{1}{2}\frac{\partial^2 f(s+1+\Delta,x,y,z,t,w)}{\partial q^2}(q-z)^2 + \frac{\partial^2 f(s+1+\Delta,x,y,z,t,w)}{\partial u \partial \vartheta}(u-x)(\vartheta-y)$$

$$+\frac{\partial^2 f(s+1+\Delta,x,y,z,t,w)}{\partial u \partial q}(u-x)(q-z) + \frac{\partial^2 f(s+1+\Delta,x,y,z,t,w)}{\partial \vartheta \partial q}(\vartheta-y)(q-z)$$

$$+\frac{1}{2}\frac{\partial^3 f(s+1+\Delta,x,y,z,t,w)}{\partial u^2 \partial \vartheta}(u-x)^2(\vartheta-y) + \frac{1}{2}\frac{\partial^3 f(s+1+\Delta,x,y,z,t,w)}{\partial u^2 \partial q}(u-x)^2(q-z)$$

$$+\frac{1}{2}\frac{\partial^3 f(s+1+\Delta,x,y,z,t,w)}{\partial \vartheta^2 \partial u}(\vartheta-y)^2(u-x) + \frac{1}{2}\frac{\partial^3 f(s+1+\Delta,x,y,z,t,w)}{\partial \vartheta^2 \partial q}(\vartheta-y)^2(q-z)$$

$$+\frac{1}{2}\frac{\partial^3 f(s+1+\Delta,x,y,z,t,w)}{\partial q^2 \partial u}(q-z)^2(u-x) + \frac{1}{2}\frac{\partial^3 f(s+1+\Delta,x,y,z,t,w)}{\partial q^2 \partial \vartheta}(q-z)^2(\vartheta-y)$$

$$+\frac{1}{6}\frac{\partial^3 f(s+1+\Delta,x,y,z,t,w)}{\partial q^3}(u-x)^3 + \frac{1}{6}\frac{\partial^3 f(s+1+\Delta,x,y,z,t,w)}{\partial \vartheta^2}(\vartheta-y)^3$$

$$+\frac{1}{6}\frac{\partial^3 f(s+1+\Delta,x,y,z,t,w)}{\partial q^2}(q-z)^3 + \frac{\partial^3 f(s+1+\Delta,x,y,z,t,w)}{\partial u \partial \vartheta \partial q}(u-x)(\vartheta-y)(q-z). \quad (5.2)$$

Substituting (5.2) into (5.1) we consider non-zero summands:

$$\int\limits_E \int\limits_E \int\limits_E \{f(s,x,y,z,s+1+\Delta,u) - f(s+\Delta,x,y,z,s+1+\Delta,u)\} \times$$

$$\frac{\partial f(s+1+\Delta,x,y,z,t,w)}{\partial x}(u-x) m_{s+1+\Delta}(d\vartheta) m_{s+1+\Delta}(dq) du$$

$$= \frac{\partial f(s+1+\Delta,x,y,z,t,w)}{\partial x} \int\limits_E \{f(s,x,y,z,s+1+\Delta,u) -$$

$$- f(s+\Delta,x,y,z,s+1+\Delta,u)\} (u-x) du.$$

Denote

$$a(s,x,y,z,\Delta)=\int\limits_E \{f(s,x,y,z,s+1+\Delta,u)\ -\ f(s+\Delta,x,y,z,s+1+\Delta,u)\}\,(u-x)du.$$

Now consider the summands with second order of derivations

$$\int\limits_E\int\limits_E\int\limits_E \{f(s,x,y,z,s+1+\Delta,u)-f(s+\Delta,x,y,z,s+1+\Delta,u)\}\times$$
$$\frac{1}{2}\frac{\partial^2 f(s+1+\Delta,x,y,z,w)}{\partial x^2}(u-x)^2 m_{s+1+\Delta}(d\vartheta)m_{s+1+\Delta}(dq)du$$
$$=\frac{1}{2}\frac{\partial^2 f(s+1+\Delta,x,y,z,w)}{\partial x^2}\int\limits_E \{f(s,x,y,z,s+1+\Delta,u)-$$
$$-\ f(s+\Delta,x,y,z,s+1+\Delta,u)\}\,(u-x)^2du.$$

Denote

$$b^2(s,x,y,z,\Delta)=\int\limits_E \{f(s,x,y,z,s+1+\Delta,u)-f(s+\Delta,x,y,z,s+1+\Delta,u)\}\,(u-x)^2du$$

Then

$$\int\limits_E\int\limits_E\int\limits_E \{f(s,x,y,z,s+1+\Delta,u)-f(s+\Delta,x,y,z,s+1+\Delta,u)\}\times$$
$$\frac{\partial^2 f(s+1+\Delta,x,y,z,w)}{\partial x\partial y}(u-x)(\vartheta-y)m_{s+1+\Delta}(d\vartheta)m_{s+1+\Delta}(dq)du$$
$$=\frac{\partial^2 f(s+1+\Delta,x,y,z,t,w)}{\partial x\partial y}\int\limits_E \{f(s,x,y,z,s+1+\Delta,u)-$$
$$-\ f(s+\Delta,x,y,z,s+1+\Delta,u)\}\,(u-x)du\int\limits_E (\vartheta-y)m_{s+1+\Delta}(d\vartheta).$$

Denote

$$\int\limits_E (\vartheta-y)m_{s+1+\Delta}(d\vartheta)=\alpha(s+1,y,\Delta).$$

Consequently

$$\int\limits_E\int\limits_E\int\limits_E \{f(s,x,y,z,s+1+\Delta,u)-f(s+\Delta,x,y,z,s+1+\Delta,u)\}\times$$
$$\frac{\partial^2 f(s+1+\Delta,x,y,z,t,w)}{\partial x\partial z}(u-x)(q-z)m_{s+1+\Delta}(d\vartheta)m_{s+1+\Delta}(dq)du$$
$$=\frac{\partial^2 f(s+1+\Delta,x,y,z,t,w)}{\partial x\partial z}\int\limits_E \{f(s,x,y,z,s+1+\Delta,u)$$

$$- f(s+\Delta, x, y, z, s+1+\Delta, u)\} \int\limits_E (u-x)du \int\limits_E (q-z)m_{s+1+\Delta}(dq).$$

$$\int\limits_E \int\limits_E \int\limits_E \{f(s,x,y,z,s+1+\Delta,u) - f(s+\Delta,x,y,z,s+1+\Delta,u)\} \times$$

$$\frac{1}{2}\frac{\partial^3 f(s+1+\Delta,x,y,z,t,w)}{\partial x^2 \partial y}(u-x)^2(\vartheta-y)m_{s+1+\Delta}(d\vartheta)m_{s+1+\Delta}(dq)du$$

$$= \frac{1}{2}\frac{\partial^3 f(s+1+\Delta,x,y,z,t,w)}{\partial x^2 \partial y} \int\limits_E \{f(s,x,y,z,s+1+\Delta,u) -$$

$$- f(s+\Delta,x,y,z,s+1+\Delta,u)\}(u-x)^2 du \int\limits_E (\vartheta-y)m_{s+1+\Delta}(d\vartheta)$$

$$= \frac{1}{2}\frac{\partial^3 f(s+1+\Delta,x,y,z,t,w)}{\partial x^2 \partial y} b^2(s,x,y,z,\Delta)\alpha(s+1,y,\Delta).$$

$$\int\limits_E \int\limits_E \int\limits_E \{f(s,x,y,z,s+1+\Delta,u) - f(s+\Delta,x,y,z,s+1+\Delta,u)\} \times$$

$$\frac{1}{2}\frac{\partial^3 f(s+1+\Delta,x,y,z,t,w)}{\partial x^2 \partial z}(u-x)^2(q-z)m_{s+1+\Delta}(d\vartheta)m_{s+1+\Delta}(dq)du$$

$$= \frac{1}{2}\frac{\partial^3 f(s+1+\Delta,x,y,z,t,w)}{\partial x^2 \partial z} \int\limits_E \{f(s,x,y,z,s+1+\Delta) -$$

$$- f(s+\Delta,x,y,z,s+1+\Delta,u)\}(u-x)^2 du \int\limits_E (q-z)m_{s+1+\Delta}(dq)$$

$$= \frac{1}{2}\frac{\partial^3 f(s+1+\Delta,x,y,z,t,w)}{\partial x^2 \partial z} b^2(s,x,y,z,\Delta)\alpha(s+1,z,\Delta).$$

$$\int\limits_E \int\limits_E \int\limits_E \{f(s,x,y,z,s+1+\Delta,u) - f(s+\Delta,x,y,z,s+1+\Delta,u)\} \times$$

$$\frac{1}{2}\frac{\partial^3 f(s+1+\Delta,x,y,z,t,w)}{\partial^2 y \partial x}(\vartheta-y)^2(u-x)m_{s+1+\Delta}(d\vartheta)m_{s+1+\Delta}(dq)du$$

$$= \frac{1}{2}\frac{\partial^3 f(s+1+\Delta,x,y,z,t,w)}{\partial^2 y \partial x} \int\limits_E \{f(s,x,y,z,s+1+\Delta,u) -$$

$$- f(s+\Delta,x,y,z,s+1+\Delta,u)\}(u-x)du \int\limits_E (\vartheta-y)^2 m_{s+1+\Delta}(d\vartheta)$$

$$= \frac{1}{2}a(s,x,y,z,\Delta)\frac{\partial^3 f(s+1+\Delta,x,y,z,t,w)}{\partial^2 y \partial x} \cdot \alpha_2(s+1,y,\Delta),$$

where

$$\alpha_2(s+1,y,\Delta) = \int\limits_E (\vartheta-y)^2 m_{s+1+\Delta}(d\vartheta).$$

$$\int\limits_E\int\limits_E\int\limits_E \{f(s,x,y,z,s+1+\Delta,u) - f(s+\Delta,x,y,z,s+1+\Delta,u)\}\times$$

$$\frac{1}{2}\frac{\partial^3 f(s+1+\Delta,x,y,z,t,w)}{\partial z^2\partial x}(q-z)^2(u-x)m_{s+1+\Delta}(d\vartheta)m_{s+1+\Delta}(dq)du$$

$$=\frac{1}{2}\frac{\partial^3 f(s+1+\Delta,x,y,z,t,w)}{\partial z^2\partial x}\int\limits_E \{f(s,x,y,z,s+1+\Delta,u)-$$

$$-f(s+\Delta,x,y,z,s+1+\Delta,u)\}(u-x)du\int\limits_E (q-z)^2 m_{s+1+\Delta}(dv)$$

$$=\frac{1}{2}\frac{\partial^3 f(s+1+\Delta,x,y,z,t,w)}{\partial z^2\partial x}a(s,x,y,z,\Delta)\alpha_2(s+1,z,\Delta).$$

$$\int\limits_E\int\limits_E\int\limits_E \{f(s,x,y,z,s+1+\Delta,u) - f(s+\Delta,x,y,z,s+1+\Delta,u)\}\times$$

$$\frac{\partial^3 f(s+1+\Delta,x,y,z,t,w)}{\partial z\partial y\partial z}(u-x)(\vartheta-y)(q-z)m_{s+1+\Delta}(d\vartheta)m_{s+1+\Delta}(dq)du$$

$$=\frac{\partial^3 f(s+1+\Delta,x,y,z,t,w)}{\partial x\partial y\partial z}\int\limits_E \{f(s,x,y,z,s+1+\Delta,u)-$$

$$-f(s+\Delta,x,y,z,s+1+\Delta,u)\}(u-x)du\int\limits_E(\vartheta-y)m_{s+1+\Delta}(d\vartheta)\cdot\int\limits_E(q-z)m_{s+1+\Delta}(dq)=$$

$$=\frac{\partial^3 f(s+1+\Delta,x,y,z,t,w)}{\partial x\partial y\partial z}a(s,x,y,z,\Delta)\alpha(s+1,y,\Delta)\alpha(s+1,z,\Delta).$$

Since other summands are equal to zero, we get

$$f(s,x,y,z,t,w)-f(s+\Delta,x,y,z,t,w)=a(s,x,y,z,\Delta)\frac{\partial f(s+1+\Delta,x,y,z,t,w)}{\partial x}$$

$$+b^2(s,x,y,z,\Delta)\cdot\frac{1}{2}\frac{\partial^2 f(s+1+\Delta,x,y,z,t,w)}{\partial x^2}+\frac{1}{2}a(s,x,y,z,\Delta)\alpha(s+1,y,\Delta)\times$$

$$\frac{\partial^2 f(s+1+\Delta,x,y,z,t,w)}{\partial x\partial y}+a(s,x,y,z,\Delta)\frac{\partial^2 f(s+1+\Delta,x,y,z,t,w)}{\partial x\partial z}\alpha(s+1,z,\Delta)$$

$$+\frac{1}{2}\frac{\partial^3 f(s+1+\Delta,x,y,z,t,w)}{\partial x^2\partial y}b^2(s,x,y,z,\Delta)\alpha(s+1,y,\Delta)+\frac{1}{2}\frac{\partial^3 f(s+1+\Delta,x,y,z,t,w)}{\partial x^2\partial z}\times$$

$$b^2(s,x,y,z,\Delta)\alpha(s+1,z,\Delta)+\frac{1}{2}a(s,x,y,z,\Delta)\frac{\partial^3 f(s+1+\Delta,x,y,z,t,w)}{\partial y^2\partial x}\times$$

$$\alpha_2(s+1,y,\Delta)+\frac{1}{2}\frac{\partial^3 f(s+1+\Delta,x,y,z,t,w)}{\partial z^2\partial x}a(s,x,y,z,\Delta)\alpha_2(s+1,z,\Delta)+a(s,x,y,z,\Delta)\times$$

$$\times\alpha(s+1,y,\Delta)\alpha(s+1,z,\Delta)\frac{\partial^3 f(s+1+\Delta,x,y,z,t,w)}{\partial x\partial y\partial z}. \qquad (5.3)$$

Dividing both sides of the equality (5.3) by Δ and passing to the limit (assuming the limits exist) as $\Delta \to 0$, and denoting

$$A(s,x,y,z) = \lim_{\Delta\to 0} \frac{a(s,x,y,z,\Delta)}{\Delta}, \quad B^2(s,x,y,z) = \lim_{\Delta\to 0} \frac{b^2(s,x,y,z,\Delta)}{2\Delta},$$

$$D(s+1,y) = \lim_{\Delta\to 0} \alpha(s+1,y,\Delta), \quad D_2(s+1,y) = \lim_{\Delta\to 0} \frac{\alpha_2(s+1,y,\Delta)}{2\Delta}$$

from (5.3) we get

$$\frac{\partial f(s,x,y,z,t,w)}{\partial s} = -A(s,x,y,z)\frac{\partial f(s+1,x,y,z,t,w)}{\partial x} - B^2(s,x,y,z)\frac{\partial^2 f(s+1,x,y,z,t,w)}{\partial x^2}$$

$$-\frac{1}{2}A(s,x,y,z)D(s+1,y)\frac{\partial^2 f(s+1,x,y,z,t,w)}{\partial x\partial y} - A(s,x,y,z)D(s+1,z)\frac{\partial^2 f(s+1,x,y,z,t,w)}{\partial x\partial z}$$

$$-B^2(s,x,y,z)D(s+1,y)\frac{\partial^3 f(s+1,x,y,z,t,w)}{\partial x^2\partial y}$$

$$-B^2(s,x,y,z)D(s+1,z)\frac{\partial^3 f(s+1,x,y,z,t,w)}{\partial x^2\partial z} - \frac{1}{2}A(s,x,y,z)D_2(s+1,y)\frac{\partial^3 f(s+1,x,y,z,t,w)}{\partial y^2\partial x}$$

$$-\frac{1}{2}A(s,x,y,z)D_2(s+1,z)\frac{\partial^3 f(s+1,x,y,z,t,w)}{\partial z^2\partial x}$$

$$-A(s,x,y,z)D(s+1,y)D(s+1,z)\frac{\partial^3 f(s+1,x,y,z,t,w)}{\partial x\partial y\partial z}. \tag{5.4}$$

Remark 5.1. *We note that the equation (5.4) is known as differential equation with advanced argument. Similarly one can reduce the equation (4.4) to a differential equation which is known as a differential equation with delay argument. For theory of such kind of equations we refer to [4] and [18].*

Example 5.2. *Let $E = \mathbb{R}$. If $m_\tau(du) = r_\tau(u)du$ then from (4.3) we get*

$$f(s,x,y,z,t,w) = \int_{-\infty}^{\infty}\int_{-\infty}^{\infty}\int_{-\infty}^{\infty} f(s,x,y,z,\tau,u)f(\tau,u,\vartheta,q,t,w)r_\tau(\vartheta)r_\tau(q)dud\vartheta dq. \tag{5.5}$$

Let $a(s,t)$ and $b(t)$ be strictly positive functions. Consider

$$f(s,x,y,z,t,w) = (\pi a(s,t))^{-1/2}\exp\left[-\frac{(w-x-y-z)^2}{a(s,t)}\right], \tag{5.6}$$

$$r_t(u) = (\pi b(t))^{-1/2}\exp\left[-\frac{u^2}{b(t)}\right].$$

For any $A, B, C \in \mathbb{R}$, with $A > 0$ it is known[3] *that*

$$\int_{-\infty}^{\infty} \exp(-Ax^2 + Bx + C)dx = \sqrt{\frac{\pi}{A}}\exp\left(\frac{B^2}{4A} + C\right).$$

[3] see https://en.wikipedia.org/wiki/Gaussian_integral

Using this formula (three times) one can see that the function (5.6) satisfies (5.5) iff functions $a(s,t) > 0$ *and* $b(t) > 0$ *satisfy the following equation*

$$a(s,t) = 2b(\tau) + a(s,\tau) + a(\tau,t), \quad \text{for any} \quad s < \tau < t \quad \text{with} \quad \tau - s \geq 1, \quad t - \tau \geq 1. \tag{5.7}$$

Thus we obtain a family of CSP generated by functions (5.6) with $a(s,t)$, $b(t)$ *satisfying (5.7). For example, take* $\epsilon \in (0,1)$, $a(s,t) = t - s - \epsilon$ *and* $b(t) = \epsilon/2$. *These functions satisfy (5.7) and the corresponding* f *and* r *are defined by*

$$f(s,x,y,z,t,w) = \frac{\exp\left[-\frac{(w-x-y-z)^2}{t-s-\epsilon}\right]}{\sqrt{(t-s-\epsilon)\pi}},$$

$$r_t(u) = \frac{\exp\left[-\frac{2u^2}{\epsilon}\right]}{\sqrt{\epsilon\pi/2}}$$

generate a CSP.

Acknowledgments. We thank the referee for careful reading of the manuscript and for useful suggestions.

References

[1] Avise JC 2014 *The Hope, Hype, and Reality of Genetic Engineering* (New York: Oxford Univ. Press).
[2] Bernshtein S N 1924 *Uch. Zap. Nauchno-Issled. kaf. Ukr. Otd. Mat.* **1** 83–115.
[3] Dohtani A 1992 *SIAM J.Appl. Math.* **52** 1707-1721.
[4] Elgolz L E and Norkin S B 1971 *Introduction to the theory of differential equations with a deviating agument* (Moscow: Nauka)
[5] Fisher M E and Goh B S 1977 *J. Math. Biol.* **4** 265-274
[6] Ganikhodjaev N N and Rozikov U A 2006 *Regul. Chaotic Dyn.* **11** 467–473
[7] Ganikhodjaev N N 2000 *Doklady Math.* **61** 321–323
[8] Ganikhodjaev N N 1991 *J. Theor. Prob.* **4** 639-653
[9] Ganikhodzhaev R N 1993 *Russian Acad. Sci. Sbornik Math.* **76** 489–506
[10] Ganikhodzhaev R N, Mukhamedov F M and Rozikov U A 2011 *Inf. Dim. Anal. Quant. Prob. Rel. Fields.* **14** 279-335
[11] Erson R C and Stewart J 1971 *Hum. Hered.* **21** 523–542
[12] Jenks R 1969 *J. Differ. Equations.* **5** 497–514
[13] Khamraev A Yu 2009 *Uzbek. Mat. Zh.* **3** 65–71
[14] Khamraev A Yu 2004 *Uzbek. Mat. Zh.* **2** 79–84.
[15] Lyubich Yu I 1992 *Mathematical structures in population genetics* (Berlin: Springer)
[16] Malevanets A and Kapral R 2000 *J. Chem. Phys.* **112** 7260–7269
[17] Mamurov B J 2010 *Uzbek. Mat. Zh.* **4** 113–117
[18] Mishkis A D 1972 *Linear differensial equations with delay of argument* (Moscow: Nauka) (Russian)
[19] Mukhamedov F and Ganikhodjaev N 2015 *Lect. Notes Math.* **2133** (Berlin: Springer)
[20] Mukhamedov F and Supar N A 2015 *Bull. Malay. Math. Sci. Soc.* **38** 1281–1296
[21] Mukhamedov F, Supar N A and Pah Ch H 2013 *J. Phys.: Conf. Ser.* **435** 012013.
[22] Rozikov U A and Khamraev A Yu 2004 *Ukraine Math. Jour.* **56** 1699–1711
[23] Rozikov U A and Khamraev A Yu 2014 *Nonlinear Dyn. Syst. Theory.* **14** 92–100
[24] Sarymsakov T A and Ganikhodzhaev N N 1989 *Soviet Math. Dokl.* **39** 369-373
[25] Sarymsakov T A and Ganikhodzhaev N N 1991 *Soviet Math. Dokl.* **43** 279-283
[26] Sarymsakov T A and Ganikhodzhaev N N 1990 *J. Theor. Prob.* **3** 51-70
[27] Udwadia F E and Raju N 1998 *Physica D* **111** 16-26

On Construction of Quantum Markov Chains on Cayley trees

Luigi Accardi[1], **Farrukh Mukhamedov**[2], **Abdessatar Souissi**[3]

[1] Centro Interdisciplinare Vito Volterra, II Università di Roma "Tor Vergata", Via Columbia 2, 00133 Roma, Italy

[2] Department of Computational & Theoretical Sciences, Faculty of Science, International Islamic University Malaysia, P.O. Box, 25200, Kuantan, Pahang, Malaysia

[3] Department of Mathematics, Marsa Preparatory Institute for Scientific and Technical Studies, Carthage University, Tunisia

E-mail: accardi@volterra.uniroma2.it, farrukh_m@iium.edu.my, s.abdessatar@hotmail.fr

Abstract. The main aim of the present paper is to provide a new construction of quantum Markov chain (QMC) on arbitrary order Cayley tree. In that construction, a QMC is defined as a weak limit of finite volume states with boundary conditions, i.e. QMC depends on the boundary conditions. Note that this construction reminds statistical mechanics models with competing interactions on trees. If one considers one dimensional tree, then the provided construction reduces to well-known one, which was studied by the first author. Our construction will allow to investigate phase transition problem in a quantum setting.

1. Introduction

One of the basic open problems in quantum probability is the construction of a theory of quantum Markov fields, that is quantum process with multi-dimensional index set. This program concerns the generalization of the theory of Markov fields (see [17],[22])) to non- commutative setting, naturally arising in quantum statistical mechanics and quantum led theory.

The quantum analogues of Markov chains were first constructed in [1], where the notion of quantum Markov chain on infinite tensor product algebras was introduced. Nowadays, quantum Markov chains have become a standard computational tool in solid state physics, and several natural applications have emerged in quantum statistical mechanics and quantum information theory. The reader is referred to [19, 24, 25, 32] and the references cited therein, for recent developments of the theory and the applications.

First attempts to construct a quantum analogue of classical Markov fields have been done in [3]-[6],[9, 20, 26]. In these papers the notion of *quantum Markov state*, introduced in [8] (see also [23, 20]), extended to fields as a sub-class of the *quantum Markov chains* (QMC) introduced in [1]. In [7] it has been proposed a definition of quantum Markov states and chains, which extend a proposed one in [31], and includes all the presently known examples. Note that in the mentioned papers quantum Markov fields were considered over multidimensional integer lattice $\mathbb{Z}$. This lattice has so-called amenability property. Moreover, there do not exist analytical solutions (for example, critical temperature) on such lattice. On the other hand, investigations of phase transitions of spin models on hierarchical lattices showed that there are exact calculations of various physical quantities (see for example, [16]). Therefore, it is natural to investigate quantum Markov fields over hierarchical lattices. For example, a Cayley tree is the simplest hierarchical lattice with non-amenable graph structure [33]. First attempts to investigate QMC over such trees was done in [15], such studies were related to investigation of thermodynamic limit of valence-bond-solid models on a Cayley tree [18]. The mentioned considerations naturally suggest the study of the following problem: the extension to fields of the notion of generalized QMC. In [14] we have introduced a hierarchy of notions of Markovianity for states on discrete infinite tensor products of C^*–algebras and for each of these notions we constructed some explicit examples. We showed that the construction of [8] can be generalized to trees. It is worth to note that, in a different context and for quite different purposes, the special role of trees was already emphasized in [26]. In [18] the VBS-model was considered on the Cayley tree. It was established the existence of the phase transition for the model in term of finitely correlated states which describe ground states of the model. Note that more general structure of finitely correlated states was studied in [19]. We stress that finitely correlated states can be considered as quantum Markov chains. In [21, 27, 28, 29, 30] noncommutative extensions of classical Markov fields, associated with Ising and Potts models on a Cayley tree, were investigated. In the classical case, Markov fields on trees are also considered in [35, 36],[37]-[40].

In the present paper, for a given Hamiltonian, we provide a more general construction (than [10, 11]) of QMC associated with the Hamiltonian. Namely, in this that construction, the Hamiltonian exhibits nearest-neighbor and next-nearest-neighbor interactions (in the previous papers [11]-[13] the Hamiltonian contained only nearest-neighbor interactions), and the corresponding QMC is defined as a weal limit of finite volume states (which depend on the Hamiltonian) with boundary conditions, i.e. QMC depends on the boundary conditions. We remark that in this construction, one can observe some similarities with Gibbs measures. We stress that all previous considered examples (in the literature) of QMC related to Hamiltonians with nearest-neighbor interactions. In classical setting, our construction, in particularly, leads to the Ising model with competing interactions, which has been rigorously investigated in many papers (see for example [29, 30, 34, 36].

2. Preliminaries

Let $\Gamma^k_+ = (L, E)$ be a semi-infinite Cayley tree of order $k \geq 1$ with the root x^0 (i.e. each vertex of Γ^k_+ has exactly $k+1$ edges, except for the root x^0, which has k edges). Here L is the set of vertices and E is the set of edges. The vertices x and y are called *nearest neighbors* and they are denoted by $l =< x, y >$ if there exists an edge connecting them. A collection of the pairs $< x, x_1 >, \dots, < x_{d-1}, y >$ is called a *path* from the point x to the point y. The distance $d(x, y), x, y \in V$, on the Cayley tree, is the length of the shortest path from x to y.

Recall a coordinate structure in Γ^k_+: every vertex x (except for x^0) of Γ^k_+ has coordinates $(i_1, \dots, i_n)$, here $i_m \in \{1, \dots, k\}$, $1 \leq m \leq n$ and for the vertex x^0 we put (0). Namely, the symbol (0) constitutes level 0, and the sites $(i_1, \dots, i_n)$ form level n (i.e. $d(x^0, x) = n$) of the lattice (see Fig. 1).

Let us set

$$W_n = \{x \in L : d(x, x_0) = n\}, \qquad \Lambda_n = \bigcup_{k=0}^{n} W_k, \qquad \Lambda_{[n,m]} = \bigcup_{k=n}^{m} W_k, \ (n < m)$$

$$E_n = \big\{ < x, y > \in E \, : \, x, y \in \Lambda_n \big\}, \qquad \Lambda_n^c = \bigcup_{k=n}^{\infty} W_k$$

For $x \in \Gamma^k_+$, $x = (i_1, \dots, i_n)$ denote

$$S(x) = \{(x, i) : \ 1 \leq i \leq k\}.$$

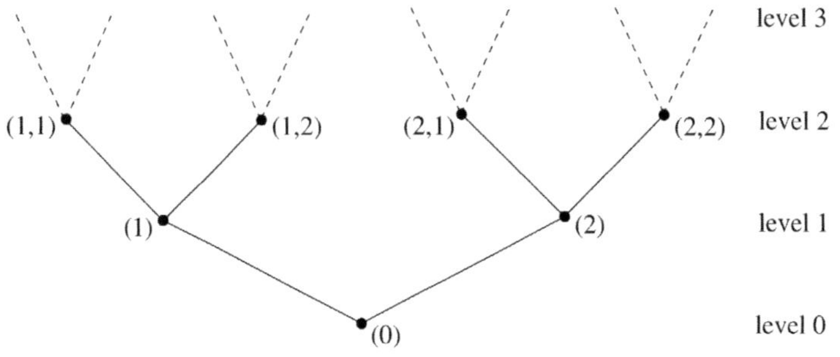

Figure 1. The first levels of Γ_+^2

Here (x,i) means that $(i_1,\dots,i_n,i)$. This set is called a set of *direct successors* of x.

Two vertices $x, y \in V$ is called *one level next-nearest-neighbor vertices* if there is a vertex $z \in V$ such that $x, y \in S(z)$, and they are denoted by $>x, y<$. In this case the vertices x, z, y was called *ternary* and denoted by $<x, z, y>$.

Let us define on Γ_+^k a binary operation $\circ : \Gamma_+^k \times \Gamma_+^k \to \Gamma_+^k$ as follows: for any two elements $x = (i_1,\dots,i_n)$ and $y = (j_1,\dots,j_m)$ put

$$x \circ y = (i_1,\dots,i_n) \circ (j_1,\dots,j_m) = (i_1,\dots,i_n,j_1,\dots,j_m) \tag{2.1}$$

and

$$x \circ x^0 = x^0 \circ x = (i_1,\dots,i_n) \circ (0) = (i_1,\dots,i_n). \tag{2.2}$$

By means of the defined operation Γ_+^k becomes a noncommutative semigroup with a unit. Using this semigroup structure one defines translations $\tau_g : \Gamma_+^k \to \Gamma_+^k$, $g \in \Gamma_+^k$ by

$$\tau_g(x) = g \circ x. \tag{2.3}$$

It is clear that $\tau_{(0)} = id$.

The algebra of observables $\mathcal{B}_x$ for any single site $x \in L$ will be taken as the algebra M_d of the complex $d \times d$ matrices. The algebra of observables localized in the finite volume $\Lambda \subset L$ is then given by $\mathcal{B}_\Lambda = \bigotimes_{x\in\Lambda} \mathcal{B}_x$. As usual if $\Lambda^1 \subset \Lambda^2 \subset L$, then $\mathcal{B}_{\Lambda^1}$ is identified as a subalgebra of $\mathcal{B}_{\Lambda^2}$ by tensoring with unit matrices on the sites $x \in \Lambda^2 \setminus \Lambda^1$. Note that, in the sequel, by $\mathcal{B}_{\Lambda,+}$ we denote the positive part of $\mathcal{B}_\Lambda$. The full algebra $\mathcal{B}_L$ of the tree is obtained in the usual manner by an inductive limit

$$\mathcal{B}_L = \overline{\bigcup_{\Lambda_n} \mathcal{B}_{\Lambda_n}}.$$

In what follows, by $\mathcal{S}(\mathcal{B}_\Lambda)$ we will denote the set of all states defined on the algebra $\mathcal{B}_\Lambda$.

Consider a triplet $\mathcal{C} \subset \mathcal{B} \subset \mathcal{A}$ of unital C^*-algebras. Recall [2] that a *quasi-conditional expectation* with respect to the given triplet is a completely positive (CP) linear map $\mathcal{E} : \mathcal{A} \to \mathcal{B}$ such that $\mathcal{E}(ca) = c\mathcal{E}(a)$, for all $a \in \mathcal{A}$, $c \in \mathcal{C}$.

Definition 2.1 ([14]). *Let φ be a state on $\mathcal{B}_L$. Then φ is called*

(i) a forward quantum d-Markov chain (QMC), associated to $\{\Lambda_n\}$, if for each Λ_n, there exist a quasi-conditional expectation $\mathcal{E}_{\Lambda_n^c}$ with respect to the triplet

$$\mathcal{B}_{\Lambda_{n+1}^c} \subseteq \mathcal{B}_{\Lambda_n^c} \subseteq \mathcal{B}_{\Lambda_{n-1}^c} \tag{2.4}$$

and a state $\hat{\varphi}_{\Lambda_n^c} \in \mathcal{S}(\mathcal{B}_{\Lambda_n^c})$ *such that for any* $n \in \mathbb{N}$ *one has*

$$\hat{\varphi}_{\Lambda_n^c}|\mathcal{B}_{\Lambda_{n+1}\setminus\Lambda_n} = \hat{\varphi}_{\Lambda_{n+1}^c} \circ \mathcal{E}_{\Lambda_{n+1}^c}|\mathcal{B}_{\Lambda_{n+1}\setminus\Lambda_n} \tag{2.5}$$

and

$$\varphi = \lim_{n\to\infty} \hat{\varphi}_{\Lambda_n^c} \circ \mathcal{E}_{\Lambda_n^c} \circ \mathcal{E}_{\Lambda_{n-1}^c} \circ \cdots \circ \mathcal{E}_{\Lambda_1^c} \tag{2.6}$$

in the weak- topology.*

(ii) a backward quantum d-Markov chain, associated to $\{\Lambda_n\}$*, if there exist a quasi-conditional expectation* $\mathcal{E}_{\Lambda_n}$ *with respect to the triple* $\mathcal{B}_{\Lambda_{n-1}} \subseteq \mathcal{B}_{\Lambda_n} \subseteq \mathcal{B}_{\Lambda_{n+1}}$ *for each* $n \in \mathbb{N}$*and an initial state* $\rho_0 \in S(B_{\Lambda_0})$ *such that*

$$\varphi = \lim_{n\to\infty} \rho_0 \circ \mathcal{E}_{\Lambda_0} \circ \mathcal{E}_{\Lambda_1} \circ \cdots \circ \mathcal{E}_{\Lambda_n}$$

in the weak- topology.*

Note that (2.5) is an analogue of the DRL equation from classical statistical mechanics [17, 22], and QMC is thus the counterpart of the infinite-volume Gibbs measure.

3. Construction of Quantum Markov Chains on Cayley tree

In this section we are going to provide a general construction of quantum Markov chains. Note that in our construction generalizes our previous works [10]-[13].

Let us rewrite the elements of W_n in the following order, i.e.

$$\overrightarrow{W_n} := \left(x_{W_n}^{(1)}, x_{W_n}^{(2)}, \cdots, x_{W_n}^{(|W_n|)}\right), \quad \overleftarrow{W_n} := \left(x_{W_n}^{(|W_n|)}, x_{W_n}^{(|W_n|-1)}, \cdots, x_{W_n}^{(1)}\right).$$

Note that $|W_n| = k^n$. Vertices $x_{W_n}^{(1)}, x_{W_n}^{(2)}, \cdots, x_{W_n}^{(|W_n|)}$ of W_n can be represented in terms of the coordinate system as follows

$$\begin{aligned} &x_{W_n}^{(1)} = (1,1,\cdots,1,1), \quad x_{W_n}^{(2)} = (1,1,\cdots,1,2), \quad \cdots \quad x_{W_n}^{(k)} = (1,1,\cdots,1,k,), \\ &x_{W_n}^{(k+1)} = (1,1,\cdots,2,1), \quad x_{W_n}^{(2)} = (1,1,\cdots,2,2), \quad \cdots \quad x_{W_n}^{(2k)} = (1,1,\cdots,2,k), \\ &\vdots \\ &x_{W_n}^{(|W_n|-k+1)} = (k,k,,\cdots,k,1), \; x_{W_n}^{(|W_n|-k+2)} = (k,k,\cdots,k,2), \quad \cdots x_{W_n}^{|W_n|} = (k,k,\cdots,k,k). \end{aligned} \tag{3.1}$$

Analogously, for a given vertex x, we shall use the following notation for the set of direct successors of x:

$$\overrightarrow{S(x)} := ((x,1),(x,2),\cdots(x,k)), \quad \overleftarrow{S(x)} := ((x,k),(x,k-1),\cdots(x,1)).$$

In what follows, by $\prod$ we mean an ordered product, i.e.

$$\prod_{k=1}^{n} a_k = a_1 a_2 \cdots a_n,$$

where elements $\{a_k\} \subset \mathcal{B}_L$ are multiplied in the indicated order. This means that we are not allowed to change this order.

Assume that we are given a sequence of operators $\{K_{[n,n+1]}\}_{n\geq 0}$, where each operator $K_{[n,n+1]}$ belongs to $\mathcal{B}_{\Lambda_{[n,n+1]}}$ $(n \geq 0)$. Moreover, we suppose that boundary conditions $w_0 \in \mathcal{B}^+_{(0)}$ and $\mathfrak{h} = \{\mathbf{h}_n \in \mathcal{B}^+_{W_n} : n \in \mathbb{N}\}$ are also given.

For each $n \in \mathbb{N}$ we denote

$$\mathbf{K}_n := w_0^{1/2} \prod_{m=0}^{n-1} K_{[m,m+1]} \mathbf{h}_n^{1/2} \tag{3.2}$$

$$\mathcal{W}^{(f)}_{n]} := \mathbf{K}_n \mathbf{K}_n^*, \quad \mathcal{W}^{(b)}_{n]} := \mathbf{K}_n^* \mathbf{K}_n. \tag{3.3}$$

One can see that $\mathcal{W}^{(f)}_{n]}$ and $\mathcal{W}^{(b)}_{n]}$ are positive.

In what follows, by $\mathrm{tr}_\Lambda : \mathcal{B}_L \to \mathcal{B}_\Lambda$ we mean normalized partial trace (i.e. $\mathrm{tr}_\Lambda(\mathbb{1}_L) = \mathbb{1}_\Lambda$, here $\mathbb{1}_\Lambda = \bigotimes_{y\in\Lambda} \mathbb{1}$), for any $\Lambda \subseteq_{\mathrm{fin}} L$. For the sake of shortness we put $\mathrm{tr}_{n]} := \mathrm{tr}_{\Lambda_n}$.

Let us define positive functionals $\varphi^{(n,f)}_{w_0,\mathfrak{h}}$, $\varphi^{(n,b)}_{w_0,\mathfrak{h}}$ on $\mathcal{B}_{\Lambda_n}$, respectively, by

$$\varphi^{(n,f)}_{w_0,\mathfrak{h}}(a) = \mathrm{tr}(\mathcal{W}^{(f)}_{n+1]}(a \otimes \mathbb{1}_{W_{n+1}})), \quad \varphi^{(n,b)}_{w_0,\mathfrak{h}}(a) = \mathrm{tr}(\mathcal{W}^{(b)}_{n+1]}(a \otimes \mathbb{1}_{W_{n+1}})), \tag{3.4}$$

for every $a \in \mathcal{B}_{\Lambda_n}$. Note that here, tr is a normalized trace on $\mathcal{B}_L$ (i.e. $\mathrm{tr}(\mathbb{1}_L) = 1$).

To get an infinite-volume state φ on $\mathcal{B}_L$ such that $\varphi\lceil_{\mathcal{B}_{\Lambda_n}} = \varphi^{(n,\varepsilon)}_{w_0,\mathfrak{h}}$, $(\varepsilon = f, b)$ we need to impose some constrains to the boundary conditions $\{w_0, \mathfrak{h}\}$ so that the functionals $\{\varphi^{(n,\varepsilon)}_{w_0,\mathfrak{h}}\}$ satisfy the compatibility condition, i.e.

$$\varphi^{(n+1,\varepsilon)}_{w_0,\mathfrak{h}}\lceil_{\mathcal{B}_{\Lambda_n}} = \varphi^{(n,\varepsilon)}_{w_0,\mathfrak{h}}. \tag{3.5}$$

In the following we need an auxiliary fact.

Lemma 3.1. *Let $\Lambda \subseteq \Lambda' \subseteq_{fin} L$, then for any $A \in \mathcal{B}_\Lambda, B \in \mathcal{B}_{\Lambda'}$ one has $\mathrm{Tr}(AB) = \mathrm{Tr}[A\,\mathrm{Tr}_{\mathcal{B}_\Lambda}(B)]$.*

Theorem 3.2. *Let a sequence $\{K_{[n,n+1]}\}_{n\geq 0}$, ($K_{[n,n+1]} \in \mathcal{B}_{\Lambda_{[n,n+1]}}$) be given. Assume that boundary conditions $w_0 \in \mathcal{B}_{(0),+}$ and $\mathfrak{h} = \{h_x \in \mathcal{B}_{x,+}\}_{x\in L}$ satisfy*

$$\mathrm{Tr}(w_0 h_0) = 1, \tag{3.6}$$

$$\mathrm{Tr}_{n]}\left(K_{[n,n+1]}\mathbf{h}_{n+1}K^*_{[n,n+1]}\right) = \mathbf{h}_n, \quad \text{for every} \quad n \in \mathbb{N}. \tag{3.7}$$

Then the functionals $\{\varphi^{(n,f)}_{w_0,\mathfrak{h}}\}$ satisfy the compatibility condition (3.5). Moreover, there is a unique forward quantum Markov chain $\varphi^{(f)}_{w_0,\mathfrak{h}}$ on $\mathcal{B}_L$ such that $\varphi^{(f)}_{w_0,\mathfrak{h}} = w - \lim_{n\to\infty} \varphi^{(n,f)}_{w_0,\mathfrak{h}}$.

Proof. We first show that a sequence $\{\mathcal{W}^{(f)}_{n]}\}$ is *projective* with respect to $\mathrm{tr}_{n]}$, i.e.

$$\mathrm{tr}_{n-1]}(\mathcal{W}^{(f)}_{n]}) = \mathcal{W}^{(f)}_{n-1]}, \quad \forall n \in \mathbb{N}. \tag{3.8}$$

It is known [8] that the projectivity implies the compatibility condition.

Now let us check the equality (3.8). From (3.2) one has

$$\mathcal{W}^{(f)}_{n]} = w_0^{1/2}\left(\prod_{m=0}^{n-1} K_{[m-1,m]}\right) K_{[n-1,n]}\mathbf{h}_n K^*_{[n-1,n]} \left(\prod_{m=0}^{n-1} K_{[m-1,m]}\right)^* w_0^{1/2}.$$

Hence, from the last equality with (3.16) we get

$$\begin{aligned}\mathrm{tr}_{n-1]}(\mathcal{W}^{(f)}_{n]}) &= w_0^{1/2}\Big(\prod_{m=0}^{n-1} K_{[m-1,m]}\Big)\,\mathrm{tr}_{n-1]}(K_{[n-1,n]}\mathbf{h}_n K^*_{[n-1,n]})\Big(\prod_{m=0}^{n-1} K_{[m-1,m]}\Big)^* w_0^{1/2}\\ &= \mathcal{W}^{(f)}_{n-1]}.\end{aligned}$$

From the above argument and (3.15), one can show that $\mathcal{W}^{(f)}_{n]}$ is density operator, i.e. $\mathrm{tr}(\mathcal{W}^{(f)}_{n]}) = 1$.

Let us show that the defined state $\varphi^{(f)}_{w_0,\mathfrak{h}}$ is a forward QMC. Indeed, define quasi-conditional expectations $\mathcal{E}_{\Lambda_n^c}$ as follows:

$$\hat{\mathcal{E}}_{\Lambda_1^c}(x_{[0}) = \mathrm{tr}_{[1}(K_{[0,1]}w_0^{1/2}x_{[0}w_0^{1/2}K^*_{[0,1]}),\quad x_{[0}\in\mathcal{B}_{\Lambda_0^c} \tag{3.9}$$
$$\mathcal{E}_{\Lambda_k^c}(x_{[k-1}) = \mathrm{tr}_{[n}(K_{[k-1,k]}x_{[k-1}K^*_{[k-1,k]}),\quad x_{[k-1}\in\mathcal{B}_{\Lambda_{k-1}^c},\quad k=1,2,\dots,n+1, \tag{3.10}$$

here $\mathrm{tr}_{[n} = \mathrm{tr}_{\Lambda_n^c}$. Then for any monomial $a_{\Lambda_1}\otimes a_{W_2}\otimes\cdots\otimes a_{W_n}\otimes \mathbb{1}_{W_{n+1}}$, where $a_{\Lambda_1}\in\mathcal{B}_{\Lambda_1}, a_{W_k}\in\mathcal{B}_{W_k}$, $(k=2,\dots,n)$, we have

$$\begin{aligned}\varphi^{(n,f)}_{w_0,\mathfrak{h}}(a_{\Lambda_1}\otimes a_{W_2}\otimes\cdots\otimes a_{W_n}) &= \mathrm{tr}\Big(\mathbf{h}_{n+1}K^*_{[n,n+1]}\cdots K^*_{[0,1]}w_0^{1/2}(a_{\Lambda_1}\otimes a_{W_2}\otimes\cdots\otimes a_{W_n})\\ &\qquad w_0^{1/2}K_{[0,1]}\cdots K_{[n,n+1]}\Big)\\ &= \mathrm{tr}_{[1}\Big(\mathbf{h}_{n+1}K^*_{[n,n+1]}\cdots K^*_{[1,2]}\hat{\mathcal{E}}_{\Lambda_1^c}(a_{\Lambda_1})a_{W_2}K_{[1,2]}\\ &\qquad\cdots a_{W_n}K_{[n,n+1]}\Big)\\ &= \mathrm{tr}_{[n+1}\big(\mathbf{h}_{n+1}\mathcal{E}_{\Lambda_{n+1}^c}\circ\mathcal{E}_{\Lambda_n^c}\circ\cdots\\ &\qquad \mathcal{E}_{\Lambda_2^c}\circ\hat{\mathcal{E}}_{\Lambda_1^c}(a_{\Lambda_1}\otimes a_{W_2}\otimes\cdots\otimes a_{W_n})\big).\end{aligned} \tag{3.11}$$

Hence, for any $a\in\Lambda\subset\Lambda_{n+1}$ from (3.4) with (3.2), (3.9)-(3.11) one can see that

$$\varphi^{(n,f)}_{w_0,\mathfrak{h}}(a) = \mathrm{tr}_{[n+1}\big(\mathbf{h}_{n+1}\mathcal{E}_{\Lambda_{n+1}^c}\circ\mathcal{E}_{\Lambda_n^c}\circ\cdots\mathcal{E}_{\Lambda_2^c}\circ\hat{\mathcal{E}}_{\Lambda_1^c}(a)\big). \tag{3.12}$$

The projectivity of $\mathcal{W}^{(f)}_{n]}$ yields the equality (2.5) for $\varphi^{(n,f)}_{w_0,\mathbf{h}}$, therefore, from (3.12) we conclude that $\varphi^{(f)}_{w_0,\mathfrak{h}}$ is a forward QMC. □

Theorem 3.3. *Let a sequence $\{K_{[n,n+1]}\}_{n\geq 0}$, ($K_{[n,n+1]}\in\mathcal{B}_{\Lambda_{[n,n+1]}}$) be given. Assume that boundary conditions $w_0\in\mathcal{B}_{(0),+}$ and $\mathfrak{h} = \{h_x\in\mathcal{B}_{x,+}\}_{x\in L}$ satisfy (3.15),(3.16). Then the functionals $\{\varphi^{(n,b)}_{w_0,\mathfrak{h}}\}$ satisfy the compatibility condition (3.5). Moreover, there is a unique backward quantum Markov chain $\varphi^{(b)}_{w_0,\mathfrak{h}}$ on $\mathcal{B}_L$ such that $\varphi^{(b)}_{w_0,\mathfrak{h}} = w-\lim_{n\to\infty}\varphi^{(n,b)}_{w_0,\mathfrak{h}}$.*

Proof. Let us check that the states $\{\varphi^{(n,b)}_{w_0,\mathfrak{h}}\}$ satisfy the compatibility condition. For every $a\in\mathcal{B}_{\Lambda_n}$

we have

$$
\begin{aligned}
& \varphi^{(n+1,b)}_{w_0,\mathfrak{h}}(a\otimes \mathbb{1}_{W_{n+1}}) = \mathrm{tr}(\mathbf{K}_n^*\mathbf{K}_n(a\otimes \mathbb{1}_{\Lambda_{[n+1,n+2]}}))\\
= \; & \mathrm{tr}\left(K^*_{[n,n+1]}\cdots K^*_{[0,1]}w_0K_{[0,1]}\cdots K_{[n+1,n+2]}\mathbf{h}^{1/2}_{n+2}(a\otimes \mathbb{1}_{\Lambda_{[n+1,n+2]}})\mathbf{h}^{1/2}_{n+2}K^*_{[n+1,n+2]}\right)\\
= \; & \mathrm{tr}\left(K^*_{[n,n+1]}\cdots K^*_{[0,1]}w_0K_{[0,1]}\cdots K_{[n,n+1]}(a\otimes \mathbb{1}_{\Lambda_{[n+1,n+2]}})K_{[n+1,n+2]}\mathbf{h}_{n+2}K^*_{[n+1,n+2]}\right)\\
= \; & \mathrm{tr}\left(K^*_{[n,n+1]}\cdots K^*_{[0,1]}w_0K_{[0,1]}\cdots K_{[n,n+1]}(a\otimes \mathbb{1}_{\Lambda_{[n+1,n+2]}})\,\mathrm{tr}_{n+1)]}\,K_{[n+1,n+2]}\mathbf{h}_{n+2}K^*_{[n+1,n+2]}\right)\\
= \; & \mathrm{tr}\left(K^*_{[n,n+1]}\cdots K^*_{[0,1]}w_0K_{[0,1]}\cdots K_{[n,n+1]}(a\otimes \mathbb{1}_{\Lambda_{[n+1,n+2]}})\mathbf{h}_{n+1}\right)\\
= \; & \mathrm{tr}\left(\mathbf{h}^{1/2}_{n+1}K^*_{[n,n+1]}\cdots K^*_{[0,1]}w_0K_{[0,1]}\cdots K_{[n,n+1]}\mathbf{h}^{1/2}_{n+1}(a\otimes \mathbb{1}_{\Lambda_{[n+1,n+2]}})\right)\\
= \; & \mathrm{tr}\left(\mathcal{W}^{(b)}_{n+1]}(a\otimes \mathbb{1}_{\Lambda_{[n+1,n+2]}})\right)\\
= \; & \varphi^{(n,b)}_{w_0,\mathfrak{h}}(a)
\end{aligned}
$$

Let us show that the defined state $\varphi^{(b)}_{w_0,\mathfrak{h}}$ is a backward QMC. Indeed, define quasi-conditional expectations $\mathcal{E}_n$ as follows:

$$
\begin{aligned}
\hat{\mathcal{E}}_n(x_{n+1]}) &= \mathrm{tr}_{n]}(K_{[n,n+1]}\mathbf{h}^{1/2}_{n+1}x_{n+1]}\mathbf{h}^{1/2}_{n+1}K^*_{[n,n+1]}), \quad x_{n+1]}\in\mathcal{B}_{\Lambda_{n+1}}\\
\mathcal{E}_k(x_{k+1]}) &= \mathrm{tr}_{k]}(K_{[k,k+1]}x_{k+1]}K^*_{[k,k+1]}), \quad x_{k+1]}\in\mathcal{B}_{\Lambda_{k+1}}, \quad k=0,1,\dots,n-1,
\end{aligned}
$$

Then using the same argument as in the proof of Theorem 3.2 we have

$$
\varphi^{(n,b)}_{w_0,\mathfrak{h}}(a) = \rho_0\big(\mathcal{E}_0\circ\cdots\circ\mathcal{E}_{n-1}\circ\hat{\mathcal{E}}_n(a)\big).
$$

This means that the limit state $\varphi^{(b)}_{w_0,\mathfrak{h}}$ is a backward QMC. This completes the proof. □

A sequence $\{K_{[n,n+1]}\}_{n\geq 0}$, ($K_{[n,n+1]}\in\mathcal{B}_{\Lambda_{[n,n+1]}}$) is called *localized* if for each $n\in\mathbb{N}$ one has

$$
K_{[n,n+1]} = \prod_{x\in\overrightarrow{W_n}} A_{x,(x,1),\dots,(x,k)} \tag{3.13}
$$

where $A_{x,(x,1),\dots,(x,k)}\in\mathcal{B}_x\otimes\mathcal{B}_{(x,1)}\otimes\cdots\otimes\mathcal{B}_{(x,k)}$.

Now we assume that the boundary condition $\mathfrak{h}$ is also localized, i.e. for every $n\in\mathbb{N}$ one has

$$
\mathbf{h}_n = \prod_{x\in\overrightarrow{W}_n} h_x \tag{3.14}
$$

where $h_x\in\mathcal{B}_x^+$, $x\in W_n$.

Corollary 3.4. *Let $\{K_{[n,n+1]}\}_{n\geq 0}$ be a localized sequence given by (3.13). Assume that boundary condition $\mathfrak{h}$ is also localized by (3.14). If one has*

$$
\mathrm{Tr}(w_0h_0) = 1, \tag{3.15}
$$

$$
\mathrm{Tr}_{x]}\big(A_{x,(x,1),\dots,(x,k)}\prod_{y\in\overrightarrow{S(x)}} h_yA^*_{x,(x,1),\dots,(x,k)}\big) = h_x, \tag{3.16}
$$

for all $x\in L\setminus\{(0)\}$. Then there is a unique backward quantum Markov chain $\varphi^{(b)}_{w_0,\mathfrak{h}}$ on $\mathcal{B}_L$.

Theorem 3.5. *Let a sequence $\{K_{[n,n+1]}\}_{n\geq 0}$, ($K_{[n,n+1]} \in \mathcal{B}_{\Lambda_{[n,n+1]}}$) be given. Assume that two boundary conditions $(w_0, \mathfrak{h})$ and $(\tilde{w}_0, \mathfrak{s})$ satisfy (3.15),(3.16), and $\varphi^{(b)}_{w_0,\mathfrak{h}}$ and $\varphi^{(b)}_{\tilde{w}_0,\mathfrak{s}}$ are the corresponding backward QMC. Then for every $\lambda \in [0,1]$ the state $\lambda\varphi^{(b)}_{w_0,\mathfrak{h}} + (1-\lambda)\varphi^{(b)}_{\tilde{w}_0,\mathfrak{s}}$ is also backward QMC.*

Proof. Without loss of generality we may assume that for every boundary condition $w_0 = \mathbb{1}$. Otherwise, we re-denote $w_0 h_0$ by h'_0. In the latter case, one gets $w'_0 = \mathbb{1}$. Therefore, $\varphi^{(b)}_{\mathbb{1},\mathfrak{h}}$ and $\varphi^{(b)}_{\mathbb{1},\mathfrak{s}}$ are QMC corresponding to $(\mathbb{1}, \mathfrak{h})$ and $(\mathbb{1}, \mathfrak{s})$.

Now define a sequence $\mathfrak{g} = \{\mathbf{g}_n\}$ as follows:

$$\mathbf{g}_n = \lambda\mathbf{h}_n + (1-\lambda)\mathbf{s}_n, \quad n \geq 0.$$

It is clear that $\mathbf{g}_n \in \mathcal{B}^+_{W_n}$

We first observe that the boundary condition $(\mathbb{1}, \mathfrak{g})$ satisfies (3.15),(3.16). Indeed, one can see that $\mathrm{tr}(g_0) = \mathrm{tr}(\lambda h_0 + (1-\lambda)s_0) = 1$ and

$$\begin{aligned} \mathrm{tr}_{n]}(K_{[n,n+1]}\mathbf{g}_{n+1}K^*_{[n,n+1]}) &= \lambda\,\mathrm{tr}_{n]}(K_{[n,n+1]}\mathbf{h}_{n+1}K_{[n,n+1]}) \\ &\quad +(1-\lambda)\,\mathrm{tr}_{n]}(K_{[n,n+1]}\mathbf{s}_{n+1}K_{[n,n+1]}) \\ &= \lambda\mathbf{h}_n + (1-\lambda)\mathbf{s}_n \\ &= \mathbf{g}_n. \end{aligned}$$

Hence, due to Theorem 3.3 there is a backward QMC $\varphi^{(b)}_{\mathbb{1},\mathfrak{g}}$ corresponding to the boundary condition $(\mathbb{1}, \mathfrak{g})$. Let us show that $\varphi^{(b)}_{\mathbb{1},\mathfrak{g}} = \lambda\varphi^{(b)}_{w_0,\mathfrak{h}} + (1-\lambda)\varphi^{(b)}_{\tilde{w}_0,\mathfrak{s}}$. Indeed, For every $a \in \mathcal{B}_{\Lambda_n}$, we have

$$\begin{aligned} \varphi^{(b)}_{\mathbb{1},\mathfrak{g}}(a) &= \mathrm{tr}\left(\mathbf{g}^{1/2}_{n+1}K^*_{[n,n+1]}\cdots K^*_{[0,1]}w_0K_{[0,1]}\cdots K_{[n,n+1]}\mathbf{g}^{1/2}_{n+1}(a\otimes \mathbb{1}_{W_{n+1}})\right) \\ &= \mathrm{tr}\left(K^*_{[n,n+1]}\cdots K^*_{[0,1]}w_0K_{[0,1]}\cdots K_{[n,n+1]}(\lambda\mathbf{h}_{n+1} + (1-\lambda)\mathbf{s}_{n+1})(a\otimes \mathbb{1}_{W_{n+1}})\right) \\ &= \lambda\,\mathrm{tr}\left(K^*_{[n,n+1]}\cdots K^*_{[0,1]}w_0K_{[0,1]}\cdots K_{[n,n+1]}\mathbf{h}_{n+1}(a\otimes \mathbb{1}_{W_{n+1}})\right) \\ &\quad +(1-\lambda)\,\mathrm{tr}\left(K^*_{[n,n+1]}\cdots K^*_{[0,1]}w_0K_{[0,1]}\cdots K_{[n,n+1]}\mathbf{s}_{n+1}(a\otimes \mathbb{1}_{W_{n+1}})\right) \\ &= \lambda\varphi^{(b)}_{\mathbb{1},\mathfrak{h}}(a) + (1-\lambda)\varphi^{(b)}_{\mathbb{1},\mathfrak{s}}(a). \end{aligned}$$

This completes the proof. □

From the proved theorem we immediately infer the following result.

Corollary 3.6. *If there are at least two backward QMC corresponding to the sequence $\{K_{[n,n+1]}\}_{n\geq 0}$, ($K_{[n,n+1]} \in \mathcal{B}_{\Lambda_{[n,n+1]}}$), then they are uncountable.*

References

[1] Accardi L 1975 *Funct. Anal. Appl.*, **9** 1–8
[2] Accardi L and Cecchini C 1982 *J. Funct. Anal.* **45** 245–273
[3] Accardi L and Fidaleo F 2005 *Annali di Matematica Pura e Applicata*, **184** 327–346
[4] Accardi L and Fidaleo F 2003 *Inf. Dim. Analysis, Quantum Probab. Related Topics* **6** 123–138
[5] Accardi L and Fidaleo F 2003 *J. Funct. Anal.* **200** 324-347
[6] Accardi L and Fidaleo F 2003 In book: *Proceedings Burg Conference 15–20 March 2001, W. Freudenberg (ed.)* (Singapore: World Scientific) 1–20

[7] Accardi L, Fidaleo F and Mukhamedov F 2007 *Inf. Dim. Analysis, Quantum Probab. Related Topics* **10** 165–183
[8] Accardi L and Frigerio A 1983 *Proc. Royal Irish Acad.* **83A** 251-263
[9] Accardi L and Liebscher V 1999 *Infin. Dimens. Anal. Quantum Probab. Relat. Top.* **2** 645-661
[10] Accardi L, Mukhamedov F and Saburov M 2011 *Math. Notes* **90** 8–20
[11] Accardi L, Mukhamedov F and Saburov M 2011 *Inf. Dim. Analysis, Quantum Probab. Related Topics* **14** 443–463
[12] Accardi L, Mukhamedov F and Saburov M 2011 *Ann. Henri Poincare* **12** 1109–1144
[13] Accardi L, Mukhamedov F and Saburov M 2014 *J. Statis. Phys.* **157** 303-329
[14] Accardi L, Ohno H and Mukhamedov F 2010 *Inf. Dim. Analysis, Quantum Probab. Related Topics* **13** 165–189
[15] Affleck L, Kennedy E, Lieb E H and Tasaki H 1988 *Commun. Math. Phys.* **115** 477–528
[16] Baxter R J 1982 *Exactly Solved Models in Statistical Mechanics* (London-New York: Acad. Press).
[17] Dobrushin R L 1968 *Probab. Theory Appl.* **13** 201–229
[18] Fannes M, Nachtergaele B amd Werner R F 1992 *J. Stat. Phys.* **66** 939–973
[19] Fannes M, Nachtergaele B and Werner R F 1992 *Commun. Math. Phys.* **144** 443–490
[20] Fidaleo F and Mukhamedov F 2004 *Probab. Math. Stat.* **24** 401–418
[21] Ganikhodzhaev N N. and Mukhamedov F M 2004 *Math. Notes* **76** 329–338
[22] Georgi H-O 1988 *Gibbs measures and phase transitions* (Berlin: Walter de Gruyter)
[23] Golodets V Y and Zholtkevich G N 1983 *Theor. Math. Phys.* **56** 686–690
[24] Ibinson B, Linden N and Winter A 2008 *Comm. Math. Phys.* **277** 289–304
[25] Kümmerer B 2006 *Lecture Notes in Math.*, **1866** 259–330
[26] Liebscher V 2003 In book: *Proceedings Burg Conference 15–20 March 2001, W. Freudenberg (ed.)* (Singapore: World Scientific) 151–159
[27] Mukhamedov F M 2004 *Rep. Math. Phys.* **53** 1–18
[28] Mukhamedov F M 2000 *Theor. Math. Phys.* **123** 489–493
[29] Mukhamedov F M and Rozikov U A 2004 *J. Stat. Phys.* **114**(2004),825–848.
[30] Mukhamedov F M and Rozikov U A 2005 *J. Stat. Phys.* **119** 427–446
[31] Ohno H. 2005 *Inf. Dim. Analysis, Quantum Probab. Related Topics*, **8** 141–152
[32] Ohya M and Petz D 1993 *Quantum entropy and its use* (Berlin: Springer)
[33] Ostilli M 2012 *Physica A* **391** 3417–3423
[34] Ostilli M, Mukhamedov F and Mendes J F F 2008 *Physica A* **387** 2777–2792
[35] Preston C 1974, *Gibbs states on countable sets* (London: Cambridge University Press)
[36] Rozikov U A 2013 *Gibbs measures on Cayley trees*, (Singapore: World Scientific)
[37] Spataru A 1990 *Advances in Math* **81** 105–116
[38] Spitzer F 1975 *Ann. Prob.* **3** 387-398
[39] Zachary S 1983 *Ann. Prob.* **11** 894–903
[40] Zachary S 1985 *Stochastic Process. Appl.* **20** 247–256

An improvement of extremality regions for Gibbs measures of the Potts model on a Cayley tree

Farhod Haydarov[1], **Rustam Khakimov**[2]

[1] National University of Uzbekistan, Tashkent, Uzbekistan
[2] Institute of Mathematics, National University of Uzbekistan, 29, Do'rmon Yo'li str., 100125, Tashkent, Uzbekistan

E-mail: haydarov_imc@mail.ru, rustam-7102@rambler.ru

Abstract. We give a condition of extemelity for translation-invariant Gibbs measures of $q-$state Potts model on a Cayley tree. We'll improve the regions of extremality for some measures considered in [14]. Moreover, some results in [14] are generalized.

1. Introduction

The Potts model is a generalization of the Ising model. In [3], [4] the q-state Potts model on a Cayley tree of order $k \geq 2$ was studied, and it has been known for a long time that at sufficiently low temperatures, there are at least $q+1$ translation-invariant Gibbs measures. This measures can be considered as tree-indexed Markov chains. Such translation-invariant tree-indexed measures are equivalently called translation- invariant splitting Gibbs measures (TISGMs).

In [6] the uniqueness of the translation-invariant Gibbs measure of the antiferromagnetic Potts model with an external field is proved. In [7] the Potts model with a countable number of states and nonzero external field on a Cayley tree was considered. In that paper, it was established that the model has a unique translation-invariant Gibbs measure.

In [13] all TISGMs (tree-indexed Markov chains) for the Potts model are found on the Cayley tree of order $k \geq 2$, and it is shown that at sufficiently low temperatures their number is $2^q - 1$. In the case $k = 2$ the explicit formulae for the critical temperatures and all TISGMs are given. Further, in [14] by means of methods and results of [10], [21], [15] it has been found some regions for the temperature parameter ensuring that a given TISGM is (non-)extreme in the set of all Gibbs measures. In particular, it was shown the existence of a temperature interval for which there are at least $2^{q-1} + q$ extreme TISGMs. In case of the order of the tree is two, it was given an explicit formulae and some numerical values of the critical temperature. Note that other properties of the Potts model on a Cayley tree were studied in [1, 5, 8, 9, 11, 12],[17]-[20].

In this paper, we consider the $q-$state Potts model on the Cayley tree of order two. Some results of [14] will be improved. Moreover, we will extend Theorem 6 of [14].

2. Definitions and known facts

A Cayley tree $\Im^k$ of order $k \geq 1$ is an infinite tree, i.e. a graph without cycles, such that exactly $k+1$ edges originate from each vertex. Let $\Im^k = (V, L, i)$, where V is the set of vertices $\Im^k$, L

the set of edges and i is the incidence function setting each edge $l \in L$ into the correspondence with its endpoints $x, y \in V$. If $i(l) = \{x, y\}$, then the vertices x and y are called the *nearest neighbors*, denoted by $l = \langle x, y \rangle$. The distance $d(x, y), x, y \in V$ on a Cayley tree is defined by

$$d(x,y) = \min \{d | \exists x = x_0, x_1, \dots, x_{d-1}, x_d = y \in V \text{ such that } \langle x_0, x_1 \rangle, \dots, \langle x_{d-1}, x_d \rangle \}.$$

For a fixed $x^0 \in V$ we set $W_n = \{x \in V \mid d(x, x^0) = n\}$,

$$V_n = \{x \in V \mid d(x, x^0) \leq n\}, \quad L_n = \{l = \langle x, y \rangle \in L \mid x, y \in V_n\}.$$

Put

$$S(x) = \{y \in W_{n+1} : d(x,y) = 1\}, \quad x \in W_n.$$

Namely, $S(x)$ is the set of *direct successors* of x.

We consider the model in which the spin variables take values in the set $\Phi = \{1, 2, \dots, q\}$, $(q \geq 2)$ and which are located at the tree vertices. For $A \subset V$ a *configuration* σ_A on A is an arbitrary function $\sigma_A : A \to \Phi$. Note that $\Omega_A = \Phi^A$ is the set of all configurations. We denote that $\Omega = \Omega_V$ and $\sigma = \sigma_V$.

A Hamiltonian of the Potts model is defined as

$$H(\sigma) = -J \sum_{\langle x,y \rangle \in L} \delta_{\sigma(x)\sigma(y)}, \tag{2.1}$$

where $J \in R$ and δ_{ij} is the Kronecker symbol.

In this paper, we restrict ourselves to the case of ferromagnetic interaction $J > 0$.

Define a finite-dimensional distribution of a probability measure μ in the volume V_n by

$$\mu_n(\sigma_n) = Z_n^{-1} \exp \left\{ -\beta H_n(\sigma_n) + \sum_{x \in W_n} h_{\sigma(x),x} \right\}, \tag{2.2}$$

where $\beta = 1/T$, $T > 0$-temperature, Z_n^{-1} is the normalizing factor, $\{h_x = (h_{1,x}, \dots, h_{q,x}) \in R^q, x \in V\}$ is a collection of vectors, and

$$H_n(\sigma_n) = -J \sum_{\langle x,y \rangle \in L_n} \delta_{\sigma(x)\sigma(y)}$$

is the restriction of Hamiltonian to V_n.

The probability distributions (2.2) are called compatible if for all $n \geq 1$ and $\sigma_{n-1} \in \Phi^{V_{n-1}}$:

$$\sum_{\omega_n \in \Phi^{W_n}} \mu_n(\sigma_{n-1} \vee \omega_n) = \mu_{n-1}(\sigma_{n-1}), \tag{2.3}$$

here $\sigma_{n-1} \vee \omega_n$ is the concatenation of configurations. In this case, by the well-known Kolmogorov's extension theorem, there exists a unique measure μ on Φ^V such that, for all n and $\sigma_n \in \Phi^{V_n}$

$$\mu(\{\sigma \in \Omega : \sigma|_{V_n} = \sigma_n\}) = \mu_n(\sigma_n).$$

This measure μ is called a *splitting Gibbs measure* corresponding to the Hamiltonian (2.1) and vector-valued function $h_x, x \in V$.

The following statement describes conditions on $\tilde{h}_x$ guaranteeing compatibility of $\{\mu_n\}$.

Theorem 2.1. *(see [3], [19, p.106]) The probability distributions μ_n, $n = 1, 2, \ldots$, in (2.2) are compatible iff for any $x \in V \setminus \{x^0\}$ the following equation holds:*

$$h_x = \sum_{y \in S(x)} F(h_y, \theta), \tag{2.4}$$

where $F : h = (h_1, \ldots, h_{q-1}) \in R^{q-1} \to F(h, \theta) = (F_1, \ldots, F_{q-1}) \in R^{q-1}$ is defined by

$$F_i = \ln\left(\frac{(\theta - 1)e^{h_i} + \sum_{j=1}^{q-1} e^{h_j} + 1}{\theta + \sum_{j=1}^{q-1} e^{h_j}}\right),$$

$\theta = \exp(J\beta)$, and $h_x = (h_{1,x}, \ldots, h_{q-1,x})$ with

$$h_{i,x} = \tilde{h}_{i,x} - \tilde{h}_{q,x}, \quad i = 1, \ldots, q-1. \tag{2.5}$$

From Theorem 2.1 it follows that for any $h = \{h_x, \quad x \in V\}$ satisfying (2.4) there exists a unique SGM μ for the Potts model.

Note that a translation-invariant splitting Gibbs measure (TISGM) corresponds to a solution h_x of (2.4) with $h_x = h = (h_1, \ldots, h_{q-1}) \in R^{q-1}$ for all $x \in V$. Then from equation (2.4) we get $h = kF(h, \theta)$, and denoting $z_i = \exp(h_i), i = 1, \ldots, q-1$, the last equation can be written as follows

$$z_i = \left(\frac{(\theta - 1)z_i + \sum_{j=1}^{q-1} z_j + 1}{\theta + \sum_{j=1}^{q-1} z_j}\right)^k, \quad i = 1, \ldots, q-1. \tag{2.6}$$

From [13] the following facts are known:

1. By solving (2.6) the full set of TISGMs is described. It is shown that any TISGM of the Potts model corresponds to a solution of the following equation

$$z = f_m(z) \equiv \left(\frac{(\theta + m - 1)z + q - m}{mz + q - m - 1 + \theta}\right)^k, \tag{2.7}$$

for some $m = 1, \ldots, [q/2]$.

2. Let $\theta_m = 1 + 2\sqrt{m(q-m)}$, $m = 1, \ldots, q-1$.. If $\theta < \theta_1$ then there exists a unique TISGM for $k \geq 2$, $J > 0$. Moreover, each θ_m is a critical value for the change of the number of TISGMs.

3. Extremity conditions

Following [14], to check the extremity of the Gibbs measure, we apply arguments of a reconstruction on trees [2], [10], [16].

From [14] it is known that for each fixed m, the equation (2.7) has up to three solutions: $z_0 = 1, z_i = z_i(\theta, q, m), i = 1, 2$, with $z_1 < z_2$ (see [13, Step 1 of the proof of Theorem 1]). Denote by $\mu_i = \mu_i(\theta, m)$ the TISGM of the Potts model which corresponds to the solution z_i.

For $l = (\underbrace{z, z, \ldots, z}_{m}, \underbrace{1, 1, \ldots, 1}_{q-m})$ a TISGM corresponding to a vector $l \in R^q$ is a tree-indexed Markov chain with states $\{1, 2, \ldots, q\}$ and transition probabilities matrix $\mathbb{P} = (P_{ij})$ with

$$P_{ij} = \frac{l_j \exp(J\beta\delta_{ij})}{\sum_{r=1}^{q} l_r \exp(J\beta\delta_{ir})}. \tag{3.1}$$

From (3.1) we get

$$P_{ij} = \begin{cases} \theta z/Z_1, & \text{if } i = j,\ i \in \{1, \dots, m\} \\ z/Z_1, & \text{if } i \neq j,\ i, j \in \{1, \dots, m\} \\ 1/Z_1, & \text{if } i \in \{1, \dots, m\}, j \in \{m+1, \dots, q\} \\ z/Z_2, & \text{if } i \in \{m+1, \dots, q\}, j \in \{1, \dots, m\} \\ \theta/Z_2, & \text{if } i = j,\ i \in \{m+1, \dots, q\} \\ 1/Z_2, & \text{if } i \neq j,\ i, j \in \{m+1, \dots, q\}, \end{cases} \tag{3.2}$$

where

$$Z_1 = (\theta + m - 1)z + q - m, \quad Z_2 = mz + \theta + q - m - 1.$$

Let us first give some necessary definitions from [15]. Considering finite complete subtrees $\mathcal{T}$ that are initial points of Cayley tree Γ^k, i.e. share the same root; if $\mathcal{T}$ has depth d (i.e. the vertices of $\mathcal{T}$ are within distance $\leq d$ from the root) then it has $(k^{d+1} - 1)/(k - 1)$ vertices, and its boundary $\partial\mathcal{T}$ consists the neighbors (in $\Gamma^k \setminus \mathcal{T}$) of its vertices, i.e., $|\partial\mathcal{T}| = k^{d+1}$. We identify subgraphs of $\mathcal{T}$ with their vertex sets and write $E(A)$ for the edges within a subset A and ∂A for the boundary of A, i.e., the neighbors of A in $(\mathcal{T} \cup \partial\mathcal{T}) \setminus A$.

For a given subtree $\mathcal{T}$ of Γ^k and a vertex $x \in \mathcal{T}$, we write $\mathcal{T}_x$ for the (maximal) subtree of $\mathcal{T}$ rooted at x. When x is not the root of $\mathcal{T}$, let $\mu^s_{\mathcal{T}_x}$ denote the (finite-volume) Gibbs measure in which the parent of x has its spin fixed to s and the configuration on the bottom boundary of $\mathcal{T}_x$ (i.e., on $\partial\mathcal{T}_x \setminus \{\text{parent of } x\}$) is specified by τ.

For two measures μ_1 and μ_2 on Ω, $\|\mu_1 - \mu_2\|_x$ denotes the variation distance between the projections of μ_1 and μ_2 onto the spin at x, i.e.,

$$\|\mu_1 - \mu_2\|_x = \frac{1}{2} \sum_{i=1}^{q} |\mu_1(\sigma(x) = i) - \mu_2(\sigma(x) = i)|.$$

Let $\eta^{x,s}$ be the configuration η with the spin at x set to s.

Following [15] we define

$$\kappa \equiv \kappa(\mu) = \sup_{x \in \Gamma^k} \max_{x,s,s'} \|\mu^s_{\mathcal{T}_x} - \mu^{s'}_{\mathcal{T}_x}\|_x;$$

$$\gamma \equiv \gamma(\mu) = \sup_{A \subset \Gamma^k} \max \|\mu^{\eta^{y,s}}_A - \mu^{\eta^{y,s'}}_A\|_x,$$

where the maximum is taken over all boundary conditions η, all sites $y \in \partial A$, all neighbors $x \in A$ of y, and all spins $s, s' \in \{1, \dots, q\}$.

We apply [15, Theorem 9.3], which says that for an arbitrary channel $\mathbb{P} = (P_{ij})_{i,j=1}^q$ on a tree reconstruction of the corresponding tree-indexed Markov chain (splitting Gibbs measure) is impossible if $k\kappa\gamma < 1$.

Note that κ has the particularly simple form (see [15])

$$\kappa = \frac{1}{2} \max_{i,j} \sum_{l} |P_{il} - P_{jl}| \tag{3.3}$$

and γ is a constant which does not have a clean general formula, but can be estimated in specific models (as Ising, Hard-Core etc.). For example, if $\mathbb{P}$ is the symmetric channel of the Potts model (i.e. corresponding to the solution $z = 1$) then $\gamma \leq \frac{\theta - 1}{\theta + 1}$ [15, Theorem 8.1].

Using (3.3) and (3.2) for $i \neq j$ we get (see [14])

$$\frac{1}{2}\sum_{l=1}^{q}|P_{il}-P_{jl}| = \begin{cases} a, & \text{if } i,j=1,\dots,m \\ b, & \text{if } i,j=m+1,\dots,q \\ c, & \text{otherwise,} \end{cases}$$

where a and b are defined by

$$a = \frac{(\theta-1)z}{Z_1}, \quad b = \frac{(\theta-1)\sqrt[k]{z}}{Z_1}, \tag{3.4}$$

$$Z_1 = (\theta+m-1)z+q-m, \quad Z_2 = mz+\theta+q-m-1,$$

$$c = \frac{1}{2Z_1}\Big(z|\theta-\sqrt[k]{z}|+|1-\theta\sqrt[k]{z}|+(z(m-1)+q-m-1)|1-\sqrt[k]{z}|\Big).$$

Clearly,

$$\kappa = \begin{cases} \max\{b,c\}, & \text{if } m=1 \\ \max\{a,b,c\} & \text{if } m\geq 2. \end{cases} \tag{3.5}$$

We consider the case $z \neq 1$ (where $z = x^2$ and x is a solution to (2.7)) and fix the solution of (2.6), which has the form $(\underbrace{z,z,\dots,z}_{m},1,\dots,1)$ and the corresponding matrix is $\mathbb{P}$.

For $p_1, p_2, u \geq 0$, $p_1+p_2+u \leq 1$, define the following functions

$$K_1(p_1,p_2,u) = \frac{\theta z p_1}{(\theta-1)zp_1+(1-z)u+z} - \frac{zp_1}{(\theta-1)zp_2+(1-z)u+z};$$

$$K_2(p_1,p_2,u) = \frac{\theta z p_1}{(\theta-1)zp_1+(1-z)u+z} - \frac{zp_1}{(\theta-1)p_2+(1-z)u+z};$$

$$K_3(p_1,p_2,u) = \frac{\theta p_1}{(\theta-z)p_1+(1-z)u+z} - \frac{p_1}{(\theta-1)zp_2+(1-z)(u+p_1)+z};$$

$$K_4(p_1,p_2,u) = \frac{\theta p_1}{(\theta-z)p_1+(1-z)u+z} - \frac{p_1}{(\theta-1)p_2+(1-z)(u+p_1)+z}.$$

Proposition 3.1. *[14]*

1) If $z \geq 1$ then

$$\gamma \leq \max_{\substack{p_1,p_2,u\geq 0:\\ p_1+p_2+u\leq 1}} \{K_1(p_1,p_2,u), K_3(p_1,p_2,u)\} \leq \frac{\theta-1}{\theta+1}. \tag{3.6}$$

2) If $z \leq 1$ then

$$\gamma \leq \max_{\substack{p_1,p_2,u\geq 0:\\ p_1+p_2+u\leq 1}} \{K_2(p_1,p_2,u), K_4(p_1,p_2,u)\} \leq \frac{\theta-1}{\theta+1}+1-z. \tag{3.7}$$

Remark 3.2. *The function $K_1(p_1,p_2,u)$ ($K_3(p_1,p_2,u)$) reaches its maximum for $u=0, p_1 = p_2 = \frac{1}{2}$, i.e. if $K_1 \leq f(\theta,z)$ ($K_3 \leq f(\theta,z)$), then $f(\theta,z) \geq \frac{\theta-1}{\theta+1}$ for $z \geq 1$.*

The following Proposition improves the part 2) of Proposition 1.

Proposition 3.3. *If $z \le 1$ then*

$$\gamma \le \max_{\substack{p_1,p_2,u\ge 0:\\ p_1+p_2+u\le 1}} \{K_2(p_1,p_2,u), K_4(p_1,p_2,u)\} \le \frac{\sqrt{\theta(\theta-1)+\theta z}-\sqrt{z}}{\sqrt{\theta(\theta-1)+\theta z}+\sqrt{z}}. \tag{3.8}$$

Proof. We shall find maximum values of functions $K_2(p_1,p_2,u)$ and $K_4(p_1,p_2,u)$. We consider

$$K_2(p_1,p_2,u) = \frac{\theta z p_1}{(\theta-1)zp_1+(1-z)u+z} - \frac{zp_1}{(\theta-1)p_2+(1-z)u+z},$$

where $p_1,p_2,u \ge 0,\ p_1+p_2+u \le 1,\ z>0,\ \theta>1$. Let be $p_1+p_2+u=\alpha\le 1$. Note that the function $K_2(p_1,p_2,u)$ is an increasing function of p_2. Then $K_2(p_1,p_2,u) \le K_2(p_1,p_2+1-\alpha,u)$. Hence, it is sufficient to find the maximum of function $K_2(p_1,p_2,u)$ for $\alpha=1$.

$$K_2(p_1,p_2,u) = \frac{\theta p_1}{\theta p_1+z_1u+p_2} - \frac{p_1}{[(\theta-1)z_1+1]p_2+z_1u+p_1},$$

where $z_1=\frac{1}{z},\ 0<z\le 1$. Let $\frac{u}{p_1}=x,\ \frac{p_2}{p_1}=y$. Then

$$K_2(p_1,p_2,u) = \frac{\theta}{\theta+z_1x+y} - \frac{1}{1+z_1x+[(\theta-1)z_1+1]y}.$$

Denote $L(x,y) = \frac{\theta}{\theta+z_1x+y} - \frac{1}{1+z_1x+[(\theta-1)z_1+1]y}$. By inequalities $(L(x,y))'_x \le 0$ (resp.$(L(x,y))'_x \ge 0$) we obtain following inequalities

$$\sqrt{\theta}-(1+(\theta+\sqrt{\theta})z_1)y \le z_1x, \tag{3.9}$$

$$(resp.\sqrt{\theta}-(1+(\theta+\sqrt{\theta})z_1)y \ge z_1x) \tag{3.10}$$

If (x,y) satisfies (3.9) for some θ and z_1, then

$$L(x,y) \le \frac{\theta}{\theta+y} - \frac{1}{1+(\theta-1)z_1y+y} = f(y) \le \max L(0,y).$$

If (3.10) holds then from $0\le y\le \frac{\sqrt{\theta}}{1+(\theta+\sqrt{\theta})z_1}$ it follows that

$$L(x,y) \le L\left(\frac{\sqrt{\theta}-(1+(\theta+\sqrt{\theta})z_1)y}{z_1}, y\right) = \frac{\sqrt{\theta}-1}{(\sqrt{\theta}+1)(1-z_1y)}.$$

Denote $g(y) = \frac{\sqrt{\theta}-1}{(\sqrt{\theta}+1)(1-z_1y)}$. It is easy to check that $g'(y)>0$ for $\theta>1$. Hence the function $g(y)$ is an increasing. Since $y=\frac{\sqrt{\theta}}{1+(\theta+\sqrt{\theta})z_1} < \frac{1}{z_1}$ we have

$$L\left(\frac{\sqrt{\theta}-(1+(\theta+\sqrt{\theta})z_1)y}{z_1}, y\right) \le L\left(0, \frac{\sqrt{\theta}}{1+(\theta+\sqrt{\theta})z_1}\right) \le \max L(0,y).$$

Consequently, $L(x,y)\le \max L(0,y)$. We consider

$$L'(0,y) = -\frac{\theta}{(\theta+y)^2} + \frac{(\theta-1)z_1+1}{(1+[(\theta-1)z_1+1]y)^2} = 0 \ \Leftrightarrow$$

$$\frac{\sqrt{\theta}}{\theta+y}=\frac{\sqrt{(\theta-1)z_1+1}}{1+[(\theta-1)z_1+1]y}\Rightarrow y=\frac{\sqrt{\theta}}{\sqrt{(\theta-1)z_1+1}}.$$

Hence

$$\max L(0,y)=L\left(0,\frac{\sqrt{\theta}}{\sqrt{(\theta-1)z_1+1}}\right)=\frac{\sqrt{\theta(\theta-1)z_1+\theta}-1}{\sqrt{\theta(\theta-1)z_1+\theta}+1}.$$

By $z_1=\frac{1}{z}$ we obtain

$$\max K_2(p_1,p_2,u)=\frac{\sqrt{\theta(\theta-1)+\theta z}-\sqrt{z}}{\sqrt{\theta(\theta-1)+\theta z}+\sqrt{z}}.$$

Analogously, we get

$$\max K_4(p_1,p_2,u)=\frac{\sqrt{\theta(\theta-1)+\theta z}-\sqrt{z}}{\sqrt{\theta(\theta-1)+\theta z}+\sqrt{z}}.$$

□

Remark 3.4. *The function $K_2(p_1,p_2,u)$ $(K_4(p_1,p_2,u))$ reaches its maximum for $p_1+p_2=1$, $\frac{p_2}{p_1}=\frac{\sqrt{\theta z}}{z+\theta+\sqrt{\theta}}$, $u=0$ $(p_1+p_2=1$, $\frac{p_2}{p_1}=\sqrt{\frac{\theta}{z(z+\theta-1)}}$, $u=0)$, i.e. if $K_2(p_1,p_2,u)\le f(\theta,z)$ $(K_4(p_1,p_2,u)\le f(\theta,z))$, then $f(\theta,z)\ge\frac{\sqrt{\theta(\theta-1)+\theta z}-\sqrt{z}}{\sqrt{\theta(\theta-1)+\theta z}+\sqrt{z}}$ for $z<1$.*

Lemma 3.5. *The following inequality holds*

$$\frac{\sqrt{\theta(\theta-1)+\theta z}-\sqrt{z}}{\sqrt{\theta(\theta-1)+\theta z}+\sqrt{z}}\le\frac{\theta-1}{\theta+1}+1-z \tag{3.11}$$

for $0<z\le 1$, $\theta>1$.

Proof. Clearly, (3.11) is equivalent to

$$\frac{-2\sqrt{z}}{\sqrt{\theta(\theta-1)+\theta z}+\sqrt{z}}\le\frac{\theta-1}{\theta+1}-z.$$

If $z\le\frac{\theta-1}{\theta+1}$, then the inequality is hold. Let $z>\frac{\theta-1}{\theta+1}$. Then

$$2\sqrt{z}\ge\left(z-\frac{\theta-1}{\theta+1}\right)(\sqrt{\theta(\theta-1)+\theta z}+\sqrt{z}).$$

From $\sqrt{\theta(\theta-1)+\theta z}<\theta$ we get

$$\left(z-\frac{\theta-1}{\theta+1}\right)(\sqrt{\theta(\theta-1)+\theta z}+\sqrt{z})\le(\theta+1)z-(\theta-1).$$

We will prove $(\theta+1)z-(\theta-1)\le 2\sqrt{z}$. Indeed

$$\theta z-\theta+z+1\le 2\sqrt{z}\ \Leftrightarrow (1-\sqrt{z})^2\le\theta(1-\sqrt{z})(1+\sqrt{z}).$$

As $1-\sqrt{z}\le 1+\sqrt{z}\le\theta(1+\sqrt{z})$ for $0<z\le 1$ and $\theta>1$, the last inequality is hold. The equality in (3.11) is only right for $z=1$. □

Theorem 3.6. *[14] If $k=2$, $m=1$ then*

(a) *There exists* $\theta^{**} > \theta_c = q+1$ *such that the measure* $\mu_1(\theta, 1)$ *is extreme for any* $\theta \in [1+2\sqrt{q-1}, \theta^{**})$, $q \geq 2$.
(b) *The measure* $\mu_2(\theta, 1)$ *is extreme for any* $\theta \geq 1+2\sqrt{q-1}$, $q \geq 2$.

From [14] it is known that if $\theta > \theta_c = q+1$ then $z_1 < 1$.

Theorem 3.7. *If* $m = 1, k = 2$ *then there exist* $\tilde{\theta} > \theta_c = q+1$ *such that the measure* $\mu_1(\theta, 1)$ *is extreme for any* $\theta \in (1+2\sqrt{q-1},\ \tilde{\theta})$, $q \geq 2$.

Proof. Let $z_1 < 1$. In this case independently on $q \geq 2$ we get $c = \frac{\theta\sqrt{z_1}-1}{\theta z_1+q-1}$. Using $z_1 \leq 1$ (i.e. $\theta \in [\theta_c, +\infty)$) we get $b \geq c$. Consequently, $\kappa = b$. Hence by Proposition 3.3 we should check

$$2\gamma\kappa \leq 2b\left(\frac{\sqrt{\theta(\theta-1)+\theta z_1} - \sqrt{z_1}}{\sqrt{\theta(\theta-1)+\theta z_1} + \sqrt{z_1}}\right) < 1, \tag{3.12}$$

here $b = \frac{(\theta-1)\sqrt{z_1}}{\theta z_1+q-1}$ for $m = 1$ and z_1 is a solution to $z_1 - (\theta-1)\sqrt{z_1} + q - 1 = 0$. Denote

$$f(\theta, q) = \frac{2(\theta-1)\sqrt{z_1}}{\theta z_1 + q - 1} \cdot \frac{\sqrt{\theta(\theta-1)+\theta z_1} - \sqrt{z_1}}{\sqrt{\theta(\theta-1)+\theta z_1} + \sqrt{z_1}} - 1.$$

It is easy to check

$$f(\theta_c, q) = \frac{\theta-1}{\theta+1} - 1 < 0.$$

Since

$$\sqrt{z_1} = \frac{2(q-1)}{\theta - 1 + \sqrt{(\theta-1)^2 - 4(q-1)}},$$

we get

$$\lim_{\theta \to +\infty} f(\theta, q) > 0.$$

Hence there exist $\theta' \in (\theta_c, +\infty)$ such that $f(\theta', q) = 0$, $\tilde{\theta} = \min\{\theta' :\ f(\theta', q) = 0\}$. Then $f(\theta, q) < 0$ for any $\theta \in [\theta_c, \tilde{\theta})$. Consequently the measure $\mu_1(\theta, 1)$ is extreme for any $\theta \in [\theta_c, \tilde{\theta})$.

For $q = 3$ ($m = 1$) a numerical analysis shows that $\theta^* - \tilde{\theta} = 6.243 - 5.079 = 1.164$ (see Fig.1), where $\theta^* = 1 + (\sqrt{2}+1)q - 2m$ is the critical value of θ above of which the measure $\mu_1(\theta, 1)$ is not extreme (see [14]). □

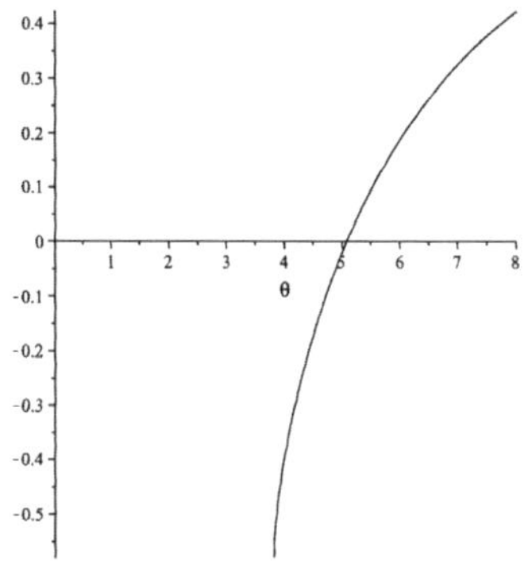

Fig. 1. The graph of the function $f(\theta, q)$ for $q = 3$. The solution of $f(\tilde{\theta}, 3) = 0$ is $\tilde{\theta} \approx 5.079$.

Remark 3.8. *1. In [14] for* $q = 3$, $z_1 < 1$ *the measure* $\mu_1(\theta, 1)$ *was extreme for* $\theta \in (4, \theta^{**})$ *and it was non-extreme for* $\theta \in (\theta^*, 2 + 3\sqrt{2})$. *The difference is* $\theta^* - \theta^{**} = 2.0149$. *In our case this difference is equal to* $\theta^* - \tilde{\theta} = 1.1636$, *i.e. it was reduced the interval where it is not known the extremality of the measure* $\mu_1(\theta, 1)$.

Theorem 3.9. *[14]*

(i) If $m = 2$ *then for each* $q = 4, 5, 6, 7, 8$ *there exists* $\breve{\theta} > q + 1$ *such that the measure* $\mu_1(\theta, 2)$ *is extreme for any* $\theta \in [\theta_2, \breve{\theta})$. *Moreover, if* $q = 9, 10, 11, 12, 13$ *then there exists* $\acute{\theta} = \acute{\theta}(q)$ *such that* $\theta_2 < \acute{\theta} < q + 1$ *and* $\mu_1(\theta, 2)$ *is extreme for* $\theta \in [\acute{\theta}, \breve{\theta})$.

(ii) If $m = 2$ *then for each* $q = 4, 5, 6, 7, 8$ *there exists* $\grave{\theta} = \grave{\theta}(q)$ *such that* $\theta_2 < \grave{\theta} \le q + 1$ *and* $\mu_2(\theta, 2)$ *is extreme for* $\theta \in [\theta_2, \grave{\theta})$ *(see Fig. 7).*

(iii) If $q < \frac{m+1}{2m}\left[3m + 1 + \sqrt{m^2 + 6m + 1}\right]$, $m \ge 2$ *then the measure* $\mu_1(\theta_m, m) = \mu_2(\theta_m, m)$ *is extreme.*

The following theorem is the generalization of Theorem 3.9

Theorem 3.10. *Let* $k = 2$, $m \ge 2$. *Then*

1. If $2m \le q < \frac{m+1}{2m}[3m + 1 + \sqrt{m^2 + 6m + 1}]$, *then there exists* $\ddot{\theta} > q + 1$ *such that the measure* $\mu_1(\theta, m)$ *is extreme for any* $\theta \in [\theta_m, \ddot{\theta})$;

2. If $q > \frac{m+1}{2m}[3m + 1 + \sqrt{m^2 + 6m + 1}]$, *then there exists* $\overline{\theta} \in (\theta_m, \theta_c)$ *such that the measure* $\mu_1(\theta, m)$ *is extreme for any* $\theta \in (\overline{\theta}, \ddot{\theta})$;

3. If $2m \le q < m + \frac{1}{4m}(m + 1 + \sqrt{m^2 + 2m + 7})^2 = \zeta(m)$, *then there exists* $\overline{\overline{\theta}} \in (\theta_m, +\infty)$ *such that the measure* $\mu_2(\theta, m)$ *is extreme for any* $\theta \in [\theta_m, \overline{\overline{\theta}})$.

Proof. 1. Denote $\frac{m+1}{2m}[3m+1+\sqrt{m^2 + 6m + 1}] = \alpha(m)$. We'll check extremality for the measure $\mu_1(\theta, m)$ with $z_1 \ge 1$. In this case $\theta \in [\theta_m, \theta_c]$. Then it is easy to check that $\kappa = a$. Consequently

$$2\kappa\gamma \le 2a\frac{\theta - 1}{\theta + 1} < 1.$$

Since

$$\sqrt{z_1} = \frac{2(q - m)}{\theta - 1 + \sqrt{(\theta - 1)^2 - 4m(q - m)}},$$

we get

$$\theta^2 - (2m + 4)\theta - 2m + 3 - (\theta - 3)\sqrt{(\theta - 1)^2 - 4m(q - m)} < 0.$$

Consequently

$$u(\theta) = \theta^3 - (q + 3)\theta^2 + (8q - 10m - 1)\theta - (9q - 8m - 12) > 0. \tag{3.13}$$

We note that $u(\theta_m) > 0$ for $q < \alpha(m)$. Moreover the function $u(\theta)$ increases in $[\theta_m^{(1)}, +\infty]$,

$$\theta_m^{(1)} = \frac{q + 3 + \sqrt{(q + 3)^2 - 3(8q - 10m - 1)}}{3}.$$

From $\left[\frac{q}{2}\right] \ge m$ it follows that $\theta_m^{(1)} \le \theta_m$. Indeed,

$$\frac{q + 3 + \sqrt{(q + 3)^2 - 3(8q - 10m - 1)}}{3} \le 1 + 2\sqrt{m(q - m)},$$

which is equivalent to

$$\xi(q) = 2q\sqrt{m(q - m)} + 5m + 2 - 3q - 6m(q - m) \le 0. \tag{3.14}$$

We compute the derivative

$$\xi'(q) = 2\sqrt{m(q-m)} + \frac{mq}{\sqrt{m(q-m)}} - 3 - 6m.$$

For $m \le \left[\frac{q}{2}\right]$ and $q < \frac{m+1}{2m}[3m+1+\sqrt{m^2+6m+1}]$ the function $\xi(q)$ is increasing. By the last inequality we get

$$q < \frac{2(m+1)^2}{m} \le 2(m+3).$$

Hence $\xi(q) < \xi(2(m+3))$. For $m \ge 2$ we can see easily $\xi(2(m+3)) < 0$ and $\theta_m^{(1)} \le \theta_m$. Thus the function $u(\theta)$ increases on the segment $[\theta_m, \theta_c]$, i.e. $u(\theta) > 0$ for any $\theta \in [\theta_m, \theta_c]$. Consequently the measure $\mu_1(\theta, m)$ is extreme for $z_1 > 1$ and $\theta \in [\theta_m, \theta_c]$.

Case: $z_1 < 1$. In this case we have, independently on the values of q and m,

$$c = \frac{1}{Z_1}\left(\theta\sqrt{z_1} - 1\right).$$

It is easy to see that $b \ge c$ for $z_1 < 1$, i.e. for $\theta > \theta_c$. Consequently, $\kappa = b$ and we should check

$$2\kappa\gamma \le \frac{2(\theta-1)\sqrt{z_1}}{\theta z_1 + q - 1} \cdot \frac{\sqrt{\theta(\theta-1)+\theta z_1} - \sqrt{z_1}}{\sqrt{\theta(\theta-1)+\theta z_1} + \sqrt{z_1}} < 1.$$

Denote

$$g(\theta, q, m) = \frac{2(\theta-1)\sqrt{z_1}}{\theta z_1 + q - 1} \cdot \frac{\sqrt{\theta(\theta-1)+\theta z_1} - \sqrt{z_1}}{\sqrt{\theta(\theta-1)+\theta z_1} + \sqrt{z_1}} - 1.$$

From

$$\sqrt{z_1} = \frac{2(q-m)}{\theta - 1 + \sqrt{(\theta-1)^2 - 4m(q-m)}},$$

we have

$$g(\theta_c, q, m) = \frac{\theta-1}{\theta+1} - 1 < 0, \quad \lim_{\theta\to+\infty} g(\theta, q, m) > 0.$$

Hence there exists $\theta'' \in (\theta_c, +\infty)$ such that $g(\theta'', q, m) = 0$, $\ddot{\theta} = \min\{\theta'' : g(\theta'', q, m) = 0\}$. Then $g(\theta, q, m) < 0$ for any $\theta \in [\theta_c, \ddot{\theta})$, i.e. the measure $\mu_1(\theta, m)$ is extreme for this condition.

Note that for $q = 6$, $m = 2$ we have $\theta^* = 3(1+2\sqrt{2}) \approx 11.485$ and a numerical analysis shows that $\ddot{\theta} \approx 8.22$ (see Fig.2). So $\theta^* - \ddot{\theta} \approx 3.265$.

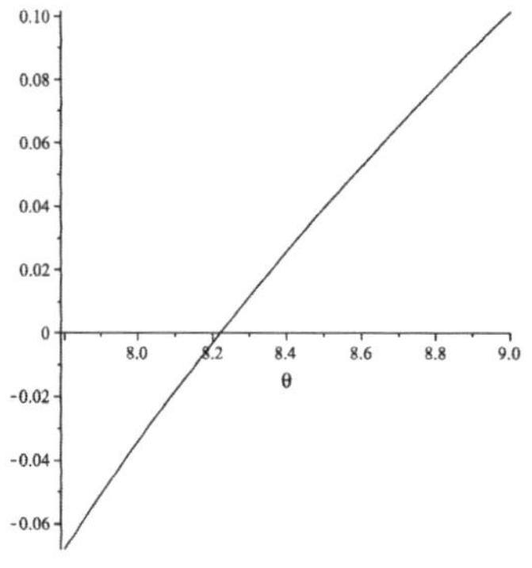

Fig. 2. The graph of the functions $g(\theta, q, m)$ for $q = 6, m = 2$. $\theta^* - \ddot{\theta} \approx 3.265$.

2. The case $z_1 \geq 1$. If $q > \frac{m+1}{2m}[3m+1+\sqrt{m^2+6m+1}]]$, then $u(\theta_m) < 0$. We note that $u(\theta_c) > 0$. Moreover the function $u(\theta)$ increases for $[\theta_m, \theta_c]$. Hence there exists a unique $\overline{\theta} \in (\theta_m, \theta_c]$ such that $u(\overline{\theta}) = 0$. From the condition of extremality (3.13) we get the measure $\mu_1(\theta, m)$ is extreme for $(\overline{\theta}, \theta_c)$.

The case $z_1 < 1$ is similar to the proof of part 1 of the theorem. That's why the measure $\mu_1(\theta, m)$ is extreme for $(\overline{\theta}, \ddot{\theta})$.

3. We have $z_2 \geq 1$ for $\theta \geq \theta_m$. We check the condition of extremality of the measure $\mu_2(\theta, m)$

$$2\kappa\gamma \leq 2c\frac{\theta-1}{\theta+1} < 1,$$

which is equivalent to

$$v(\theta) = \theta^2 - (2m+4)\theta - 2m + 3 + (\theta - 3)\sqrt{(\theta-1)^2 - 4m(q-m)} < 0. \tag{3.15}$$

The inequality (3.15) has a solution if the inequality

$$\theta^2 - (2m+4)\theta - (2m-3) < 0,$$

has a solution. The solution of the last inequality has following form

$$\theta < m + 2 + \sqrt{m^2 + 2m + 7}.$$

Since $\theta \geq \theta_m = 1 + 2\sqrt{m(q-m)}$ we obtain

$$1 + 2\sqrt{m(q-m)} < m + 2 + \sqrt{m^2 + 2m + 7}.$$

Hence

$$2m \leq q < m + \frac{1}{4m}(m + 1 + \sqrt{m^2 + 2m + 7})^2.$$

For this condition it is easy to check that $\lim_{\theta \to +\infty} v(\theta) = +\infty$ for this condition. Consequently there exist $\overline{\overline{\theta}} \in (\theta_m, +\infty)$ such that $v(\overline{\overline{\theta}}) = 0$. So the measure $\mu_2(\theta, m)$ is extreme for $\theta \in (\theta_m, \overline{\overline{\theta}})$. □

Remark 3.11. *1. If $q > m + \frac{1}{4m}(m + 1 + \sqrt{m^2 + 2m + 7})^2$, then the condition of extremality (3.15) does not satisfy for any θ, i.e. in this case it is not known the extremality of the measure $\mu_2(\theta, m)$.*

2. Using a graphs we can see that derivatives $f'(\theta, q) > 0$, $g'(\theta, q, m) > 0$ for finite q and m (see Fig 3.), i.e. functions $f(\theta, q)$ and $g(\theta, q, m)$ are increasing (by analytical method the proof of this statement is very difficult). So values θ', θ'' are uniquely determined by $\tilde{\theta} = \theta'$, $\ddot{\theta} = \theta''$.

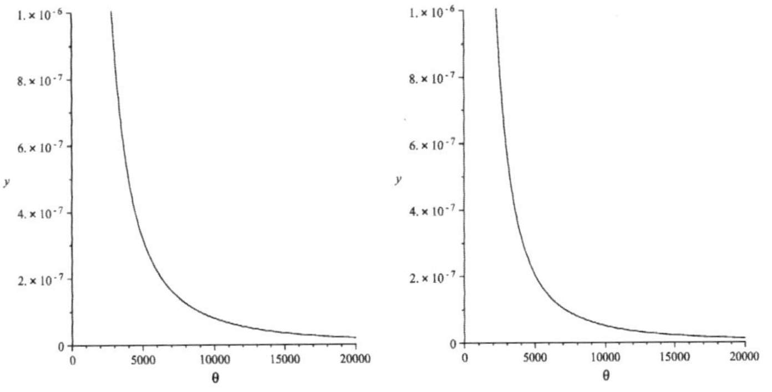

Fig. 3. The graph of the functions $f'(\theta, q)$ for $q = 5, m = 1$ (on the left). The graph of the functions $g'(\theta, q, m) > 0$ for $q = 8, m = 3$ (on the right).

References

[1] Brightwell G R and Winkler P 1999 *J. Combin. Theory Ser.B* **77**(2) 221–262
[2] Formentin M and Külske C 2009 *Electron. Commun. Probab.* 14 587–596
[3] Ganikhodjaev N N 1990 *Theor. Math. Phys.* **85**(2) 1125–1134
[4] Ganikhodjaev N N 1992 *Dokl. AN Uzbekistan* No. 6-7 4–7
[5] Ganikhodjaev N N, Mukhamedov F M and Mendes J F F 2006 *Jour. Stat. Mech.* P08012
[6] Ganikhodzhaev N N and Rozikov U A 1997 *Theor. Math. Phys.* **111**(1) 480–486
[7] Ganikhodzhaev N N and Rozikov U A 2006 *Lett. Math. Phys.* **75**: 2 99–109
[8] Häggström O 1996 *Probab. Theory and Relat. Fields* **104** 231–253
[9] Häggström O and Külske C 2004 *Markov Proc. Rel. Fields* **10**(3) 477–506
[10] Kesten H and Stigum B P 1966 *Ann. Math. Statist.* **37** 1463–1481
[11] Khakimov R M 2014 *Journal of Siberian Federal University. Mathematics and Phisics* **7**(3) 297–304
[12] Khakimov R M 2014 *Theor. and Math. Phys.* **179**(1) 405–415
[13] Külske C, Rozikov U A and Khakimov R M *Jour. Stat. Phys.* **156**(1) 189–200
[14] Külske C and Rozikov U A 2014 arXiv:1403.5775v1
[15] Martinelli F, Sinclair A and Weitz D 2007 *Random Structures and Algoritms* **31** 134–172
[16] Mossel E 2004 In book: *Graphs, morphisms and statistical physics* **63** (New-York: AMS) 155–170
[17] Mukhamedov F M and Rozikov U A 2001 *Theor. Math. Phys.* **126** 169–174.
[18] Mukhamedov F M and Rozikov U A 2000 *Theor. Math. Phys.* **124** 1202–1209.
[19] Rozikov U A 2013 *Gibbs measures on Cayley trees* (Singapore: World Scientific)
[20] Rozikov U A and Khakimov R M 2013 *Theor. Math. Phys.* **175**(2) 699–709
[21] Sly A 2011 *Ann. Probab.* **39** 1365–1406

On new weakly periodic Gibbs measures of the Ising model on the Cayley tree of order ≤ 5

Muzaffar Rahmatullaev

Institute of Mathematics, National University of Uzbekistan, 29, Do'rmon Yo'li str., 100125, Tashkent, Uzbekistan.

E-mail: mrahmatullaev@rambler.ru

Abstract. For the Ising model on the Cayley tree, we find new weakly periodic Gibbs measures corresponding to normal subgroups of index two in the group representation of the Cayley tree of order $k \leq 5$.

1. Introduction

One of the main problems for the Ising model is to describe all limiting Gibbs measures corresponding to this model. It is well known that for the Ising model, such measures form a nonempty, convex, and compact subset in the set of all probability measures. The problem of completely describing the element of this set is far from being completely solved. Some translation-invariant (see, e.g., [1],[8],[14]), periodic [3], [9], and continuum sets of non-periodic [8], [10] Gibbs measures for the Ising model on a Cayley tree have already been described.

To extend the set of periodic Gibbs measures corresponding to the Ising model, the notion of periodic measures was generalized to that of weakly periodic Gibbs measures in [6],[11]-[13]. In the mentioned paper under some conditions on parameters for the Ising model (on the Cayley tree of order ≥ 6), the existence of such new measures was proved. But the used methods did not allowed us to find conditions for the existence of weakly periodic Gibbs measures. Therefore, main aim of this paper to study existence of weakly periodic (non-periodic) Gibbs measures for the mentioned model on the Cayley tree of order less than 5.

This paper is organized as follows. In Section 2 we give necessary definitions and formulate the problem. Section 3 is devoted to weakly periodic (non-periodic) Gibbs measures corresponding to the normal subgroups of index two in the group representation of the Cayley tree.

2. Definitions and the main problem.

Let $\tau^k = (V, L), k \geq 1$, be the Cayley tree of order k, i.e. an infinity graph every vertex of which is incident to exactly $k+1$ edges. Here V is the set of all vertices, L is the set of all edges of the tree τ^k. It is known that τ^k can be represented as G_k, which is the free product of $k+1$ cyclic groups of the second order [4], [8].

For an arbitrary point $x^0 \in V$ we set $W_n = \{x \in V | d(x^0, x) = n\}$, $V_n = \bigcup_{m=0}^{n} W_m$, $L_n = \{< x, y >\in L | x, y \in V_n\}$, where $d(x, y)$ is the distance between the vertices x and y in the Cayley tree, i.e. the number of edges in the shortest path joining the vertices x and y.

Let $\Phi = \{-1, 1\}$ and let $\sigma \in \Omega = \Phi^V$ be a configuration, i.e. $\sigma = \{\sigma(x) \in \Phi : x \in V\}$. Let $A \subset V$. We let Ω_A denote the space of configurations defined on the set A and taking values in Φ.

We consider the Hamiltonian of the Ising model:

$$H(\sigma) = -J \sum_{<x,y>\in L} \sigma(x)\sigma(y), \tag{2.1}$$

where $J \in R$, $\sigma(x) \in \Phi$ and $< x, y >$ are nearest neighbors.

Let $h_x \in R, x \in V$. For every n, we then define a measure μ_n on Ω_{V_n} setting

$$\mu_n(\sigma_n) = Z_n^{-1} \exp\{-\beta H(\sigma_n) + \sum_{x\in W_n} h_x\sigma(x)\}, \tag{2.2}$$

where $\beta = \frac{1}{T}$ (T is temperature, $T > 0$), $\sigma_n = \{\sigma(x), x \in V_n\} \in \Omega_{V_n}$, Z_n^{-1} is the normalizing factor, and

$$H(\sigma_n) = -J \sum_{<x,y>\in L_n} \sigma(x)\sigma(y).$$

The compatibility condition for the measures $\mu_n(\sigma_n), n \geq 1$, is

$$\sum_{\sigma^{(n)}} \mu_n(\sigma_{n-1}, \sigma^{(n)}) = \mu_{n-1}(\sigma_{n-1}), \tag{2.3}$$

where $\sigma^{(n)} = \{\sigma(x), x \in W_n\}$.

Let $\mu_n, n \geq 1$ be a sequence of measures on the sets Ω_{V_n} that satisfy compatibility condition (2.3). By the Kolmogorov theorem, we then have a unique limit measure μ on $\Omega_V = \Omega$ (called the limit Gibbs measure) such that

$$\mu(\sigma_n) = \mu_n(\sigma_n)$$

for every $n = 1, 2, \ldots$. It is known that measures (2.2) satisfies the condition (2.3) if and only if the set $h = \{h_x, x \in G_k\}$ of quantities satisfies the condition

$$h_x = \sum_{y\in S(x)} f(h_y, \theta), \tag{2.4}$$

where $S(x)$ is the set of direct successors (children) of the point $x \in V$ (see.[8], [1],[14]). Here, $f(x, \theta) = arctanh(\theta \tanh x)$, $\theta = \tanh(J\beta)$, $\beta = \frac{1}{T}$.

Let $G_k/\widehat{G}_k = \{H_1, ..., H_r\}$ be a factor group, where $\widehat{G}_k$ is a normal subgroup of index $r \geq 1$.

Definition 2.1. *A set $h = \{h_x,\ x \in G_k\}$ of quantities is called $\widehat{G}_k$-periodic if $h_{xy} = h_x$, for all $x \in G_k$ and $y \in \widehat{G}_k$.*

For $x \in G_k$ we denote by $x_\downarrow$ the unique point of the set $\{y \in G_k : \langle x, y\rangle\} \setminus S(x)$.

Definition 2.2. *A set of quantities $h = \{h_x,\ x \in G_k\}$ is called $\widehat{G}_k$-weakly periodic, if $h_x = h_{ij}$, for any $x \in H_i, x_\downarrow \in H_j$.*

We note that the weakly periodic h coincides with an ordinary periodic one (see Definition 2.1) if the quantity h_x is independent of $x_\downarrow$.

Definition 2.3. *A Gibbs measure μ is said to be $\widehat{G}_k$-(weakly) periodic if it corresponds to the $\widehat{G}_k$-(weakly) periodic h. We call a G_k-periodic measure a translation-invariant measure.*

In this paper, we study weakly periodic Gibbs measures and demonstrate that such measures exist for the Ising model on a Cayley tree of order ≤ 5.

3. Weakly periodic measures

The level of difficulty in the describing of weakly periodic Gibbs measures is related to the structure and index of the normal subgroup relative to which the periodicity condition is imposed. It is known (see Chapter 1 of [8]) that in the group G_k, there is no normal subgroup of odd index different from one. Therefore, we consider normal subgroups of even indices. Here, we restrict ourself to the case of index two.

We describe $\widehat{G}_k$-weakly periodic Gibbs measures for any normal subgroup $\widehat{G}_k$ of index two. We note (see Chapter 1 of [8]) that any normal subgroup of index two of the group G_k has the form

$$H_A = \left\{ x \in G_k : \sum_{i \in A} \omega_x(a_i) - \text{even} \right\},$$

where $\emptyset \neq A \subseteq N_k = \{1, 2, \ldots, k+1\}$, and $\omega_x(a_i)$ is the number of letters a_i in a word $x \in G_k$.

Let $A \subseteq N_k$ and H_A be the corresponding normal subgroup of index two. We note that in the case $|A| = k+1$, i.e., in the case $A = N_k$, weak periodicity coincides with ordinary periodicity. Therefore, we consider $A \subset N_k$ such that $A \neq N_k$. Then, in view of (2.4), the H_A-weakly periodic set of h has the form

$$h_x = \begin{cases} h_1, & x \in H_A,\ x_\downarrow \in H_A, \\ h_2, & x \in H_A,\ x_\downarrow \in G_k \setminus H_A, \\ h_3, & x \in G_k \setminus H_A,\ x_\downarrow \in H_A, \\ h_4, & x \in G_k \setminus H_A, x_\downarrow \in G_k \setminus H_A, \end{cases} \tag{3.1}$$

where $h_i, i = 1, 2, 3, 4$, satisfy the following equations:

$$\begin{cases} h_1 = |A| f(h_3, \theta) + (k - |A|) f(h_1, \theta), \\ h_2 = (|A| - 1) f(h_3, \theta) + (k + 1 - |A|) f(h_1, \theta), \\ h_3 = (|A| - 1) f(h_2, \theta) + (k + 1 - |A|) f(h_4, \theta), \\ h_4 = |A| f(h_2, \theta) + (k - |A|) f(h_4, \theta). \end{cases} \tag{3.2}$$

Consider operator $W : R^4 \to R^4$, defined by

$$\begin{cases} h_1' = |A| f(h_3, \theta) + (k - |A|) f(h_1, \theta) \\ h_2' = (|A| - 1) f(h_3, \theta) + (k + 1 - |A|) f(h_1, \theta) \\ h_3' = (|A| - 1) f(h_2, \theta) + (k + 1 - |A|) f(h_4, \theta) \\ h_4' = |A| f(h_2, \theta) + (k - |A|) f(h_4, \theta). \end{cases} \tag{3.3}$$

Note that the system of equations (3.2) is the equation $h = W(h)$.

It is obvious that the following sets are invariant with respect to operator W:

$$I_1 = \{h \in R^4 : h_1 = h_2 = h_3 = h_4\}, \quad I_2 = \{h \in R^4 : h_1 = h_4; h_2 = h_3\},$$

$$I_3 = \{h \in R^4 : h_1 = -h_4; h_2 = -h_3\}.$$

Set $\alpha = \frac{1-\theta}{1+\theta}$.

Theorem 3.1. *([7]). The following assertions hold:*

1) *For the Ising model, all H_A-weakly periodic Gibbs measures on I_1 and I_2 are translation invariant.*
2) *For $|A| = k$, all H_A-weakly periodic Gibbs measures are translation invariant.*

3) For $|A| = 1$ and $k = 4$, there exists a critical value $\alpha_{\rm cr}$ ($\approx 0,1569$) such that there exist five H_A-weakly periodic Gibbs measures $\mu_0, \mu_1^-, \mu_1^+, \mu_2^-, \mu_2^+$; for $0 < \alpha < \alpha_{\rm cr}$, three H_A-weakly periodic Gibbs measures μ_0, μ_1^-, μ_1^+, for $\alpha = \alpha_{\rm cr}$, and only one H_A-weakly periodic Gibbs measure μ_0 for $\alpha > \alpha_{\rm cr}$.

4) For $|A| = 1$, $k \geq 6$, and $\theta \in (\theta_1, \theta_2)$, where $\theta_{1,2} = \frac{k-1\pm\sqrt{k^2-6k+1}}{2k}$, there exist three H_A-weakly periodic Gibbs measures μ_0 and μ^+, μ^- on I_3.

Remark 3.2. *The measures $\mu^-, \mu^+, \mu_i^-, \mu_i^+, i = 1, 2$ are H_A-weakly periodic Gibbs measures for the Ising model. All the other measures constructed in Theorem 3.1 are translation invariant.*

In parts 3) and 4) of Theorem 3.1 the cases $k = 4$ and $k \geq 6$ are considered (see [11]). Therefore remain the cases $k = 2, k = 3, k = 5$. In this paper we consider this cases.

Using the fact that

$$f(h, \theta) = \operatorname{arctanh}(\theta \tanh h) = \frac{1}{2} \ln \frac{(1+\theta)e^{2h} + (1-\theta)}{(1-\theta)e^{2h} + (1+\theta)},$$

and introducing the notation $z_i = e^{2h_i}$, $i = 1, 2, 3, 4$ we obtain the following system of equations instead of (3.3):

$$\begin{cases} z_1 = \left(\frac{z_3+\alpha}{\alpha z_3+1}\right)^{|A|} \cdot \left(\frac{z_1+\alpha}{\alpha z_1+1}\right)^{(k-|A|)} \\ z_2 = \left(\frac{z_3+\alpha}{\alpha z_3+1}\right)^{|A|-1} \cdot \left(\frac{z_1+\alpha}{\alpha z_1+1}\right)^{(k+1-|A|)} \\ z_3 = \left(\frac{z_2+\alpha}{\alpha z_2+1}\right)^{|A|-1} \cdot \left(\frac{z_4+\alpha}{\alpha z_4+1}\right)^{(k+1-|A|)} \\ z_4 = \left(\frac{z_2+\alpha}{\alpha z_2+1}\right)^{|A|} \cdot \left(\frac{z_4+\alpha}{\alpha z_4+1}\right)^{(k-|A|)}. \end{cases} \tag{3.4}$$

We consider the case $|A| = 1$. Introduce the notation $f(x) = \frac{x+\alpha}{\alpha x+1}$. Then the system of equations (3.4) takes the form

$$\begin{cases} z_1 = f(z_3) \cdot (f(z_1))^{k-1} \\ z_2 = (f(z_1))^k \\ z_3 = (f(z_4))^k \\ z_4 = f(z_2) \cdot (f(z_4))^{k-1}. \end{cases} \tag{3.5}$$

Theorem 3.3. *Let $|A| = 1$.*
1) For $k = 2$ and $k = 3$ there is not any H_A-weakly periodic (non translation-invariant) Gibbs measures corresponding to a set of quantities from I_3.
2) Let $k = 5$. For the weakly periodic Gibbs measures corresponding to the set of quantities from I_3 there exists a critical value $\alpha_{cr}(\approx 0,377)$ such that there is not any H_A- weakly periodic (non translation-invariant) Gibbs measures for $\alpha > \alpha_{cr}$, there are two H_A- weakly periodic Gibbs measures for $\alpha = \alpha_{cr}$, and there are four H_A- weakly periodic (non translation-invariant) Gibbs measures for $0 < \alpha < \alpha_{cr}$.

Proof. System of equations (3.5) on the invariant set I_3 has the following form:

$$\begin{cases} z_1 = f(\frac{1}{z_2}) \cdot (f(z_1))^{k-1} \\ z_2 = (f(z_1))^k. \end{cases} \tag{3.6}$$

Thus the system (3.6) can be reduced to the equation

$$z_1 = \left(\frac{z_1+\alpha}{\alpha z_1+1}\right)^{k-1} \frac{\alpha(\alpha+z_1)^k + (1+\alpha z_1)^k}{(\alpha+z_1)^k + \alpha(1+\alpha z_1)^k}. \tag{3.7}$$

Denoting $u = \frac{z_1+\alpha}{\alpha z_1+1}$ we reduce equation (3.7) to the equation

$$\alpha^2 u^{2k} - \alpha u^{2k-1} + u^{k+1} - u^{k-1} + \alpha u - \alpha^2 = 0. \tag{3.8}$$

Equation (3.8) is equivalent to

$$(u^2 - 1)P_{2k-2}(u) = 0, \tag{3.9}$$

where $P_{2k-2}(u)-$ is a symmetric polynomial of degree $2k-2$. It is well known that setting $u + 1/u = \xi$, we can decrease the degree of the equation $P_{2k-2}(u) = 0$ twofold, i.e., reduce this equation to the equation $P_{k-1}(\xi) = 0$, where $P_{k-1}(\xi)$ is a nonsymmetric polynomial of degree $k-1$ in the general case. But for $k \geq 6$, the equation $P_{k-1}(\xi) = 0$ cannot be solved in radicals.

Now we prove the first part of Theorem 3.3. Let $k = 3$. Then the equation (3.9) has the following form

$$(u^2 - 1)(\alpha^2 u^4 - \alpha u^3 + (\alpha^2 + 1)u^2 - \alpha u + \alpha^2) = 0. \tag{3.10}$$

In this case one solution of the equation (3.10) is $u = 1$. Therefore we assume that $u \neq 1$. From (3.10) we have

$$\alpha^2 u^4 - \alpha u^3 + (\alpha^2 + 1)u^2 - \alpha u + \alpha^2 = 0.$$

Denoting $\xi = u + \frac{1}{u}$, we obtain the equation

$$\alpha\xi^2 - \alpha\xi + 1 - \alpha^2 = 0.$$

It is easy to show that the equation for $\xi > 2$ does not have a solution, so the equation (3.10) has no solution other than $u = 1$.

By similar way one can show that for $k = 2$ the equation (3.8) has no solution other than $u = 1$.

We consider the case $k = 5$, where equation (3.9) has the form

$$(u^2 - 1)\left(\alpha^2 u^8 - \alpha u^7 + \alpha^2 u^6 - \alpha u^5 + (\alpha^2 + 1)u^4 - \alpha u^3 + \alpha^2 u^2 - \alpha u + \alpha^2\right) = 0. \tag{3.11}$$

From (3.11) we have $u^2 - 1 = 0$ or

$$\alpha^2 u^8 - \alpha u^7 + \alpha^2 u^6 - \alpha u^5 + (\alpha^2 + 1)u^4 - \alpha u^3 + \alpha^2 u^2 - \alpha u + \alpha^2 = 0. \tag{3.12}$$

Since $u > 0$, we have for equation (3.11), $u = 1$ is a solution. We assume that $u \neq 1$. Setting $\xi = u + \frac{1}{u} > 2$, from (3.12) we obtain the equation

$$\alpha^2\xi^4 - \alpha\xi^3 - 3\alpha^2\xi^2 + 2\alpha\xi + \alpha^2 + 1 = 0. \tag{3.13}$$

It is well known (see [5], p.28) that the number of positive roots of the polynomial (3.13) does not exceed the number of sign changes of the sequence of its coefficients. Therefore, the equation (3.13) has at most two positive solutions. From equation (3.13) we find the parameter α:

$$\alpha_1 = \frac{\xi^3 - 2\xi + \sqrt{\xi^6 - 8\xi^4 + 16\xi^2 - 4}}{2(\xi^4 - 3\xi^2 + 1)} := \gamma_1(\xi), \tag{3.14}$$

$$\alpha_2 = \frac{\xi^3 - 2\xi - \sqrt{\xi^6 - 8\xi^4 + 16\xi^2 - 4}}{2(\xi^4 - 3\xi^2 + 1)} := \gamma_2(\xi). \tag{3.15}$$

Denoting $v = \xi^2$ and $\varphi(v) = v^3 - 8v^2 + 16v - 4$. We consider $\varphi'(v) = 3v^2 - 16v + 16$. It is clear that $\varphi'(v) > 0$ for $v > 4$. On the other hand $\varphi(4) < 0$, $\varphi(+\infty) > 0$. It follows that $\varphi(v) = 0$ has a unique solution v_0 for $v > 4$. Therefore, the system of inequalities

$$\begin{cases} \xi^6 - 8\xi^4 + 16\xi^2 - 4 \geq 0 \\ \xi > 2, \end{cases}$$

true for $\xi \in [\xi_0, +\infty)$, where $\xi_0 = \sqrt{v_0}(\approx 2.214)$.

Note that $\gamma_1(\xi_0) = \gamma_2(\xi_0)$.

One can check that

$$\lim_{\xi \to +\infty} \alpha_i(\xi) = 0, i = 1, 2. \tag{3.16}$$

Note that if $\xi \in [\xi_0, +\infty)$ then the equation $\gamma_1'(\xi) = 0$ has a unique solution, which is $\xi_1 \approx 2.3841$, and also we get $\gamma_1(\xi_0) \approx 0.311$, $\gamma_1(\xi_1) \approx 0.377$.

Hence it is clear that the function $\gamma_1(\xi)$ reaches its maximum in $[\xi_0, +\infty)$ at ξ_1. Consequently, for $\alpha \in (0, \gamma_1(\xi_0)) \cup \{\gamma_1(\xi_1)\}$ there exists a unique $\xi > 2$ satisfying the equation (3.14), for $\alpha \in [\gamma_1(\xi_0), \gamma_1(\xi_1))$ there exist two $\xi > 2$ satisfying the equation (3.14), moreover if $\alpha > \gamma_1(\xi_1)$ then there is not $\xi > 2$ satisfying the equation (3.14).

Now consider the function $\gamma_2(\xi)$. For $\xi \in [\xi_0, +\infty)$ the equation $\gamma_2'(\xi) = 0$ has not a solution. Since $\gamma_2(\xi_0) \approx 0.311$, by (3.16) we have that the function $\gamma_2(\xi)$ is decreasing. Then we get following: for $\alpha \in (0, \gamma_2(\xi_0)]$ there exists a unique $\xi > 2$ which satisfies the equation (3.15), for $\alpha > \gamma_2(\xi_0)$ there is not $\xi > 2$ satisfying the equation (3.15).

Let n_α be the number of solutions of the equation (3.13). Then n_α has the following form

$$n_\alpha = \begin{cases} 2, & \text{if } \alpha \in (0, \alpha_{cr}) \\ 1, & \text{if } \alpha = \alpha_{cr} \\ 0, & \text{if } \alpha \in (\alpha_{cr}, +\infty). \end{cases}$$

For $\alpha \in (0, \alpha_{cr})$ from $u + \frac{1}{u} = \xi$ we get four solutions of the equation (3.12). In this case the equation (3.11) has five solutions. For $\alpha = \alpha_{cr}$ from $u + \frac{1}{u} = \xi$ we get that the equation (3.12) has two solutions. Consequently equation (3.11) has three solutions. In the case $\alpha > \alpha_{cr}$ the equation (3.11) has a unique solution $u = 1$. Theorem is proved.

Remark 3.4. *The new Gibbs measures described in Theorem 3.3 allow describing of a continuum set of non-periodic Gibbs measures that differ from the well-known ones.*

Remark 3.5. *In the paper [2] for $k = 2, 3$ the authors construct many 'new' Gibbs measures which are similar to the weakly periodic measures of the Ising model on the Cayley tree. Comparing their result with our Theorem 3.3 and with [7, Theorem 3] one can see that in the case $k = 2, 3$ we did not obtain H_A- weakly periodic measures under conditions of these theorems. If $k = 3$ and some conditions of Theorem 3.3 are not satisfied then we do not know about existence of any type of weakly periodic Gibbs measures.*

Aknowledgments. The author thanks Professor U.A.Rozikov for useful discussions.

References

[1] Bleher P M and Ganikhodjaev N N 1990 *Theor. Probab. Appl.* **35** 216-227
[2] Gandolfo D, Ruiz J and Shlosman S 2015 *Commun. Math. Phys.* **334** 313-330
[3] Ganikhodjaev N N and Rozikov U A 1997 *Theor. Math. Phys.* **111** 480-486
[4] Ganikhodjaev N N 1994 *Dokl.AN Uzbekistan* **4** 3–5
[5] Prasolov V V 2004 *Polinomials* (Berlin: Springer. Algorithms and Computation in Mathematics).
[6] Rahmatullaev M M 2009 *Uzbek. Math. J.* **2**. 144–152
[7] Rahmatullaev M M 2015 *Russian mathematics* **59** 45-53
[8] Rozikov U A 2013 *Gibbs measures on Cayley trees*, (Singapore: World Scientific)
[9] Rozikov U A 1997 *Theor. Math. Phys.* **112** 929-933
[10] Rozikov U A 1999 *Theor. Math. Phys.* **118** 77-84
[11] Rozikov U A and Rahmatullaev M M 2008 *Theor. Math. Phys.* **156** 1218-1227
[12] Rozikov U A and Rahmatullaev M M 2009 *Theor. Math. Phys.* **160** 1291-1299
[13] Rozikov U A and Rahmatullaev M M 2008 *Dokl.AN Uzbekistan* **4** 12-15
[14] Zachary S 1983 *Ann. Prob.* **11**, 894–903

Fredholm partial integral equations of second type with degenerate kernel

Ramziddin Kucharov[1], **Yusup Eshkabilov**[2]

[1,2] Faculty of Mechanics and Mathematics, National University of Uzbekistan, Vuzgorodok, Tashkent, Uzbekistan, 100174

E-mail: ramz3364647@yahoo.com, yusup62@mail.ru

Abstract. In this paper we study the solvability of the Fredholm partial integral equations of second type with degenerate kernels.

1. Introduction

Various problems of quantum mechanics [1], quantum field theory [2], partial differential equations [3], mathematical physics [4], and a number of other problems are reduced to special cases of the following integral equation

$$
\begin{aligned}
f(x,y) &= \int_{\Omega_1} k_1(x,s,y)f(s,y)d\mu_1(s) + \int_{\Omega_2} k_2(x,t,y)f(x,t)d\mu_2(t) \\
&\quad + \int_{\Omega_1}\int_{\Omega_2} k(x,s;y,t)f(s,t)d\mu_1(s)d\mu_2(t) + g(x,y). \qquad (1.1)
\end{aligned}
$$

where Ω_1 and Ω_2 are sets with a finite Lebesgue measure in $\mathbb{R}^{\nu_1}$ and $\mathbb{R}^{\nu_2}$, respectively and $k_1 : \Omega_1^2 \times \Omega_2 \to \mathbb{C}$, $k_2 : \Omega_1 \times \Omega_2^2 \to \mathbb{C}$, $k : \Omega_1^2 \times \Omega_2^2 \to \mathbb{C}$, $g : \Omega_1 \times \Omega_2 \to \mathbb{C}$ are given measurable functions, $\mu_1(\cdot), \mu_2(\cdot)$ are the Lebesgue measures on $\sigma-$ algebras of subsets Ω_1 and Ω_2, respectively.

The equation (1.1) contains a partial integrals, i.e. integrals in which an unknown function $f(x,y)$ is integrated by parts of variables. Therefore, this kind of equations are called *partial integral equations (PIE)*.

Solvability and properties of the solutions of PIE (1.1) depend on the spaces in which it is considered. The solvability of PIE (1.1) on the space of continuous functions were investigated in [5]-[8].

In 1975, Likhtarnikov and Vitova [9] studied spectral properties of partial integral operators. In [9], the following restrictions were imposed: $k_1(x,s) \in L_2(\Omega_1 \times \Omega_1)$, $k_2(y,t) \in L_2(\Omega_2 \times \Omega_2)$ and $T_0 = K = 0$. In [10], spectral properties of partial integral operators (PIO) with positive kernels were studied (under restriction $T_0 = K = 0$)). In [11] spectral properties of PIO with kernels of two variables in $L_p, p \geq 1$ are studied. In [8], [12]-[14], for more general PIO's with continuous kernels or kernels in $C(L_1)$, spectral properties of the PIO and solvability of partial integral equations in the space of $C([a,b] \times [c,d])$ were studied. In [3] some applications of partial integral equations and operators in solving problems of continuous mechanics, elasticity problems and other problems were considered. Until now, a solvability of PIE in the space L_2 is left open. The present paper is devoted to this problem. Namely, we investigate a solvability of problem for PIE with degenerate kernels from L_2.

Let $(\Omega, \varepsilon, \mu)$ be a measurable space with a finite measure, $L_0(\Omega)$ be the $*$- algebra of equivalence classes of all complex measurable functions on Ω and $L_\infty(\Omega) \subset L_0(\Omega)$ be a subalgebra of equivalence classes of all bounded measurable functions on Ω. By $[f]$ we denote an equivalence class containing a function $f \in L_0(\Omega)$.

By $L_\infty[L_2(\Omega_1), \Omega_2]$ we denote the set of all complex measurable functions $f(x, y)$ on $\Omega_1 \times \Omega_2$ satisfying the following condition: the integral

$$\varphi_0(y) = \int_{\Omega_1} |f(x,y)|^2 d\mu_1(x)$$

exists for almost all $y \in \Omega_2$ and $\varphi_0 \in L_\infty(\Omega_2)$.

In the $L_\infty[L_2(\Omega_1), \Omega_2]$ we define $L_\infty(\Omega_2)$ - valued inner product $\langle f, g\rangle$ by

$$\langle f, g\rangle = \langle f, g\rangle(y) = \int_{\Omega_1} f(x,y)\overline{g(x,y)} d\mu_1(x).$$

Let $\varphi_i, \psi_i \in L_\infty\left[L_2(\Omega_1), \Omega_2\right], \; i = \overline{1,m}$ and

$$k_1(x,s,y) = \sum_{i=1}^{m} \varphi_i(x,y)\psi_i(s,y), \quad (x,s,y) \in \Omega_1^2 \times \Omega_2.$$

Then the partial integral operator (PIO) T_1 defined by

$$T_1 f(x,y) = \int_{\Omega_1} k_1(x,s,y) f(s,y) d\mu_1(s)$$

is linear and bounded on $L_2(\Omega_1 \times \Omega_2)$.

In this paper we study the solvability of the partial integral equation

$$f - T_1 f = g, \tag{1.2}$$

in space $L_2(\Omega_1 \times \Omega_2)$, where $g = g(x,y) \in L_2(\Omega_1 \times \Omega_2)$ is a given function.

2. Solvability of nonhomogeneous partial integral equation with degenerate kernel

If $g(x,y) \neq \theta$, then PIE (1.2) is called *nonhomogeneous Fredholm PIE (NPIE)* of second type with degenerate kernel. The homogeneous partial integral equation (HPIE) corresponding the NPIE (1.2) has the following form

$$h - T_1 h = \theta. \tag{2.1}$$

We define measurable functions τ_{ij} on Ω_2 by

$$\tau_{ij}(\omega) = \int_{\Omega_1} \psi_i(s,\omega)\varphi_j(s,\omega) d\mu_1(s), \quad , \; i,j = 1,...,m.$$

One can see that $\tau_{ij} \in L_\infty(\Omega_2)$.

Let $\mathbf{e}$ be an identity element of the algebra $L_\infty(\Omega_2)$, i.e. $\mathbf{e}(\omega) = 1$ for almost all $\omega \in \Omega_2$.

Now we define $m \times m$ matrices $\mathbb{T}$ and $\mathbb{I}$, respectively, whose entries are elements of $L_0(\Omega_2)$, as follows

$$\mathbb{T} = \mathbb{T}(\omega) = (\tau_{ij}(\omega))_{i,j=\overline{1,m}}, \quad \mathbb{I} = \mathbb{I}(\omega) = (\delta_{ij}(\omega))_{i,j=\overline{1,n}},$$

where $\delta_{ii}(\omega) = \mathbf{e}(\omega)$ and $\delta_{ij}(\omega) = \theta(\omega)$ at the $i \neq j$.

Let $\mathbb{D}_1(\omega)$ be a function on Ω_2 given by

$$\mathbb{D}_1(\omega) = \det(\mathbb{T}(\omega) - \mathbb{I}(\omega)), \quad \omega \in \Omega_2.$$

One can see that $\mathbb{D}_1(\omega)$ is a measurable function. Moreover, we have $\mathbb{D}_1 = \mathbb{D}_1(\omega) \in L_\infty(\Omega_2)$. The function $\mathbb{D}_1$ is called a *determinant Fredholm of PIE* (1.2).

Let $\phi \in L_0(\Omega)$. We define its support by the equality $s(\phi) = s_\phi = [\chi_{\Omega(\phi\neq 0)}]$.

Theorem 2.1. *If $s(\mathbb{D}_1) = \mathbf{e}$ (i.e. $\mathbb{D}_1(\omega) \neq 0$ for almost all $\omega \in \Omega_2$), then the HPIE (2.1) has a trivial solution in $L_2(\Omega_1 \times \Omega_2)$, and the NPIE (1.2) has a unique solution in $L_0[L_2(\Omega_1)]$.*

Proof. Suppose that $h(x, y)$ is a solution HPIE (2.1). Let us denote

$$b_i(\omega) = \int_{\Omega_1} \psi_i(s, \omega) h(s, \omega) d\mu_1(s), \quad i = 1, \ldots, m. \tag{2.2}$$

Obviously, $b_i \in L_2(\Omega_2)$ and

$$h(x, y) = \sum_{i=1}^{m} b_i(y) \varphi_i(x, y). \tag{2.3}$$

From the equality (2.2) and (2.3) we obtain the following system of equations for the unknown functions $b_1(y), \ldots, b_m(y)$:

$$b_i(y) = \sum_{j=1}^{m} \tau_{ij}(y) b_j(y), \quad i = 1, \ldots, m.$$

Consequently,

$$\begin{cases} (\tau_{11}(y) - 1)b_1(y) + \tau_{12}(y)b_2(y) + \cdots + \tau_{1m}(y)b_m(y) = \theta(y), \\ \tau_{21}(y)b_1(y) + (\tau_{22}(y) - 1)b_2(y) + \cdots + \tau_{2m}(y)b_m(y) = \theta(y), \\ \cdots\cdots\cdots\cdots\cdots\cdots\cdots\cdots\cdots\cdots\cdots\cdots \\ \tau_{m1}(y)b_1(y) + \tau_{m2}(y)b_2(y) + \cdots + (\tau_{mm}(y) - 1)b_m(y) = \theta(y). \end{cases} \tag{2.4}$$

We can rewrite the last system (2.4) in matrix form as follows:

$$(\mathbb{T} - \mathbb{I})\mathbf{b} = \theta, \tag{2.5}$$

where $\mathbf{b} = \mathbf{b}(\omega)$ columns of the matrix which consisting of functions $b_{11} = b_1(y), b_{21} = b_2(y)$, ..., $b_{m1} = b_m(y)$.

Let $s_{\mathbb{D}_1} = \mathbf{e}$. Then $\mathbb{D}_1(\omega) \neq 0$ for almost all $\omega \in \Omega_2$. It follows that, for almost all $y \in \Omega_2$ equations (2.4) has only a trivial solution, i.e. the equation (2.5) has only zero solution: $\mathbf{b} = \theta$.

An arbitrary solution of the NPIE (1.2) has the form

$$f(x, y) = g(x, y) + \sum_{i=1}^{m} b_i(y) \varphi_i(x, y),$$

where $b_i(y) = \int_{\Omega_1} \psi_i(s, y) f(s, y) d\mu_1(s)$.

Let

$$a_i(\omega) = \int_{\Omega_1} \psi_i(s, \omega) g(s, \omega) d\mu_1(s).$$

It is clear that $a_i \in L_2(\Omega_2)$. For unknown functions $b_1(y), ..., b_m(y)$ we obtain the system of equations

$$\begin{cases} (1-\tau_{11}(\omega))b_1(\omega) - \tau_{12}(\omega)b_2(\omega) - \cdots - \tau_{1m}(\omega)b_m(\omega) = a_1(\omega), \\ -\tau_{21}(\omega)b_1(\omega) - (1-\tau_{22}(\omega))b_2(\omega) + \cdots + \tau_{2m}(\omega)b_m(\omega) = a_2(\omega), \\ \cdots\cdots\cdots\cdots\cdots\cdots\cdots\cdots\cdots\cdots\cdots\cdots\cdots\cdots\cdots\cdots\cdots\cdots \\ -\tau_{m1}(\omega)b_1(\omega) - \tau_{m2}(\omega)b_2(\omega) - \cdots - (1-\tau_{mm}(\omega))b_m(\omega) = a_m(\omega). \end{cases}$$

Put $\widetilde{\mathbb{D}_1}(\omega) = \det(\mathbb{I}(\omega) - \mathbb{T}(\omega)) = -\mathbb{D}_1(\omega)$.

a) if $s(a_i) = \theta$ for all $i \in \{1, ..., m\}$, then $g(x, y) = \theta$ is a solution of the HPIE (1.2) and there is not another solution of the equation (1.2).

b) Assume that, for some $i_0 \in \{1, ..., m\}$ one has $s(a_{i_0}) \neq \theta$. We define measurable functions $\Delta_1(\omega),\ \Delta_2(\omega),\ ..., \Delta_m(\omega)$ on Ω_2 as follows: the elements in k – th column of the determinant $\det(\mathbb{I}(\omega) - \mathbb{T}(\omega))$ we replace by the functions $a_1(\omega), a_2(\omega), ..., a_m(\omega)$ and the resulting determinant is denoted by $\Delta_k(\omega)$. For example,

$$\Delta_1(\omega) = \begin{vmatrix} a_1(\omega) & -\tau_{12}(\omega) & \ldots & -\tau_{1m}(\omega) \\ a_2(\omega) & 1-\tau_{22}(\omega) & \ldots & -\tau_{2m}(\omega) \\ \ldots & \ldots & \ldots & \ldots \\ a_m(\omega) & -\tau_{m2}(\omega) & \ldots & 1-\tau_{mm}(\omega) \end{vmatrix}.$$

It is easy to see that $\Delta_k \in L_2(\Omega_2),\ \ k \in \{1, ..., m\}$.

Let $\omega \in \Omega_2$ is a fixed element. In $L_2(\Omega_1)$ we consider the Fredholm second type equation

$$\varphi(x) - (K_\omega\varphi)(x) = g(x, \omega), \tag{2.6}$$

where

$$K_\omega\varphi(x) = \int_{\Omega_1} k_1(x, s, \omega)\varphi(s)d\mu_1(s).$$

The equation (2.6) for every $\omega \in \Omega_2$ has a unique solution

$$\varphi(x) = \varphi_\omega(x) = g(x, \omega) - \sum_{i=1}^{m} \frac{\Delta_i(\omega)}{\mathbb{D}_1(\omega)}\varphi_i(x, \omega).$$

Clearly, the function

$$f_0(x, y) = g(x, y) - \sum_{i=1}^{m} \frac{\Delta_i(y)}{\mathbb{D}_1(y)}\varphi_i(x, y) \tag{2.7}$$

belongs to $L_0[L_2(\Omega_1), \Omega_2]$ and

$$f_0 - T_1 f_0 = g,$$

i.e. f_0 is a solution of (1.2).

Now we prove the uniqueness of the solution f_0 for the equation (1.2). Let $f_1 \in L_0[L_2(\Omega_2)]$ be a solution of (1.2) and $f_1 \neq f_0$. Then from the equality $f_1 - T_1 f_1 = g$ for a.e. $\omega \in \Omega_2$ we get

$$f_1(x, \omega) - (T_1 f_1)(x, \omega) = g(x, \omega),$$

which means

$$f_1(x,\omega) - (K_\omega f_1)(x,\omega) = g(x,\omega) \quad \text{a.e.} \quad \omega \in \Omega_2.$$

By the uniqueness of the solution of the Fredholm equation (2.6), we obtain

$$f_1(x,\omega) = g(x,\omega) - \sum_{i=1}^{m} \frac{\Delta_i(\omega)}{\mathbb{D}_1(\omega)} \varphi_i(x,\omega)$$

for a.e. $\omega \in \Omega_2$, i.e. $f_1(x,y) = f_0(x,y)$.

□

3. Solvability of homogeneous partial integral equation with degenerate kernel

We define linear operators $A_k f(x,y) = \Delta_{f,k}(y)$, $k = \overline{1,m}$, on the space $L_2(\Omega_1 \times \Omega_2)$, here $\Delta_{f,k}(\omega)$ corresponds to the determinant, which instead of k-th column of the determinant $\det(\mathbb{I}(\omega) - \mathbb{T}(\omega))$ for the following functions:

$$a_{f,1}(\omega) = \int_{\Omega_1} \psi_1(s,\omega) f(s,\omega) d\mu_1(s), \quad \ldots, \quad a_{f,m}(\omega) = \int_{\Omega_1} \psi_m(s,\omega) f(s,\omega) d\mu_1(s),$$

respectively.

By Theorem 2.1 and equality (2.7), we get the following theorem

Theorem 3.1. *Let $s(\mathbb{D}_1) = \mathbf{e}$. If $\mathbb{D}_1^{-1} \in L_\infty(\Omega_2)$, then the operator $I - T_1$ (I is the identity operator) is invertible and*

$$(I - T_1)^{-1} f(x,y) = f(x,y) + \frac{1}{\mathbb{D}_1(y)} (S_1 f)(x,y),$$

where

$$S_1 f(x,y) = \sum_{i=1}^{m} \varphi_i(x,y) A_i f(x,y), \quad f \in L_2(\Omega_1 \times \Omega_2).$$

Remark 3.2. *Note that each A_i is a linear bounded operator on $L_2(\Omega_1 \times \Omega_2)$ and it is PIO as well.*

Theorem 3.3. *Let $s(\mathbb{D}_1) \neq \mathbf{e}$. Then HPIE (2.1) has a nontrivial solution in the $L_2(\Omega_1 \times \Omega_2)$ and moveover, any solution $h(x,y)$ of the equation (2.1) has the form*

$$h(x,y) = \sum_{j=1}^{m} (\boldsymbol{e}(y) - s_{\mathbb{D}_1}(y)) b_j(y) \varphi_j(x,y),$$

where $b_k \in L_2(\Omega_2)$ is arbitrary, $k \in \{1, ..., m\}$.

Proof. Let $h \in L_2(\Omega_1 \times \Omega_2)$ be a solution of the equation (2.1). Then the function h has a form

$$h(x,y) = \sum_{i=1}^{m} d_i(y) \varphi_i(x,y),$$

where $d_i \in L_2(\Omega_2)$ is arbitrary. Let $\mathcal{D}_0 = \{\omega \in \Omega_2 : \ \mathbb{D}_1(\omega) = 0\}$ and $\mathcal{D}_1 = \Omega_2 \setminus \Omega_0$. For each $\omega \in \Omega_2$ we consider the homogeneous Fredholm's equation of the second type in $L_2(\Omega_1)$:

$$\varphi(x) - (K_\omega \varphi)(x) = \theta. \tag{3.1}$$

By the Fredholm's theorem for all $\omega \in \mathcal{D}_1$ the equation (3.1) has only a trivial solution $\varphi(x) = \varphi_\omega(x) = \theta$, and for all $\omega \in \mathcal{D}_0$ the equation (3.1) has not trivial solution in $L_2(\Omega_1)$. For $\omega \in \mathcal{D}_0$ the solution of the equation (3.1) has a form

$$\varphi(x) = \varphi_\omega(x) = \sum_{i=1}^{m} d_i \varphi_i(x, y),$$

where $d_i(\omega) \in \mathbb{C}$ is an arbitrary number.

Put

$$h(x, y) = \sum_{i=1}^{m} (\mathbf{e}(y) - \mathbb{D}_1(y)) b_i(y) \varphi_i(x, y). \tag{3.2}$$

It is easy to verify that

$$h(x, y) = \begin{cases} \theta, & \text{if} \quad \omega \in \mathcal{D}_1, \\ \sum\limits_{i=1}^{m} b_i(\omega)\varphi_i(x, \omega), & \text{if} \quad \omega \in \mathcal{D}_0. \end{cases}$$

Hence, $h(x, \omega) \in L_2(\Omega_1)$ is a solution of (3.1). It is easy to see that $h(x, y) \in L_2(\Omega_1 \times \Omega_2)$ and the function $h(x, y)$ (3.2) is a solution of the HPIE (2.1). □

References

[1] Aleksandrov V M and Kovalenko E V 1986 *Problems of continuum mechanics with mixed Boundary conditions* (Moscow: Nauka)
[2] Kakichev V A and Kovalenko N V 1973 *Ukr. Mat. Zhurnal*, **25** 302–312
[3] Appell J, Kalitvin A S and Zabrejko P P 2000 *Partial Integral operators and Integro-differential Equations* (Basel: Birkhäuser)
[4] Eshkabilov Yu Kh 2006 *Theor. Math. Phys.*, **149** 1497–1511
[5] Okolelov O.P. 1967 *Investigation of Equations with Partial Integral*, (PhD Thesis:Irkutsk Univ.)
[6] Kalitvin A S and Frolova E V 1998 *Pontryagin Readings - IX* (Voronezh)
[7] Frolova E V 1998 *Dep. VINITI* No. 2576-V98
[8] Zabrejko P P, Kalitvin A S and Frolova E V 2002 *Diff. Eqs.* **38** 538–546
[9] Likhtarnikov L M nad Vitova L Z 1975 *Litov. Mat. Sbornik*, **15** 41–47
[10] Kalitvin A S 1988 *Pribl. Funk. Spektr. Theor.* (Leningrad), 43-50.
[11] Kalitvin A S and Zabrejko P P 1991 *J. Integral Eqs Appl.*, **3** 351-382
[12] Eshkabilov Yu Kh 2008 *Central European J. Math.* **6** 149–157
[13] Kalitvin A S and Frolova EV 2004 *Linear equations with partial integrals* (Lipeck: LGPU Press)
[14] Kalitvin A S 2006 *Diff. Eqs* **42** 1262–1268

Bound states of the system of two fermions on the three-dimensional lattice

Janikul Abdullayev[1], **Berdiyor Mamirov**[2]
[1,2]Samarkand State University, Samarkand, Uzbekistan
E-mail: jabdullaev@mail.ru, berdiyor@mail.ru

Abstract. In this paper, the Hamiltonian of the system of two fermions on the three-dimensional lattice $\mathbb{Z}^3$ is considered. Under some conditions to the potential, it is proven that $H(\mathbf{k})$ has four invariant subspaces: in one of them $H(\mathbf{k})$ coincides with $H_0(\mathbf{k})$ and the restrictions of $H(\mathbf{k})$ to the rests three subspaces are unitarily equivalent to each other. Moreover, for the eigenvalues and eigenfunctions of these restricted operators explicit expressions are found.

1. Introduction

The nature of appearance the bound states of two-particle cluster operators for small values of parameters are investigated in detail by Minlos and Mamatov [1], and for more general case by Minlos and Mogilner [2].

The investigation of bound states of Hamiltonian $\hat{H}$ associated with a system of two particles moving on the $d-$ dimensional lattice $\mathbb{Z}^d$ is reduced to the study of the family of Schrödinger operators $H(\mathbf{k}), \mathbf{k} \in \mathbb{T}^d = (-\pi, \pi]^d$. The eigenfunctions and eigenvalues of $H(\mathbf{k})$ are treated as the bound states and their energies of $\hat{H}$, respectively (see [3] and references therein). Bound states of $\hat{H}$ of the system of two fermions and perturbations of eigenvalues of two-particle Schrödinger operators in one-dimensional lattice are studied in [4] and [5], respectively.

In this paper, we study the bound states of two fermions in the three-dimensional lattice $\mathbb{Z}^3$ (see (2.1)), i.e. the discrete spectra of the family of Schrödinger operators $H(\mathbf{k}), \mathbf{k} = (k_1, k_2, k_3) \in \mathbb{T}^3$ associated with the Hamiltonian (see (2.3)). If we assume that the potential $\hat{v}$ is of the form (4.1) and $k_1 = k_2 = k_3 = \pi$ then it can be shown that the spectrum of the operator $H(\pi, \pi, \pi) = 6I - V$ consists of only eigenvalues $6 - \alpha$ and 6. Moreover, $6 - \alpha$ is threefold eigenvalue and 6 is an eigenvalue of infinite multiplicity. In Theorems 4.2-4.4 we show that the results will be preserved under some perturbations.

2. Description of two-particle Hamiltonian

Let $\mathbb{Z}^3$ be the three-dimensional lattice, $\ell_2(\mathbb{Z}^3 \times \mathbb{Z}^3)$ be the Hilbert space of square-summable functions, defined on $\mathbb{Z}^3 \times \mathbb{Z}^3$, $\ell_2^{as}(\mathbb{Z}^3 \times \mathbb{Z}^3) := \{f \in \ell_2(\mathbb{Z}^3 \times \mathbb{Z}^3) : f(\mathbf{x}, \mathbf{y}) = -f(\mathbf{y}, \mathbf{x})\}$ be the subspace of antisymmetric functions in $\ell_2(\mathbb{Z}^3 \times \mathbb{Z}^3)$.

In the coordinate representation the total Hamiltonian $\hat{H}$ of the system acts on the Hilbert space $\ell_2^{as}(\mathbb{Z}^3 \times \mathbb{Z}^3)$ and consists of the difference of free Hamiltonian $\hat{H}_0$ and interaction potential $\hat{V}_2$ of two-particles

$$\hat{H} = \hat{H}_0 - \hat{V}_2, \tag{2.1}$$

where free Hamiltonian $\hat{H}_0$ acts on $\ell_2^{as}(\mathbb{Z}^3 \times \mathbb{Z}^3)$ by

$$\hat{H}_0 = -\frac{1}{2m}\Delta \otimes I - \frac{1}{2m} I \otimes \Delta.$$

Here m denotes the mass of the fermion, which we further assume to be equal to 1, I is the identity operator in $\ell_2(\mathbb{Z}^3)$, the lattice Laplacian Δ is a difference operator describing the transfer of particle from one site of the lattice to the nearest neighbor site, i.e.

$$(\Delta\hat{\psi})(\mathbf{x}) = \sum_{j=1}^{3}[\hat{\psi}(\mathbf{x}+\mathbf{e}_j) - 2\hat{\psi}(\mathbf{x}) + \hat{\psi}(\mathbf{x}-\mathbf{e}_j)], \quad \mathbf{x} \in \mathbb{Z}^3, \quad \hat{\psi} \in \ell_2(\mathbb{Z}^3),$$

where $\{\mathbf{e}_1 = (1,0,0),\ \mathbf{e}_2 = (0,1,0),\ \mathbf{e}_3 = (0,0,1)\}$ is the standard basis on $\mathbb{Z}^3$. The interaction $\hat{V}_2$ of two particles are described by

$$(\hat{V}_2\hat{\psi})(\mathbf{x},\mathbf{y}) = \hat{v}(\mathbf{x}-\mathbf{y})\hat{\psi}(\mathbf{x},\mathbf{y}), \quad \hat{\psi} \in \ell_2^{as}(\mathbb{Z}^3 \times \mathbb{Z}^3).$$

Further we assume that the function $\hat{v}$ satisfies

$$\hat{v} \in \ell_1(\mathbb{Z}^3) \qquad \text{and} \qquad \hat{v}(\mathbf{x}) = \hat{v}(-\mathbf{x}) \geq 0, \quad \mathbf{x} \in \mathbb{Z}^3. \tag{2.2}$$

Under hypothesis (2.2) the operator $\hat{V}_2$ is a bounded, self-adjoint operator in $\ell_2^{as}(\mathbb{Z}^3 \times \mathbb{Z}^3)$. It then follows the boundedness and self-adjointness of the energy operator $\hat{H}$.

Let $\mathbb{T}^3$ be the three-dimensional torus, i.e. dual group to $\mathbb{Z}^3$, and $L_2^{as}(\mathbb{T}^3 \times \mathbb{T}^3)$ be the Hilbert space of square-integrable antisymmetric functions.

The momentum representation is obtained via the standard Fourier transform

$$F : \ell_2^{as}(\mathbb{Z}^3 \times \mathbb{Z}^3) \to L_2^{as}(\mathbb{T}^3 \times \mathbb{T}^3),$$

where $L_2^{as}(\mathbb{T}^3 \times \mathbb{T}^3) := \{f \in L_2(\mathbb{T}^3 \times \mathbb{T}^3) : f(\mathbf{x},\mathbf{y}) = -f(\mathbf{y},\mathbf{x})\}$. The Hamiltonian $H = H_0 - V = F\hat{H}F^{-1}$ in the momentum representation commutes with the Abelian group of unitary operators $U_{\mathbf{s}}, \mathbf{s} \in \mathbb{Z}^3$:

$$(U_{\mathbf{s}}f)(\mathbf{k}_1,\mathbf{k}_2) = \exp\big(-i(\mathbf{s},\mathbf{k}_1+\mathbf{k}_2)\big)f(\mathbf{k}_1,\mathbf{k}_2), \quad f \in L_2^{as}(\mathbb{T}^3 \times \mathbb{T}^3).$$

In [6], it is shown the existence of expansions of the Hilbert space $L_2^{as}(\mathbb{T}^3 \times \mathbb{T}^3)$ and the operators $U_{\mathbf{s}}$, H into the following direct integrals

$$L_2^{as}(\mathbb{T}^3 \times \mathbb{T}^3) = \int_{\mathbb{T}^3} \oplus L_2(F_{\mathbf{k}})\,d\mathbf{k}, \quad U_{\mathbf{s}} = \int_{\mathbb{T}^3} \oplus U_{\mathbf{s}}(\mathbf{k})\,d\mathbf{k}, \quad H = \int_{\mathbb{T}^3} \oplus \tilde{H}(\mathbf{k})\,d\mathbf{k},$$

where

$$F_{\mathbf{k}} = \{(\mathbf{k}_1,\mathbf{k}_2) \in \mathbb{T}^3 \times \mathbb{T}^3 :\ \mathbf{k}_1 + \mathbf{k}_2 = \mathbf{k}\}, \quad \mathbf{k} \in \mathbb{T}^3$$

and $U_{\mathbf{s}}(\mathbf{k})$ is the multiplication operator by the function $\exp(-i(\mathbf{s},\mathbf{k}))$ in $L_2(F_{\mathbf{k}})$. The fiber operator $\tilde{H}(\mathbf{k})$ of H acting in the Hilbert space $L_2(F_{\mathbf{k}})$ is unitarily equivalent to the Schrödinger operator $H(\mathbf{k}) := H_0(\mathbf{k}) - V$, acting on the Hilbert space

$$L_2^{o}(\mathbb{T}^3) := \{f \in L_2(\mathbb{T}^3) : f(-\mathbf{q}) = -f(\mathbf{q}), \text{ for almost all } \mathbf{q} \in \mathbb{T}^3\}$$

of odd square-integrable functions in $\mathbb{T}^3$, as follows

$$(H(\mathbf{k})f)(\mathbf{q}) = \varepsilon_{\mathbf{k}}(\mathbf{q})f(\mathbf{q}) - \frac{1}{(2\pi)^{3/2}}\int_{\mathbb{T}^3} v(\mathbf{q}-\mathbf{s})f(\mathbf{s})\,d\mathbf{s}. \tag{2.3}$$

Unperturbed operator $H_0(\mathbf{k})$ is the multiplication operator by the function

$$\varepsilon_{\mathbf{k}}(\mathbf{q}) = \varepsilon(\frac{\mathbf{k}}{2} + \mathbf{q}) + \varepsilon(\frac{\mathbf{k}}{2} - \mathbf{q}) = 2\sum_{j=1}^{3}\left[1 - \cos\frac{k_j}{2}\cos q_j\right]. \qquad (2.4)$$

The perturbation operator V is an integral operator in $L_2^o(\mathbb{T}^3)$ with the kernel

$$v(\mathbf{q}-\mathbf{s}) = \frac{1}{(2\pi)^{3/2}}\sum_{\mathbf{n}\in\mathbb{Z}^3}\hat{v}(\mathbf{n})e^{i((\mathbf{q}-\mathbf{s}),\mathbf{n})}, \quad \mathbf{q},\mathbf{s}\in\mathbb{T}^3$$

and belongs to the trace class Σ_1.

3. Spectrum of the operator $H(\mathbf{k})$

For any $\mathbf{k}\in\mathbb{T}^3$ we denote by $M(\mathbf{k})$ and $m(\mathbf{k})$ the maximum and minimum of the function $\varepsilon_{\mathbf{k}}(\mathbf{p})$:

$$M(\mathbf{k}) = \max_{\mathbf{k}\in\mathbb{T}^3}\varepsilon_{\mathbf{k}}(\mathbf{p}) = 2\sum_{j=1}^{3}\left[1+\cos\frac{k_j}{2}\right], \; m(\mathbf{k}) = \min_{\mathbf{p}\in\mathbb{T}^3}\varepsilon_{\mathbf{k}}(\mathbf{p}) = 2\sum_{j=1}^{3}\left[1-\cos\frac{k_j}{2}\right].$$

According to the Weil Theorem [6] the continuous spectrum of $H(\mathbf{k})$ coincides with the essential spectrum of $H_0(\mathbf{k})$. The spectrum of unperturbed operator $H_0(\mathbf{k})$ is purely continuous and coincides with the segment $[m(\mathbf{k}), M(\mathbf{k})]$. The width $\omega(\mathbf{k}) = M(\mathbf{k}) - m(\mathbf{k})$ of this segment is called the width of the continuous spectrum of $H(\mathbf{k})$ and it is equal to

$$\omega(\mathbf{k}) = 4\cos\frac{k_1}{2} + 4\cos\frac{k_2}{2} + 4\cos\frac{k_3}{2}.$$

Therefore,

$$\min_{\mathbf{k}\in\mathbb{T}^3}\omega(\mathbf{k}) = \omega(\pi,\pi,\pi) = 0, \quad \max_{\mathbf{k}\in\mathbb{T}^3}\omega(\mathbf{k}) = \omega(0,0,0) = 12.$$

The width of continuous spectrum of $H(\mathbf{k})$ in direction $e_j, j = 1,2,3$ is defined as

$$\omega_j(\mathbf{k}) = \max_{p_j\in\mathbb{T}^3}\varepsilon_{\mathbf{k}}(\mathbf{p}) - \min_{p_j\in\mathbb{T}^3}\varepsilon_{\mathbf{k}}(\mathbf{p}) = 4\cos\frac{k_j}{2}. \qquad (3.1)$$

Note that

$$\omega(\mathbf{k}) = \omega_1(\mathbf{k}) + \omega_2(\mathbf{k}) + \omega_3(\mathbf{k}).$$

Moreover,

$$\omega_j(\mathbf{k}) \equiv 0 \quad \text{for} \quad \mathbf{k}\in S(j) := \{\mathbf{k}\in\mathbb{T}^3 : k_j = \pi\}. \qquad (3.2)$$

If $\mathbf{k} = (\pi,\pi,\pi)$, the width of the essential spectrum is 0, i.e. $\omega(\pi,\pi,\pi) = 0$ and the spectrum of the operator $H(\pi,\pi,\pi) = 6I - V$ consists of eigenvalues of the form $6 - \hat{v}(\mathbf{n})$, $\mathbf{n}\in\mathbb{Z}^3$ and of the essential spectrum $\{6\}$. With decreasing of width of continuous spectrum the number of eigenvalues of the Schrödinger operator $H(\mathbf{k})$ increases (see [7]). If $k_j = \pi$ for some $j\in\{1,2,3\}$, then $\omega_j(\mathbf{k}) = 0$ and there exists a potential $\hat{v}$ such that the operator $H(\mathbf{k})$ has infinitely many eigenvalues outside the essential spectrum [8].

Let us recall some well-known facts and notations from the Operator and Perturbation Theory. For any self-adjoint operator B acting in Hilbert space $\mathcal{H}$ and not having essential spectrum above the point $\mu\in\mathbb{R}$, by $n(\mu, B)$ we denote the number of eigenvalues lying above μ.

By $N(\mathbf{k}, z)$ we denote the number of eigenvalues of $H(\mathbf{k})$, lying below $z \leq m(\mathbf{k})$, i.e. $N(\mathbf{k}, z) = n(-z, \ -H(\mathbf{k}))$. In fact, the number $N(\mathbf{k}, m(\mathbf{k}))$ coincides with the number of eigenvalues of $H(\mathbf{k})$ outside the essential spectrum (see [9], [10]). For any $\mathbf{k} \in \mathbb{T}^3$ and $z < m(\mathbf{k})$ we define an integral operator by

$$G(\mathbf{k}, z) = V^{\frac{1}{2}} r_0(\mathbf{k}, z) V^{\frac{1}{2}}, \tag{3.3}$$

where $r_0(\mathbf{k}, z)$ is the resolvent of unperturbed operator $H_0(\mathbf{k})$. Under the condition (2.2) the operator V is positive, and we denote by $V^{\frac{1}{2}}$ the positive square-root of V.

From the self-adjointness of $H(\mathbf{k}) = H_0(\mathbf{k}) - V$ and from the positivity of V it follows that

$$\sigma(H(\mathbf{k})) \cap (M(\mathbf{k}), \infty) = \emptyset, \tag{3.4}$$

hence, $\sigma_{disc}(H(\mathbf{k})) \subset (-\infty, m(\mathbf{k})]$, which implies that it is enough to investigate the eigenvalues for $z < m(\mathbf{k})$.

The solution f of the Schrödinger operator

$$H(\mathbf{k}) f = z f$$

and fixed points φ of the operators $G(\mathbf{k}, z)$ are related as

$$f = r_0(\mathbf{k}, z) V^{\frac{1}{2}} \varphi, \qquad \varphi = V^{\frac{1}{2}} f.$$

Moreover, the Birman-Schwinger principle also holds [9],[10].

Lemma 3.1. *If the limit operator* $\lim\limits_{z \to m(\mathbf{k})-} G(\mathbf{k}, z) = G(\mathbf{k}, m(\mathbf{k}))$ *exists in operator topology, then it is compact and*

$$N(\mathbf{k}, m(\mathbf{k})) = n(1, G(\mathbf{k}, m(\mathbf{k}))). \tag{3.5}$$

Equality (3.5) means that the number of eigenvalues of $H(\mathbf{k})$, lying below $m(\mathbf{k})$ coincides with the eigenvalues of $G(\mathbf{k}, m(\mathbf{k}))$, greater than 1.

4. Invariant subspaces of the operator $H(\mathbf{k})$

In what follows, we consider the operator $H(\mathbf{k})$ with the potential $\hat{v}$ of the form:

$$\hat{v}(\mathbf{x}) = \begin{cases} 1, & \text{if } \ \mathbf{x} = 0 \\ \alpha, & \text{if } \ |\mathbf{x}| = 1 \\ 0, & \text{if } \ |\mathbf{x}| \geq 2. \end{cases} \tag{4.1}$$

Here $\alpha > 0$ and $\mathbf{x} \in \mathbb{Z}^3$, $|\mathbf{x}| = |x_1| + |x_2| + |x_3|$. The perturbation operator V is an integral operator $L_2^o(\mathbb{T}^3)$ with the kernel

$$\frac{1}{(2\pi)^{3/2}} v(\mathbf{p} - \mathbf{q}) = \frac{2\alpha}{(2\pi)^3} \Big[\sin p_1 \sin q_1 + \sin p_2 \sin q_2 + \sin p_3 \sin q_3 \Big]. \tag{4.2}$$

If the potential $\hat{v}$ has the form (4.1) and $k_1 = k_2 = k_3 = \pi$, then the number $z_1(\pi, \pi, \pi) = 6 - \alpha$, is threefold eigenfunctions of $H(\pi, \pi, \pi)$ with normed eigenfunctions

$$\phi^{-++}_{(1,0,0)}(\mathbf{p}) = \frac{1}{2\ \pi^{3/2}} \sin p_1, \quad \phi^{+-+}_{(0,1,0)}(\mathbf{p}) = \frac{1}{2\ \pi^{3/2}} \sin p_2, \quad \phi^{++-}_{(0,0,1)}(\mathbf{p}) = \frac{1}{2\ \pi^{3/2}} \sin p_3. \tag{4.3}$$

The number $z_\infty(\pi,\pi,\pi) = 6$ is an eigenvalue of infinite multiplicity, and associated eigenfunctions are

$$\phi^{-++}_{(n,m,l)}(\mathbf{p}) = \sin np_1 \cos mp_2 \cos lp_3, \; \phi^{+-+}_{(n,m,l)}(\mathbf{p}) = \cos np_1 \sin mp_2 \cos lp_3,$$

$$\phi^{++-}_{(n,m,l)}(\mathbf{p}) = \cos np_1 \cos mp_2 \sin lp_3, \quad n+m+l \geq 2.$$

Denote

$$L_2^+(\mathbb{T}) = \{f \in L_2(\mathbb{T}) : f(p) = f(-p)\}, \quad L_2^-(\mathbb{T}) = \{f \in L_2(\mathbb{T}) : f(-p) = -f(p)\}$$

which are the subspaces of even and odd functions, respectively. Moreover, by Riesz Lemma one gets

$$L_2^o(\mathbb{T}^3) = L_2^{-++}(\mathbb{T}^3) \oplus L_2^{+-+}(\mathbb{T}^3) \oplus L_2^{++-}(\mathbb{T}^3) \oplus L_2^{---}(\mathbb{T}^3), \tag{4.4}$$

where

$$L_2^{-++}(\mathbb{T}^3) = L_2^-(\mathbb{T}) \otimes L_2^+(\mathbb{T}) \otimes L_2^+(\mathbb{T}), \qquad L_2^{+-+}(\mathbb{T}^2) = L_2^+(\mathbb{T}) \otimes L_2^-(\mathbb{T}) \otimes L_2^+(\mathbb{T})$$

and

$$L_2^{++-}(\mathbb{T}^2) = L_2^+(\mathbb{T}) \otimes L_2^+(\mathbb{T}) \otimes L_2^-(\mathbb{T}), \qquad L_2^{---}(\mathbb{T}^3) = L_2^-(\mathbb{T}) \otimes L_2^-(\mathbb{T}) \otimes L_2^-(\mathbb{T}).$$

Lemma 4.1. *The subspaces $L_2^{-++}(\mathbb{T}^3)$, $L_2^{+-+}(\mathbb{T}^3)$, $L_2^{++-}(\mathbb{T}^3)$ and $L_2^{---}(\mathbb{T}^3)$ are invariant with respect to $H(\mathbf{k})$.*

Proof. First, we prove the invariance of these spaces with respect to $H_0(\mathbf{k})$, and then for V. From the representation (2.4) it follows that $\varepsilon_{\mathbf{k}}$ belongs to the subspace

$$L_2^{+++}(\mathbb{T}^3) = \{f \in L_2(\mathbb{T}^3) : f(p_1,p_2,p_3) = f(-p_1,p_2,p_3) = f(-p_1,-p_2,p_3) = f(-p_1,-p_2,-p_3)\}.$$

Hence from $f \in L_2^{-++}(\mathbb{T}^3)$ it follows that $\varepsilon_{\mathbf{k}} f \in L_2^{-++}(\mathbb{T}^3)$. This implies that $L_2^{-++}(\mathbb{T}^3)$ is invariant with respect to $H_0(\mathbf{k})$.

If $f \in L_2^{-++}(\mathbb{T}^3)$, it then follows from (4.2) that

$$(Vf)(p_1,p_2,p_3) = \frac{1}{(2\pi)^{3/2}} \int\limits_{\mathbb{T}^3} v(p_1-s_1, p_2-s_2, p_3-s_3) f(s_1,s_2,s_3)\, \mathrm{d}s_1\, \mathrm{d}s_2\, \mathrm{d}s_3 =$$
$$\frac{2\alpha \sin p_1}{(2\pi)^3} \int\limits_{\mathbb{T}^3} \sin q_1 f(s_1,s_2,s_3)\, \mathrm{d}s_1\, \mathrm{d}s_2\, \mathrm{d}s_3$$

which belongs to $L_2^{-++}(\mathbb{T}^3)$. This proves the invariance of the subspace $L_2^{-++}(\mathbb{T}^3)$ with respect to V, and hence with respect to operator $H(\mathbf{k}) = H_0(\mathbf{k}) - V$. Analogously, one can prove the invariance of subspaces $L_2^{+-+}(\mathbb{T}^3)$, $L_2^{++-}(\mathbb{T}^3)$ and $L_2^{---}(\mathbb{T}^3)$ with respect to $H(\mathbf{k})$. This completes the proof. □

From the proof of Lemma 4.1 it follows that the restriction $H^{-++}(\mathbf{k}) = H_0(\mathbf{k}) - V^{-++}$ of the operator $H(\mathbf{k})$ acts as follows

$$H^{-++}(\mathbf{k})f(\mathbf{p}) = \varepsilon_{\mathbf{k}}(\mathbf{p})f(\mathbf{p}) - \frac{2\alpha}{(2\pi)^3} \int_{\mathbb{T}^3} \sin p_1 \sin q_1 f(\mathbf{q})\, \mathrm{d}\mathbf{q}, \;\; f \in L_2^{-++}(\mathbb{T}^3). \tag{4.5}$$

Denote by $H^{---}(\mathbf{k})$, $H^{+-+}(\mathbf{k})$ and $H^{++-}(\mathbf{k})$ the restrictions of $H(\mathbf{k})$ respectively in the Hilbert spaces $L_2^{---}(\mathbb{T}^3)$, $L_2^{+-+}(\mathbb{T}^3)$ and $L_2^{++-}(\mathbb{T}^3)$. Note that

$$(H^{---}(\mathbf{k})f)(p) = H_0(\mathbf{k})f(p,) \qquad f \in L_2^{---}(\mathbb{T}^2).$$

Notice that the spectrum of $H_0(\mathbf{k})$ is purely continuous and consists of the segment $[m(\mathbf{k}), M(\mathbf{k})]$. Hence, it is enough to study $H^{-++}(\mathbf{k})$, $H^{+-+}(\mathbf{k})$ and $H^{++-}(\mathbf{k})$. Since they are unitarily equivalent to each other, i.e.

$$H^{-++}(\mathbf{k}) = U_2 H^{+-+}(\mathbf{k}) U_2^{-1} = U_3 H^{++-}(\mathbf{k}) U_3^{-1},$$

where $(U_2 f)(p_1, p_2, p_3) = f(p_2, p_1, p_3)$, $(U_3 f)(p_1, p_2, p_3) = f(p_3, p_2, p_1)$, it suffices to consider just one of them, for example $H^{-++}(\mathbf{k}) = H^{-++}(k_1, k_2, k_3)$.

Theorem 4.2. *Let $k_1 = k_2 = \pi$. Then for any $k_3 \in (-\pi, \pi)$ the operator $H^{-++}(\pi, \pi, k_3)$ has a unique nondegenerated eigenvalue*

$$z_1^{-++}(k_3) = 6 - \sqrt{\alpha^2 + 4\cos^2\frac{k_3}{2}} \tag{4.6}$$

with an associated eigenfunction

$$f_1^{-++}(\mathbf{p}) = \frac{C \sin p_1}{6 - 2\cos\frac{k_3}{2}\cos p_3 - z_1^{-++}(k_3)} \in L_2^{-++}(\mathbb{T}^3),$$

here C is a constant.

Proof. Since $H^{-++}(\pi, \pi, k_3)$ is a self-adjoint operator, the spectrum is subset of real axis. The eigenvalue equation $H^{-++}(\pi, \pi, k_3) f(\mathbf{p}) = z f(\mathbf{p})$ is equivalent to the equation

$$(6 - 2\cos\frac{k_3}{2}\cos q_3) f(\mathbf{p}) - \frac{2\alpha}{(2\pi)^3}\int\limits_{\mathbb{T}^3} \sin q_1 \sin s_1 f(\mathbf{s})\,\mathrm{d}\mathbf{s} = z\, f(\mathbf{p}). \tag{4.7}$$

Introducing the notation

$$a = \frac{1}{(2\pi)^3}\int\limits_{\mathbb{T}^3} \sin s_1 f(\mathbf{s})\,\mathrm{d}\mathbf{s} \tag{4.8}$$

from (4.7) we get the following representation for the eigenfunction $f(\mathbf{p})$:

$$f(\mathbf{p}) = \frac{2\alpha\, a\, \sin p_1}{6 - 2\cos\frac{k_3}{2}\cos p_3 - z}. \tag{4.9}$$

Putting (4.9) and (4.8) we get the equation

$$a\left(1 - \frac{2\alpha}{(2\pi)^3}\int\limits_{\mathbb{T}^3} \frac{\sin^2 s_1\ \mathrm{d}\mathbf{s}}{6 - 2\cos\frac{k_3}{2}\cos s_3 - z}\right) = 0.$$

If $a = 0$, then it is clear that $f(\mathbf{p}) \equiv 0$. This contradicts that f is an eigenfunction. Thus,

$$\Delta(z) = 1 - \frac{2\alpha}{(2\pi)^3}\int\limits_{\mathbb{T}^3} \frac{\sin^2 s_1\ \mathrm{d}\mathbf{s}}{6 - 2\cos\frac{k_3}{2}\cos s_3 - z} = 0. \tag{4.10}$$

If $z > 6 - 2\cos\frac{k_3}{2}$ then $\Delta(z) > 1$. Hence we assume that $z < 6 - 2\cos\frac{k_3}{2}$. Let us calculate the following integral:

$$\frac{2\alpha}{(2\pi)^3}\int\limits_{\mathbb{T}^3}\frac{\sin^2 s_1\,\mathrm{ds}}{6-2\cos\frac{k_3}{2}\cos s_3 - z} = \frac{\alpha}{2\pi}\int\limits_{\mathbb{T}}\frac{\mathrm{d}s_3}{6-2\cos\frac{k_3}{2}\cos s_3 - z}. \tag{4.11}$$

Using the change of variables $\xi = e^{is_3}$, the integral in the right-hand-side of (4.11) is represented as

$$\frac{\alpha}{2\pi}\int\limits_{\mathbb{T}}\frac{\mathrm{d}s_3}{6-2\cos\frac{k_3}{2}\cos s_3 - z} = \frac{-\alpha}{\cos\frac{k_3}{2}}\frac{1}{2\pi i}\int\limits_{|\xi|=1}\frac{\mathrm{d}\xi}{\xi^2 - \frac{6-z}{\cos\frac{k_3}{2}}\xi + 1}$$

$$= \frac{-\alpha}{\cos\frac{k_3}{2}}\frac{1}{2\pi i}\int\limits_{|\xi|=1}\frac{d\xi}{(\xi-\xi_1)(\xi-\xi_2)}.$$

Hence we end up with Cauchy type integral, where

$$\xi_1 = \frac{6-z-\sqrt{(6-z)^2-4\cos^2\frac{k_3}{2}}}{2\cos\frac{k_3}{2}},\quad \xi_2 = \frac{6-z+\sqrt{(6-z)^2-4\cos^2\frac{k_3}{2}}}{2\cos\frac{k_3}{2}}$$

are zeros of $\xi^2 - \frac{6-z}{\cos\frac{k_3}{2}}\xi + 1$. If $z < 6 - 2\cos\frac{k_3}{2}$, then $|\xi_1| < 1$, $|\xi_2| > 1$. Then by Cauchy theorem on residues we get

$$\frac{1}{2\pi i}\int\limits_{|\xi|=1}\frac{d\xi}{(\xi-\xi_1)(\xi-\xi_2)} = \frac{1}{\xi_1-\xi_2} = -\frac{\cos\frac{k_3}{2}}{\sqrt{(6-z)^2-4\cos^2\frac{k_3}{2}}}. \tag{4.12}$$

Inserting this into (4.10) we get the following equation

$$1 - \frac{\alpha}{\sqrt{(6-z)^2-4\cos^2\frac{k_3}{2}}} = 0.$$

Solving it we get

$$z_1 = 6 - \sqrt{\alpha^2 + 4\cos^2\frac{k_3}{2}},\qquad z_2 = 6 + \sqrt{\alpha^2 + 4\cos^2\frac{k_3}{2}}.$$

Since $z_2 > 6 - 2\cos\frac{k_3}{2}$, z_1 is a unique eigenvalue of $H^{-++}(\pi,\pi,k_3)$. □

Theorem 4.3. *Let $k_1 = k_3 = \pi$. Then for any $k_2 \in (-\pi,\pi)$ the operator $H^{-++}(\pi,k_2,\pi)$ has a unique nondegenerated eigenvalue*

$$z_1^{-++}(k_2) = 6 - \sqrt{\alpha^2 + 4\cos^2\frac{k_2}{2}} \tag{4.13}$$

with associated eigenfunction

$$f_2^{-++}(\mathbf{p}) = \frac{C\sin p_1}{6 - 2\cos\frac{k_2}{2}\cos p_2 - z_1^{-++}(k_2)} \in L_2^{-++}(\mathbb{T}^3),$$

where C is a constant.

Theorem 4.4. *Let $k_2 = k_3 = \pi$ and $k_1 \in \mathbb{T}$. Then for any $\alpha > \cos\frac{k_1}{2}$ the operator $H^{-++}(k_1, \pi, \pi)$ has a unique nondegenerated eigenvalue*

$$z_2^{-++}(k_1) = 6 - \alpha - \frac{1}{\alpha}\cos^2\frac{k_1}{2} \tag{4.14}$$

with associated eigenfunction

$$f_3^{-++}(\mathbf{p}) = \frac{C\ \sin p_1}{6 - 2\cos\frac{k_1}{2}\cos p_1 - z_2^{-++}(k_1)} \in L_2^{-++}(\mathbb{T}^3),$$

where C is any constant.

Proof. The eigenvalue equation $H^{-++}(k_1, \pi, \pi)f(\mathbf{p}) = zf(\mathbf{p})$ is equivalent to

$$(6 - 2\cos\frac{k_1}{2}\cos p_1)f(\mathbf{p}) - \frac{2\alpha}{(2\pi)^3}\int\limits_{\mathbb{T}^3} \sin p_1 \sin s_1 f(\mathbf{s})\,\mathrm{d}\,\mathbf{s} = z\,f(\mathbf{q}), \tag{4.15}$$

which have nontrivial solutions if and only if $z < 6 - 2\cos\frac{k_1}{2}$ is the solution of the equation

$$1 - \frac{2\alpha}{(2\pi)^3}\int\limits_{\mathbb{T}^3} \frac{\sin^2 s_1\ \mathrm{d}\mathbf{s}}{6 - 2\cos\frac{k_1}{2}\cos s_1 - z} = 1 - \frac{2\alpha}{2\pi}\int\limits_{\mathbb{T}} \frac{\sin^2 s_1\ \mathrm{d}s_1}{6 - 2\cos\frac{k_1}{2}\cos s_1 - z} = 0. \tag{4.16}$$

Using

$$\frac{-\sin^2 s}{6 - 2\cos\frac{k_1}{2}\cos s - z} = \frac{\cos^2 s - 1}{6 - 2\cos\frac{k_1}{2}\cos s - z} =$$
$$-\frac{\cos s}{2\cos\frac{k_1}{2}} - \frac{6 - z}{4\cos^2\frac{k_1}{2}} + \left[\frac{(6 - z)^2}{4\cos^2\frac{k_1}{2}} - 1\right]\frac{1}{6 - 2\cos\frac{k_1}{2}\cos s - z}$$

(4.16) is rewritten as

$$1 + \frac{\alpha(6 - z)}{2\cos^2\frac{k_1}{2}} - \left[\frac{(6 - z)^2}{4\cos^2\frac{k_1}{2}} - 1\right]\frac{2\alpha}{2\pi}\int\limits_{\mathbb{T}} \frac{\mathrm{d}s}{6 - 2\cos\frac{k_1}{2}\cos s_1 - z} = 0. \tag{4.17}$$

As in the proof of Theorem 4.3 it is easy to check

$$\frac{2\alpha}{2\pi}\int\limits_{\mathbb{T}} \frac{\mathrm{d}s_1}{6 - 2\cos\frac{k_1}{2}\cos s_1 - z} = \frac{2\alpha}{\sqrt{(6 - z)^2 - 4\cos^2\frac{k_1}{2}}}.$$

Now using (4.17) and last relation we get

$$1+\frac{\alpha(6-z)}{2\cos^2\frac{k_1}{2}}-\Big[\frac{(6-z)^2}{4\cos^2\frac{k_1}{2}}-1\Big]\frac{2\alpha}{\sqrt{(6-z)^2-4\cos^2\frac{k_1}{2}}}=$$

$$1+\frac{\alpha(6-z)}{2\cos^2\frac{k_1}{2}}-\frac{\alpha\sqrt{(6-z)^2-4\cos^2\frac{k_1}{2}}}{2\cos^2\frac{k_1}{2}}$$

Solving this equation with respect to z we get $z=6-\alpha-\frac{1}{\alpha}\cos^2\frac{k_1}{2}$, which proves (4.14).

Now let us prove that for $\alpha\le\cos\frac{k_1}{2}$ the discrete spectrum of $H^{-++}(k_1,\pi,\pi)$ is empty outside the essential spectrum. The limit Birman-Schwinger operator

$$G^{-++}((k_1,\pi,\pi),m(k_1,\pi,\pi))=\big(V^{-++}\big)^{\frac{1}{2}}r_0((k_1,\pi,\pi),m(k_1,\pi,\pi))\big(V^{-++}\big)^{\frac{1}{2}}$$

corresponding to $H^{-++}(k_1,\pi,\pi)$ is of the form

$$(G^{-++}((k_1,\pi,\pi),m(k_1,\pi,\pi))f)(\mathbf{p})=\frac{\alpha}{\cos(k_1/2)}\int\limits_{-\pi}^{\pi}\sin p_1\sin s_1 f(\mathbf{s})d\mathbf{s}.$$

Thus,

$$\|G^{-++}((k_1,\pi,\pi),m(k_1,\pi,\pi))\|=(G^{-++}((k_1,\pi,\pi),m(k_1,\pi,\pi))\varphi_1^-,\varphi_1^-)=\frac{\alpha}{\cos(k_1/2)},$$

$$(\varphi_1^-,\varphi_1^-)=1,\ \ \varphi_1^-(\mathbf{p})=\frac{\sin p_1}{2\pi\sqrt{\pi}}.$$

From here if $\alpha\le\cos\frac{k_1}{2}$ then $||G^{-++}((k_1,\pi,\pi),m(k_1,\pi,\pi))||\le 1$. So, by Lemma 3.1 the operator $H^{-++}(k_1,\pi,\pi)$ has empty discrete spectrum outside the essential spectrum. □

References

[1] Mamatov Sh S and Minlos R A 1989 *Theor. Math. Phys.* **79** 163-179
[2] Minlos R A and Mogilner A I 1989 In book: *Schodinger operators: Standard and Nonstandard* (eds. P.Exner, P.Seba) (Simgapore: World Scientific)
[3] Albeverio S, Lakaev S N, Makarov K A and Muminov Z I 2006 *Commun. Math. Phys.* **262** 91–115
[4] Abdullaev J I 2006 *Theor. Math. Phys.* **147** 36–47
[5] Abdullaev J I 2005 *Theor. Math. Phys.* **145** 212-220
[6] Reed M and Simon B 1978 *Methods of modern mathematical physics.* Vol. 4, (New York: Acad. Press)
[7] Abdullaev J I 2005 *Dokl. Akad. Nauk R. Uzb.* 4 11–16
[8] Abdullaev J I and Muminov Z. E 2004 *Dokl. Akad. Nauk R. Uzb.* 3 4–8
[9] Albeverio S, Lakaev S and Muminov Z 2004 *Annales Henri Poincare* **5** 743–772
[10] Sobolev A V 1993 *Commun. Math. Phys.* **156** 127–168

Scattering of a two-soliton molecule by Gaussian potential barriers and wells

B A Umarov[1], **N A B Aklan**[2], **B B Baizakov**[3], **F Kh Abdullaev**[1]

[1] Department of Physics, Kulliyyah of Science, International Islamic University Malaysia, 25200 Kuantan, Malaysia

[2] Department of Computational and Theoretical Sciences, International Islamic University Malaysia, Kulliyyah of Science, 25710, Kuantan, Malaysia

[3] Physical-Technical Institute, 2-b, Bodomzor str., 100084, Tashkent, Uzbekistan

E-mail: `bakhram@iium.edu.my`

Abstract. Two anti-phase bright solitons in a dipolar Bose-Einstein condensate can form stable bound states, known as soliton molecules. In this paper we study the scattering of a two-soliton molecule by external potential, using the simplest and analytically tractable Gaussian potential barriers and wells, in one spatial dimension. Theoretical model is based on the variational approximation for the nonlocal Gross-Pitaevskii equation (GPE). At sufficiently low velocity of the incident molecule we observe quantum reflection from the potential well. Predictions of the mathematical model are compared with numerical simulations of the GPE, and good qualitative agreement between them is demonstrated.

1. Introduction

Manifestation of quantum behavior by macroscopic objects always attracts great interest. One of the recent findings in this direction has been the observation of quantum reflection of matter-wave solitons from attractive potential wells and downward potential steps [1, 2, 3]. Experimental study of the quantum reflection of a matter-wave soliton, formed from ^{85}Rb Bose-Einstein condensate, was reported in [4]. In the present context, the quantum reflection is understood as a classically forbidden reflection, because it occurs without reaching the classical turning point.

Gaining new knowledge by studying collisions of particles with each-other, or their interaction with external potentials, represents the leading approach in particle physics. Similar method is applicable in studies of nonlinear localized wave packets or solitons. By definition, solitons are robust localized wave packets, that can propagate along wave-guides preserving their shape and velocity. Solitons emerge from a fine balance between dispersive spreading and nonlinear self focusing of the wave packet. During the interactions solitons exhibit both particle-like and wave-like properties. In particular, two colliding solitons exchange not only their velocities, as classical particles do, but also their physical locations, as quantum particles do via the tunnel phenomenon. Experimental demonstration [5] and theoretical confirmation [6] of the last property has been a significant achievement in the physics of solitons. The discovery of stable bound states of optical solitons, known as soliton molecules [7, 8, 9], represents another significant advance in the field. Recently the existence of similar soliton complexes in dipolar Bose-Einstein condensates was reported in [10, 11].

In this work we study the scattering of a soliton molecule by potential barriers and wells to find out, whether the quantum reflection is observed in such a setting. Of particular interest is the excitation of internal modes of soliton molecules, as a result of scattering process. We address the problem using the analytical methods and numerical simulations. The analytical method is based on the variational approximation, which is briefly described below.

2. The model and variational approach

The governing equation of our model is the one-dimensional nonlocal Gross-Pitaevskii equation (GPE), which in dimensionless units has the following form

$$i\frac{\partial\psi}{\partial t}+\frac{1}{2}\frac{\partial^2\psi}{\partial x^2}-V(x)\psi+q|\psi|^2\psi+g\psi\int\limits_{-\infty}^{+\infty}R(|x-\xi|)\,|\psi(\xi,t)|^2d\xi=0, \tag{1}$$

where $\psi(x,t)$ is the mean-field wave function of the condensate, $V(x)$ is the external potential, q and g are, respectively, the coefficients of nonlinearity, responsible for the short-range contact and long-range dipole-dipole interactions between atoms in the condensate. The wave function is normalized to the reduced number of atoms in the condensate $N=\int_{-\infty}^{+\infty}|\psi(x)|^2dx$, which is a conserved quantity of Eq. (1). The response function $R(x)$ characterizes the degree of non-locality of the medium. It shows how strongly the properties of the medium at given location depends on the properties at neighboring regions. For the contact interactions, when the particles influence each-other only when they are at the same spatial point, the response function is equal to the Dirac delta function. For the long-range dipole-dipole interactions, we approximate the response function by a Gaussian

$$R(x)=\frac{1}{\sqrt{2\pi}w}\exp\left(-\frac{x^2}{2w^2}\right), \tag{2}$$

which is normalized to one $\int_{-\infty}^{+\infty}R(x)dx=1$. This response function is simple for analytical treatment and sufficient for understanding of the problem at hand. In fact $R(x)$ is qualitatively similar to the response function, obtained from rigorous calculations [12].

To develop the variational approximation (VA) we note that the governing GPE (1) can be generated from the following Lagrangian density

$$\mathcal{L}=\frac{i}{2}(\psi\psi_t^*-\psi^*\psi_t)+\frac{1}{2}|\psi_x|^2+V(x)|\psi|^2-\frac{1}{2}q|\psi|^4-\frac{1}{2}g|\psi(x,t)|^2\int\limits_{-\infty}^{\infty}R(x-\xi)|\psi(\xi,t)|^2d\xi. \tag{3}$$

An important step in development of the VA is selection of suitable trial function. As a trial function for the two-soliton molecule we use the first Gauss-Hermite function [13, 14, 15]

$$\psi(x,t)=A(t)\cdot(x-\xi)\cdot\exp\left[-\frac{(x-\xi)^2}{2a(t)^2}+ib(t)(x-\xi)^2+iv(t)(x-\xi)+i\phi(t)\right], \tag{4}$$

where $A(t),a(t),b(t),v(t),\phi(t)$ are time-dependent variational parameters, meaning the amplitude, width, chirp parameter, velocity and phase of the soliton molecule, respectively. The velocity is defined as a time derivative of the molecule's center-of-mass position $v=\xi_t$. The norm $N=A^2a^3\sqrt{\pi}/2$, which is a conserved quantity of the governing equation, is proportional to the number of atoms in the condensate. In fact the waveform (4) can be approximated quite accurately by two anti-phase Gaussian functions (single solitons). The spatial separation between two soltion's center-of-mass positions (molecular bond length) is defined as $\Delta = 4a/\sqrt{\pi}$. Below we consider the Gaussian $V(x) = V_0 \cdot \exp\left[-x^2/(2d^2)\right]$ and Pöschl-Teller $V(x) = V_0 \cdot \mathrm{sech}^2(\alpha x)$ potentials as examples. For $V_0 > 0$ we have the potential barrier, while $V_0 < 0$ corresponds to the potential well.

Now using the response function (2) and ansatz (4), we evaluate the Lagrangian density (3). Subsequent integration over the space variable $L = \int \mathcal{L} dx$ yields the averaged Lagrangian

$$\frac{L}{N} = \frac{3}{2}a^2 b_t - \frac{1}{2}\xi_t^2 + \phi_t + \frac{3}{4a^2} + 3\,a^2 b^2 - \frac{3qN}{8\sqrt{2\pi}a} - \frac{gN}{8\sqrt{2\pi}}F(a) + 2\sqrt{2}V_0 d G(a,\xi), \tag{5}$$

where

$$F(a) = \frac{4w^2(w^2+a^2)+3a^4}{(w^2+a^2)^{5/2}}, \quad G(a,\xi) = \frac{d^2(a^2+2d^2)+a^2\xi^2}{(a^2+2d^2)^{5/2}} \exp\left(-\frac{\xi^2}{a^2+2d^2}\right).$$

In the case of a Pöschl-Teller potential the hyperbolic secant can be approximated by a Gaussian function $\mathrm{sech}^2(\alpha x) \simeq \exp(-\pi\alpha^2 x^2/4)$, which was shown to be quite accurate [16]. Therefore, for the Pöschl-Teller potential in the above and subsequent equations we can substitute $d = \sqrt{2/\pi}/\alpha$.

The Euler-Lagrange equations $d/dt(\partial L/\partial \nu_t) - \partial L/\partial \nu = 0$ for variational parameters $\nu \to a,\ b,\ \xi,\ \phi$ give the following coupled system for the width and center-of-mass position of the soliton molecule

$$a_{tt} = \frac{1}{a^3} - \frac{qN}{4\sqrt{2\pi}a^2} - \frac{gN}{4\sqrt{2\pi}} \cdot \frac{a(a^4+4w^4)}{(a^2+w^2)^{7/2}} - \frac{4\sqrt{2}V_0 d}{3} \cdot \frac{\partial G(a,\xi)}{\partial a}, \tag{6}$$

$$\xi_{tt} = -4\sqrt{2}V_0 d\,\xi \cdot \frac{a^4+a^2(d^2-\xi^2)-2d^4}{(a^2+2d^2)^{7/2}} \cdot \exp\left(-\frac{\xi^2}{a^2+2d^2}\right). \tag{7}$$

From these equations it is evident, that when the molecule is far from the location of the potential ($\xi \gg d$), the dynamics of the center-of-mass and width become independent. Namely, the center-of-mass freely moves with a constant velocity ($\xi_{tt} = 0$), and the width remains fixed in equilibrium state $a = a_0$, or oscillates with a constant frequency ω_0, if perturbed. The frequency of small amplitude oscillations can be found by expanding Eq.(6) around the fixed point $a(t) = a_0 + a_1(t),\quad a_1 \ll a_0$. Then we get the equation $a_{1tt} + \omega_0^2 a_1 = 0$, where

$$\omega_0 = \left(\frac{3}{a_0^4} - \frac{qN}{2\sqrt{2\pi}a_0^3} - \frac{gN}{2\sqrt{2\pi}} \cdot \frac{a_0^4(2a_0^2-5w^2)+4w^4(6a_0^2-w^2)}{(a_0^2+w^2)^{9/2}}\right)^{1/2}. \tag{8}$$

Decoupling of equations (6)-(7) occurs due to vanishing of the action of the potential at large distances ($\xi \gg d,\ a$), since $\exp[-\xi^2/(a^2+2d^2)] \to 0$, and as a consequence, $G(a,\xi) \to 0$.

In the following sections we shall consider the scattering of a soliton molecule by weak and strong potential barriers and wells. The sketch of numerical experiments is illustrated in Fig. 1. Comparing the predictions of the VA with the results of numerical simulations of the GPE (1) will show the accuracy of the developed mathematical model.

3. Weak potential barriers and wells

The external potential is considered to be weak if its strength V_0 is much less than the amplitude of individual solitons A_s, forming the molecule. The maxima of the molecule's waveform (4) are located at $x = \pm a$, which can be found from the condition $d|\psi|/dx = 0$ at $\xi = 0$. Then the amplitude of individual soliton is

$$A_s = Aae^{-1/2} = \sqrt{\frac{2N}{ae\sqrt{\pi}}}. \tag{9}$$

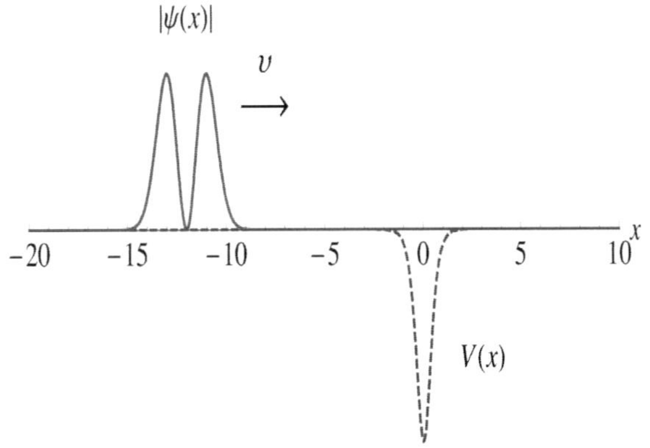

Figure 1. The sketch of numerical experiments. A two-soliton molecule is set in motion towards the external potential $V(x)$ with some initial velocity v. Depending on the strength of the potential and initial velocity, the molecule can be reflected, trapped or transmitted through the potential. If the potential is sufficiently strong and the velocity is high, the molecule can break up into individual solitons after the scattering event.

For instance, the amplitude of solitons, forming the molecule with norm $N = 2$ and parameter value $a = 1.87$, is equal to $A_s = 0.67$. These values are found from the fixed point ($a_{tt} = 0$) of Eq. (6) at $V_0 = 0$, $w = 5$, for pure dipolar condensate ($q = 0$, $g = 20$). Then the external potential with strength $|V_0| \ll A_s$ will be considered as a weak potential.

In Fig. 2 we illustrate the scattering of a two soliton molecule by weak potential barrier located at the origin ($x = 0$). At low velocity the molecule exhibits full reflection, while at greater velocity it is transmitted through the potential. As can be seen from Fig. 2b, after the interaction with the potential barrier the periodic exchange of matter between solitons takes place, which is caused by the collision induced pase shift. Deviation of the relative phase between solitons from π is responsible for the exchange of matter between them [11]. This is not accounted by the trial function (4), and therefore matter exchange cannot be described by the developed VA. Comparison with predictions of VA shows, that scattering by weak potential slightly perturbs the molecule, by exciting its internal modes. The agreement between VA and GPE is quite well for the center-of-mass dynamics, while for the width the agreement is only qualitative. These parameters are retrieved from the results of GPE simulation $\psi(x,t)$ as follows

$$\xi(t) = \frac{1}{N}\int_{-\infty}^{\infty} x\,|\psi(x,t)|^2 dx, \qquad a(t) = \left(\frac{2}{3N}\int_{-\infty}^{\infty} (x-\xi(t))^2\,|\psi(x,t)|^2 dx\right)^{1/2}. \tag{10}$$

It should be noted, that small amplitude oscillations of the width in GPE simulations before the scattering event is due to the approximate waveform of the molecule, obtained from VA. When the numerically exact waveform for the molecule is employed [11], these oscillations do not show up.

Extensive numerical experiments have shown that, scattering on weak potential barriers and wells mainly result in perturbation of the molecule, by exciting its internal modes and inducing the matter exchange between solitons. The same pertains to smooth potential barriers and wells, when the variation of the potential on the spatial scale of the molecule is small. In these limits the VA gives reliable predictions for the dynamics of soliton molecules. In particular, the period

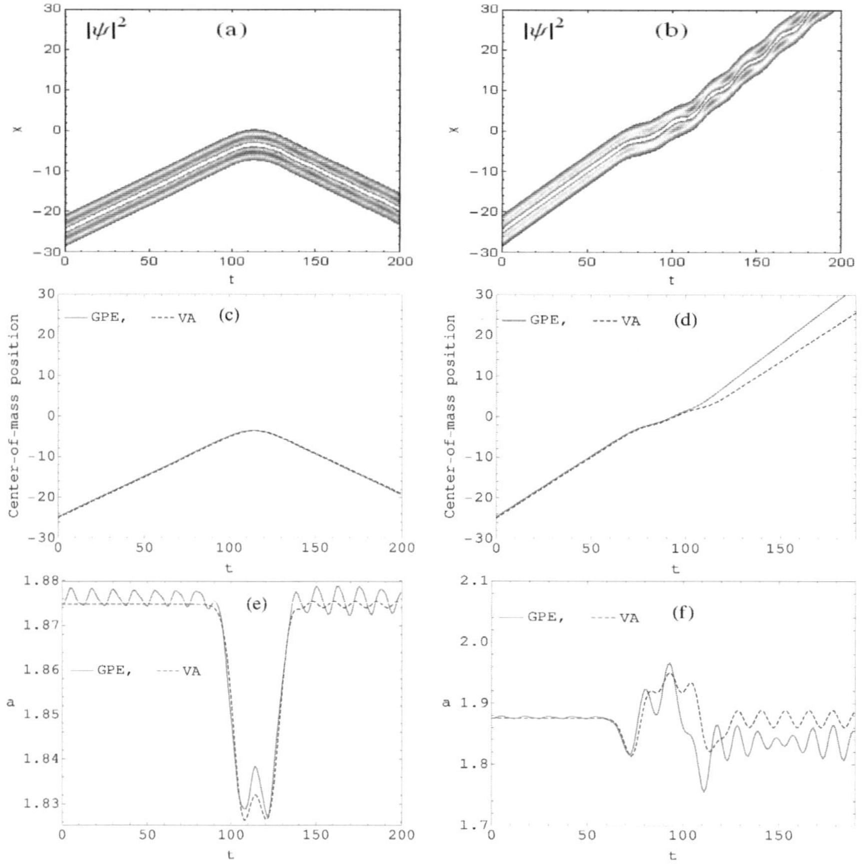

Figure 2. Scattering of a two-soliton molecule by weak potential barrier with $V_0 = 0.1$ at different velocities. Full reflection at low velocity $v = 0.2$ (a) and transmission at greater velocity $v = 0.3$ (b), according to GPE (1). Comparison GPE vs. VA for the center-of-mass position (c,d). At low velocity the two curves overlap (c), while at greater velocity there is a notable deviation after the scattering event (d). Agreement is only qualitative for the width (e,f). Initial parameter values $A = 0.585$, $a = 1.875$ are predicted by VA for $q = 0$, $g = 20$, $w = 5$, $N = 2$, $\xi_0 = -25$, $d = 1$.

of small amplitude oscillations of the width, shown in Fig. 2 (e) and (f), is very close to the VA predicted value in Eq. (8) $T = 2\pi/\omega_0 \simeq 12$.

4. Strong potential barriers and wells

In accordance with the definition, given in the previous section, by strong potential barriers and wells we presume the settings, when the strength of the potential is comparable or greater than the amplitude of individual solitons $|V_0| \gtrsim A_s$. In addition, the spatial extent of the potential is commensurate with that of the molecule $d \sim a$.

At low velocity of the soliton molecule, incident upon the strong potential barrier, the character of scattering is similar to the case of weak potential considered in the previous section. Namely, the molecule preserves its integrity after the reflection from the potential, while oscillations of solitons near their equilibrium position is excited, and matter exchange between them is induced. The dynamics of the center-of-mass and width still can be described by VA on a qualitative level. In the case of a potential well and sufficiently low velocity of the molecule, the phenomenon of quantum reflection can be observed, which will be considered in the next section. Below we study scattering of a soliton molecule by strong potential barriers and wells, at moderately large velocities. Since the molecule undergoes significant deformation during the interaction process, and can be eventually disintegrated, the agreement between GPE and VA is not expected in these conditions. Therefore, our studies in this section will be mainly based on numerical experiments.

In Fig. 3 we illustrate the main characteristic features of scattering of a two-soliton molecule by Gaussian potential wells and barriers.

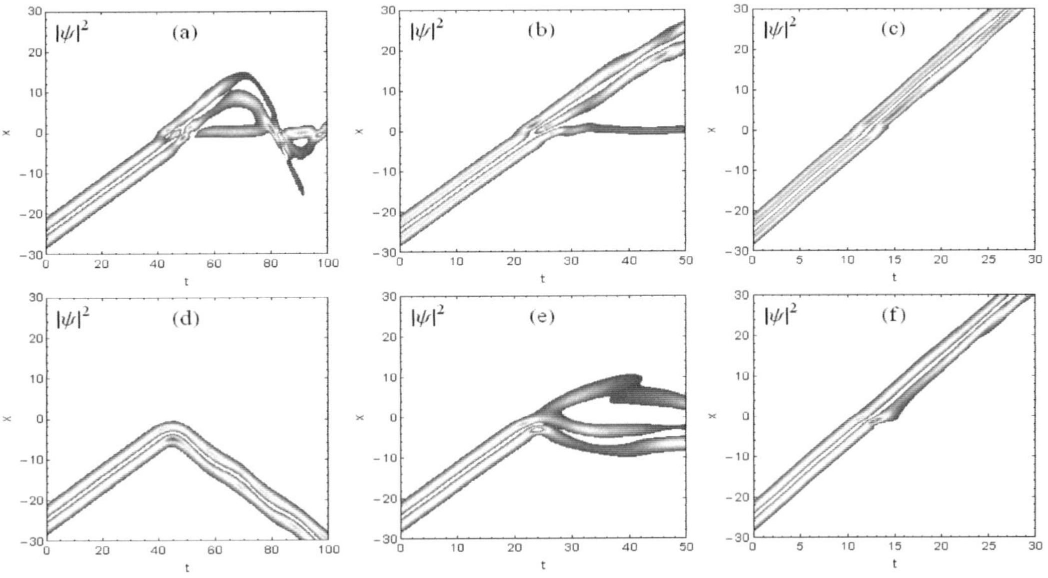

Figure 3. Scattering of a two-soliton molecule by strong potential well $V_0 = -1$ (upper row) and barrier $V_0 = 1$ (lower row) at different velocities, according to GPE (1). The molecule is destroyed by the potential well at moderately low velocity $v = 0.5$, comparable with $v_c = a_0\omega_0/\pi \simeq 0.3$ (a). At greater velocity $v = 1$ partial trapping of matter by the potential well occurs, and the remaining part traverses the well keeping its molecular structure (b). At even greater velocity $v = 2$ the molecule preserves its shape and almost intact transmission occurs (c). Low velocity collision at $v = 0.5$ with a strong potential barrier slightly perturbs the molecule (d). At greater velocity $v = 1$ the molecule is destroyed (e). At $v = 2$ the molecule can pass the barrier preserving its shape (f). Initial parameter values are similar to previous figure.

The length scales, associated with the soliton molecule and external potential are equal to $l_s = 2a$ and $l_p = d$, respectively. The greater of these two parameters defines the interaction time of the soliton molecule with the potential. In numerical simulations we use the narrow potential ($l_p < l_s$). The result of scattering depends on the ratio between the interaction time of the soliton molecule with a narrow potential $t_{int} \sim 2a_0/v$ and nonlinear relaxation time $T = 2\pi/\omega_0$. The characteristic velocity follows from the relation $t_{int} \sim T$, and is given by $v_c = a_0\omega_0/\pi$. When collision occurs at a velocity range, exceeding the quantum reflection threshold, but comparable with v_c, the molecule is destroyed by the potential well, as shown in Fig. 3a. In both cases of potential wells and barriers, collisions at high velocity ensure short interaction time, and soliton molecule can retain its coherence after the scattering event, as shown in Fig. 3 (c) and (f).

5. Quantum reflection of soliton molecules

A counterintuitive phenomenon of quantum reflection of matter-wave solitons from attractive potential wells was reported for the first time in [1]. Later the phenomenon was investigated in more details, theoretically [17] and experimentally [4]. The effect of accelerated motion of matter-wave solitons and two-soliton compounds, traversing the potential well, alongside with the quantum reflection phenomenon, was studied in [3]. Below we consider the effect of quantum reflection for two-soliton molecules.

The setting for numerical simulation of the quantum reflection is illustrated in Fig. 1. For the attractive potential well we consider the reflectionless Pöschl-Teller potential

$$V(x) = V_0 \, \mathrm{Sech}^2(\alpha x), \quad V_0 < 0. \tag{11}$$

The molecule with initial position x_0 is set in motion with velocity v towards the potential well (11). We assume that the process of interaction starts at time $t_0 \simeq x_0/v$, and finishes at $t_1 \simeq 2t_0$. Then the coefficients of reflection, transmission and trapping are calculated as follows

$$R = \frac{1}{N} \int_{-\infty}^{-h} |\psi(x,t_1)|^2 dx, \qquad T = \frac{1}{N} \int_{h}^{\infty} |\psi(x,t_1)|^2 dx, \qquad L = \frac{1}{N} \int_{-h}^{h} |\psi(x,t_1)|^2 dx, \tag{12}$$

where h denotes the position on the x-axis, at which the effect of the potential vanishes $V(h) \sim 0$, and N is the 1D norm of the soliton molecule.

The results of numerical simulations are presented in Fig. 4. For specified parameters the

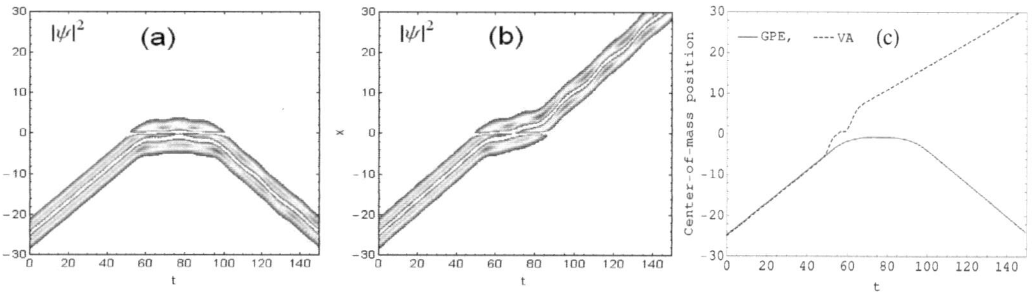

Figure 4. Quantum reflection at under-critical velocity $v = 0.40$ (a) and transmission at over-critical velocity $v = 0.42$ (b) of a two-soliton molecule, incident on the Pöschl-Teller potential well $V(x) = V_0 \, \mathrm{Sech}^2(\alpha x)$ with $V_0 = -4$ and $\alpha = 2$, according to GPE (1). Quantum reflection on panel (a) cannot be described by VA with a trial function (4), which is evident from the discrepancy after the scattering event (c).

quantum reflection of a two-soliton molecule from attractive potential takes place at velocities, below the critical value $v_q = 0.41$. At slower collision with $v = 0.40 < v_q$ we observe the full reflection ($R \simeq 1, \quad T \simeq 0$), while at slightly greater velocity $v = 0.42 > v_q$ the full transmission ($R \simeq 0, \quad T \simeq 1$) occurs. The physical mechanism behind the phenomenon of quantum reflection is linked to population of quantum states in the potential well by approaching soliton [1], as clearly observed in Fig. 4a. If the emerging localized state in the potential well has anti-phase configuration with respect to approaching soliton, they repel each-other, which results in reflection of the soliton.

Since the VA with a trial function (4) leads to equation of motion for the molecule's center-of-mass position, which is similar to equation for a particle in some external potential, it cannot describe the quantum reflection, as indicated by discrepancy after the scattering event, shown in

Fig. 4c. Incorporation of the quantum reflection into the frame of VA is possible by taking into regard the immobile trapped mode in the potential well [1, 18]. In all numerical experiments and calculations according to Eq. (12), the radiative and trapped parts of the field were negligible ($L \simeq 0$). This is the manifestation of the reflectionless property of the Pöschl-Teller potential, held in the nonlinear regime.

It should be pointed out, that the reflection from the potential well, shown in Fig. 4a would not be possible if the soliton molecule was ordinary particle, because there is no classical turning point for it in this potential. Therefore, the quantum reflection in the present context is understood as a classically forbidden reflection.

6. Conclusions

The scattering of a two-soliton molecule by Gaussian potential barriers and wells, described by the nonlocal Gross-Pitaevskii equation, has been studied by variational approximation and numerical simulations. Interaction with weak potentials gives raise to excitation of internal modes of soliton molecules and leads to matter exchange between solitons. When the potential is strong and collision velocity is moderately high, the molecule can break up and partially trapped by the potential well. At sufficiently high collision velocity the molecule again preserves its coherence, because during short interaction period with the potential the nonlinear relaxation and rearrangement processes cannot fully develop. In numerical experiments we observed the quantum reflection of a two-soliton molecule from the Pöschl-Teller potential well and found the critical velocity, at which the full reflection changes to full transmission. Predictions of the developed model, based on the variational approach, corroborate with the results of numerical simulations of the governing equation. Advantages and limitations of the variational method, applied to soliton scattering problems are discussed.

7. Acknowledgements

This work has been supported by the project FRGS 13-092-0333 of the International Islamic University Malaysia.

References

[1] Lee C and Brand J 2006 *Europhys. Lett.* **73** 321
[2] Cornish S L, Parker N G, Martin A M, Judd T E, Scott R G, Fromhold T M and Adams C S 2009 *Physica* D **238** 1299
[3] Al-Marzoug S M, Al-Amoudi S M, Al Khawaja U, Bahlouli H and Baizakov B B 2011 *Phys. Rev.* E **83** 026603
[4] Marchant A L, Billam T P, Yu M M H, Rakonjac A, Helm J L, Polo J, Weiss C, Gardiner S A and Cornish S L 2015 "Quantum reflection of bright solitary matter-waves from a narrow attractive potential" arXiv: 1507.04639
[5] Nguyen J H V, Dyke P, Luo D, Malomed B A and Hulet R G 2014 *Nature Phys.* **10** 918
[6] Rag H S and Gea-Banacloche J 2015 *Am. J. Phys.* **83** 305
[7] Stratmann M, Pagel T and Mitschke F 2005 *Phys. Rev. Lett.* **95** 143902
[8] Rohrmann P, Hause A and Mitschke F 2012 *Sci. Rep.* **2** 866
[9] Rohrmann P, Hause A and Mitschke F 2013 *Phys. Rev.* A **87** 043834
[10] Turmanov B Kh, Baizakov B B, Umarov B A and Abdullaev F Kh 2015 *Phys. Lett.* A **379** 1828
[11] Baizakov B B, Al-Marzoug S M and Bahlouli H 2015 *Phys. Rev.* A **92** 033605
[12] Sinha S and Santos L 2007 *Phys. Rev. Lett.***99** 140406
[13] Pare C and Belanger P A 1999 *Opt. Commun.* **168** 103
[14] Feng B F and Malomed B A 2004 *Opt. Commun.* **229** 173
[15] Alamoudi S M, Al Khawaja U and Baizakov B B 2014 *Phys. Rev.* A **89** 053817
[16] Umarov B A, Messikh A, Regaa N and Baizakov B B 2013 *J. Phys. Conf. Ser.* **435** 012024
[17] Ernst Th and Brand J 2010 *Phys. Rev.* A **81** 033614
[18] Goodman R H, Holmes P J and Weinstein M I 2004 Physica D **192** 215

Journal of Physics: Conference Series **697** (2016) 012024 doi:10.1088/1742-6596/697/1/012024

On one system of the Burgers equations arising in the two-velocity hydrodynamics

Kholmatzhon Imomnazarov[1], **Baxtier Mamasoliyev**[2] **Georgy Vasiliev**[3]

[1,3] Institute of Computational Mathematics and Mathematical Geophysics, Siberian Branch of the RAS, Novosibirsk,

[2] National University of Uzbekistan, Mechanics and Mathematics Faculty, Tashkent

E-mail: `E-mail: imom@omzg.sscc.ru`

Abstract. A system of the Burgers equations of the two-velocity hydrodynamics is constructed. We consider the Cauchy problem in the case of a one-dimensional system. We have obtained a formula for solving the Cauchy problem and the estimate of the stability of this solution. It is shown that with disappearance of the kinetic friction coefficient, which is responsible for the energy dissipation, this formula turns to the famous Cauchy problem for the one-dimensional Burgers equation. The existence and uniqueness of solutions to the Cauchy problem for the one-dimensional systems of the Burgers type are proved using the method of weak approximation.

1. Introduction

In recent decades, mathematicians have become increasingly interested in the problems associated with the behavior of solutions to systems of partial differential equations with a small parameter in high derivatives with allowance for the kinetic parameters. These problems arose from the physical applications, mostly from contemporary hydrodynamics (compressible multiphase fluids with low viscosity). An analog to the Burgers equation arises, for example, in studying a weak nonlinear one-dimensional acoustic wave moving with a linear velocity of sound. In this case, nonlinear velocity terms in the system of the Burgers equations come from the sound velocity depending on the amplitude of the sound wave, on the second derivative terms and on the difference in the velocities representing the attenuation of sound waves associated with energy dissipation. In other words, these terms provide the continuity of solutions and are dissipative processes associated with the production of entropy. These terms in turn provide non-roll waves [9]. The system under consideration is a special case of a two-velocity system of hydrodynamics equations [4]-[6],[10, 13].

A one-dimensional analog of the Navier-Stokes equations for a compressible fluid can be considered as a system of the Burgers equations which is a system of nonlinear convection-diffusion equations [8].

$$u_t + u\,u_x = \nu\,u_{xx} - \tilde{b}\,(u - \tilde{u}), \tag{1}$$

$$\tilde{u}_t + \tilde{u}\,\tilde{u}_x = \tilde{\nu}\,\tilde{u}_{xx} + b\,(u - \tilde{u}), \tag{2}$$

where the quantities u and $\tilde{u}$ can be regarded as velocity subsystems with the dimension $[x]/[t]$, constituting two-velocity components with the corresponding partial continuum densities ρ and

$\tilde{\rho}$, $\bar{\rho} = \tilde{\rho} + \rho$ is the common density of the continuum, $\tilde{b} = \frac{\tilde{\rho}}{\rho} b$, b is the coefficient of friction with the dimension $1/[t]$, which is the analog to the Darcy factor for porous media. The positive constants ν and $\tilde{\nu}$ play the role of kinematic viscosities subsystems of the dimension $[x]^2/[t]$.

A two-velocity system of hydrodynamic equations and a system of the Burgers equations have much in common. For example, the quadratic nonlinearity with respect to the terms u and $\tilde{u}$ with advective terms corresponding to the sound, depending on the amplitude of the sound waves and linear viscosities ν, $\tilde{\nu}$, the coefficient of friction b of the right-hand side is responsible for the attenuation of the sound waves [9]. With regard to the properties of the solutions, they are totally different. The system of the Burgers equations in the vanishing coefficients ν, $\tilde{\nu}$, b is formed of both strong (shock waves), and weak discontinuities, while the solution of the system of two-velocity hydrodynamics does not have such features. However, the range of applicability of this system is by no means limited to the above example. These systems arise in many problems, thus confirming their significance.

2. The Cauchy problem for the system of the Burgers type equations

For system (1),(2) in the domain $\Gamma_{[0,T]} = \{(t,x) : 0 \le t \le T,\ x \in R\}$ let us consider the Cauchy problem with the following initial data

$$u|_{t=0} = u_0(x), \qquad \tilde{u}|_{t=0} = \tilde{u}_0(x), \quad x \in R. \tag{3}$$

We are interested in the classical solution of the Cauchy problem for the system of the Burgers equations (1),(2), namely, $u(t,x)$, $\tilde{u}(t,x) \in C^{1,2}(\Gamma_{[0,T]})$ is the class of functions once continuously differentiable with respect to t and twice continuously differentiable with respect to x. Further assume that the inequality $\tilde{\rho} < \rho$.

Theorem 2.1 *Let $u_0(x)$, $\tilde{u}_0(x) \in C^2(R)$. Then the Cauchy problem (1) - (3) has in the class $C^{1,2}(\Gamma_{[0,T]})$ a unique solution, and the following estimate of stability holds*

$$\int_\Omega (u^2(t,x) + \tilde{u}^2(t,x))\, dx dt \le \frac{1}{2\tilde{b}}\, m(0)\, \mathrm{e}^{2bT}. \tag{4}$$

Proof. Multiply both sides of equations (1) and (2) by u and $\tilde{u}$, respectively. After integration with respect to x and after simple transformations we obtain

$$\frac{1}{2}\frac{\partial}{\partial t}\int_{-\infty}^{\infty} u^2\, dx = -\nu \int_{-\infty}^{\infty} (u_x)^2\, dx - \tilde{b}\int_{-\infty}^{\infty} (u^2 - u\,\tilde{u})\, dx, \tag{5}$$

$$\frac{1}{2}\frac{\partial}{\partial t}\int_{-\infty}^{\infty} \tilde{u}^2\, dx = -\tilde{\nu} \int_{-\infty}^{\infty} (\tilde{u}_x)^2\, dx + b\int_{-\infty}^{\infty} (u\,\tilde{u} - \tilde{u}^2)\, dx. \tag{6}$$

Hence, multiplying (5) by ρ, (6) by $\tilde{\rho}$ and summing the result obtained, after simple transformations we arrive at

$$m(t) \le \frac{\rho}{\tilde{\rho}}\, m(0) + 2\,b \int_0^t m(\tau) d\tau,$$

where $m(t) = \int_{-\infty}^{\infty} (u^2(t,x) + \tilde{u}^2(t,x))\, dx$.

From this and the Gronwall-Bellman inequality [2] follows

$$m(t) \le \frac{\rho}{\tilde{\rho}}\, m(0)\, \mathrm{e}^{2bt}.$$

Hence, integrating from 0 to t, obtain (4). The uniqueness is proved in a standard way.

3. The formula for solving the Cauchy problem for the Burgers type system of equations

It is convenient to use the Florin-Hopf-Cole transformation [9]

$$\phi = \exp\left[-\frac{1}{2\nu}\int u\,dx\right], \qquad \psi = \exp\left[-\frac{1}{2\tilde{\nu}}\int \tilde{u}\,dx\right],$$

herewith the functions u and v are expressed in terms of the functions ϕ and ψ by the formulas

$$u = -2\,\nu\frac{\phi_x}{\phi}, \qquad \tilde{u} = -2\,\tilde{\nu}\frac{\psi_x}{\psi}.$$

In terms of the functions ϕ and ψ the system of dynamic equations (1) and (2) takes the form

$$\left(\frac{\phi_t}{\phi}\right)_x = \left(\nu\,\frac{\phi_{xx}}{\phi}\right)_x - \frac{\tilde{b}}{\nu}\left(\ln\frac{\phi^{\nu}}{\psi^{\tilde{\nu}}}\right)_x,$$

$$\left(\frac{\psi_t}{\psi}\right)_x = \left(\tilde{\nu}\,\frac{\psi_{xx}}{\phi}\right)_x + \frac{b}{\tilde{\nu}}\left(\ln\frac{\phi^{\nu}}{\psi^{\tilde{\nu}}}\right)_x.$$

Hence, after the integration with respect to x, we obtain

$$\phi_t = \nu\,\phi_{xx} - \frac{\tilde{b}}{\nu}\,\phi\,(\nu\,\ln\phi - \tilde{\nu}\,\ln\psi) + C_1(t)\,\phi, \tag{7}$$

$$\psi_t = \tilde{\nu}\,\psi_{xx} + \frac{b}{\nu}\,\psi\,(\nu\,\ln\phi - \tilde{\nu}\,\ln\psi) + C_2(t)\,\psi, \tag{8}$$

where $C_1(t)$ and $C_2(t)$ are arbitrary temporal functions.

Introducing the new functions $\tilde{\phi}$ and $\tilde{\psi}$ by the formulas

$$\phi = \tilde{\phi}\,\exp\left[\int C_1\,dt\right], \qquad \psi = \tilde{\psi}\,\exp\left[\int C_2\,dt\right].$$

from (7) and (8) we obtain

$$\tilde{\phi}_t = \nu\,\tilde{\phi}_{xx} - \frac{\tilde{b}}{\nu}\,\tilde{\phi}\left(\nu\,\ln\left[\tilde{\phi}\,\exp\left(\int C_1\,dt\right)\right] - \tilde{\nu}\,\ln\left[\tilde{\psi}\,\exp\left(\int C_2\,dt\right)\right]\right), \tag{9}$$

$$\tilde{\psi}_t = \tilde{\nu}\,\tilde{\psi}_{xx} + \frac{b}{\tilde{\nu}}\,\tilde{\psi}\left(\nu\,\ln\left[\tilde{\phi}\,\exp\left(\int C_1\,dt\right)\right] - \tilde{\nu}\,\ln\left[\tilde{\psi}\,\exp\left(\int C_2\,dt\right)\right]\right). \tag{10}$$

For simplicity, assume the functions $C_k(t)\,(k = 1, 2)$ to be equal to zero.

The Cauchy problem for system (9), (10) with the data

$$\phi|_{t=0} = \phi_0(x), \quad \psi|_{t=0} = \psi_0(x)$$

has the form

$$\phi(t, x) = \int_{-\infty}^{\infty} G^{\nu}(x, \xi, t)\,\phi_0(\xi)d\xi$$

$$-\frac{\tilde{b}}{\nu}\int_0^t\int_{-\infty}^{\infty} G^{\nu}(x, \xi, t-\tau)\,\phi(\xi, \tau)\;(\nu\,\ln\phi(\xi, \tau) - \tilde{\nu}\,\ln\psi(\xi, \tau))\;d\xi\,d\tau, \tag{11}$$

Journal of Physics: Conference Series **697** (2016) 012024 doi:10.1088/1742-6596/697/1/012024

$$\psi(t,x) = \int_{-\infty}^{\infty} G^{\tilde{\nu}}(x,\xi,t)\,\psi_0(\xi)d\xi$$

$$+\frac{b}{\nu}\int_0^t\int_{-\infty}^{\infty} G^{\tilde{\nu}}(x,\xi,t-\tau)\,\psi(\xi,\tau)\,(\nu\,\ln\phi(\xi,\tau)-\tilde{\nu}\,\ln\psi(\xi,\tau))\,d\xi\,d\tau, \tag{12}$$

where

$$G^a(x,\xi,t) = \frac{1}{\sqrt{4\pi\,a\,t}}\mathrm{e}^{-\frac{(x-\xi)^2}{4\,a\,t}}$$

is the fundamental solution of the one-dimensional heat equation.

Next, assume as in [7] that the Cauchy data $u_0(x)$, $\tilde{u}_0(x)$ for large $|x|$ satisfy the following conditions

$$\int_0^x u_0(\xi)d\xi = o(x^2), \qquad \int_0^x \tilde{u}_0(\xi)d\xi = o(x^2). \tag{13}$$

Let us introduce the following functions

$$F(u,x,y,t) = \frac{(x-y)^2}{2t} + \int_0^y u(t,\eta)d\eta. \tag{14}$$

$$F_1(u,x,y,t,\tau) = \frac{(x-y)^2}{2(t-\tau)} + \int_0^y u(\tau,\eta)d\eta. \tag{15}$$

Note that the following equalities for any x, y, t, τ are valid

$$F(u_0,x,y,t) - F(\tilde{u}_0,x,y,t) = \int_0^y [u_0(\eta)-\tilde{u}_0(\eta)]d\eta.$$

$$F_1(u,x,y,t,\tau) - F_1(\tilde{u},x,y,t,\tau) = \int_0^y [u(\tau,\eta)-\tilde{u}(\tau,\eta)]d\eta.$$

Differentiate both sides of equalities (11) and (12) with respect to x. After simple transformations we obtain

$$\phi_x(t,x) = -\frac{1}{2\nu}\int_{-\infty}^{\infty} u_0(\xi)G^{\nu}(x,\xi,t)\,\exp\left[-\frac{1}{2\nu}\int_0^{\xi} u_0(\eta)d\eta\right]d\xi$$

$$+\frac{\tilde{b}}{2\nu}\int_0^t\int_{-\infty}^{\infty}\frac{x-\xi}{t-\tau}G^{\nu}(x,\xi,t-\tau)\,\exp\left[-\frac{1}{2\nu}\int_0^{\xi} u(\tau,\eta)d\eta\right]\int_0^{\xi}[u(\tau,\eta)-\tilde{u}(\tau,\eta)]d\eta\,d\xi\,d\tau,$$

$$\psi_x(t,x) = -\frac{1}{2\tilde{\nu}}\int_{-\infty}^{\infty} \tilde{u}_0(\xi)G^{\tilde{\nu}}(x,\xi,t)\,\exp\left[-\frac{1}{2\tilde{\nu}}\int_0^{\xi} \tilde{u}_0(\eta)d\eta\right]d\xi$$

$$-\frac{b}{2\tilde{\nu}}\int_0^t\int_{-\infty}^{\infty}\frac{x-\xi}{t-\tau}G^{\tilde{\nu}}(x,\xi,t-\tau)\,\exp\left[-\frac{1}{2\tilde{\nu}}\int_0^{\xi} \tilde{u}(\tau,\eta)d\eta\right]\int_0^{\xi}[u(\tau,\eta)-\tilde{u}(\tau,\eta)]d\eta\,d\xi\,d\tau.$$

Hence, taking into account (14), (15) and the definition of the fundamental solution of the operator conductivity we obtain

$$\phi_x(t,x) = -\frac{1}{2\nu\sqrt{4\pi\nu\,t}}\int_{-\infty}^{\infty}\frac{x-\xi}{t}\,\exp\left[-\frac{1}{2\nu}F(u_0,x,\xi,t)\right]d\xi$$

$$+\frac{\tilde{b}}{2\nu}\int_0^t\int_{-\infty}^{\infty}\frac{1}{\sqrt{4\pi\nu\,(t-\tau)}}\,\frac{x-\xi}{t-\tau}\,F_2(u,\tilde{u},x,\xi,t-\tau,\tau)\,d\xi\,d\tau, \tag{16}$$

$$\psi_x(t,x) = -\frac{1}{2\tilde{\nu}\sqrt{4\pi\tilde{\nu}\,t}} \int_{-\infty}^{\infty} \frac{x-\xi}{t}\, Exp\left[-\frac{1}{2\tilde{\nu}}\, F(\tilde{u}_0, x, \xi, t)\right] d\xi$$

$$-\frac{b}{2\tilde{\nu}} \int_0^t \int_{-\infty}^{\infty} \frac{1}{\sqrt{4\pi\tilde{\nu}\,(t-\tau)}} \frac{x-\xi}{t-\tau}\, G_2(u, \tilde{u}, x, \xi, t-\tau, \tau)\, d\xi\, d\tau. \tag{17}$$

In formulas (16) and (17) the following notation is used

$$F_2(u, v, x, \xi, t-\tau, \tau) = \exp\left[-\frac{1}{2\nu}\, F_1(u, x, \xi, t-\tau, \tau)\right] \int_0^{\xi} [u(\tau,\eta) - v(\tau,\eta)] d\eta$$

$$G_2(u, v, x, \xi, t-\tau, \tau) = \exp\left[-\frac{1}{2\tilde{\nu}}\, F_1(v, x, \xi, t-\tau, \tau)\right] \int_0^{\xi} [u(\tau,\eta) - v(\tau,\eta)] d\eta.$$

Thus, we come to the following theorem.

Theorem 3.1 *Let $u_0(x)$, $\tilde{u}_0(x)$ be measurable functions that satisfy relation (13). Then for the solution of the Cauchy problem (1) - (3) we have the formula*

$$u(t,x) = \frac{\int_{-\infty}^{\infty} \frac{x-\xi}{t}\, Exp\left[-\frac{1}{2\nu}\, F(u_0, x, \xi, t)\right] d\xi - \int_0^t \int_{-\infty}^{\infty} \frac{\tilde{b}\sqrt{t}}{\sqrt{t-\tau}} \frac{x-\xi}{t-\tau}\, F_2(u, \tilde{u}, x, \xi, t-\tau, \tau)\, d\xi\, d\tau}{\int_{-\infty}^{\infty} Exp\left[-\frac{1}{2\nu}\, F(u_0, x, \xi, t)\right] d\xi - \int_0^t \int_{-\infty}^{\infty} \frac{\tilde{b}\sqrt{t}}{\sqrt{t-\tau}}\, F_2(u, \tilde{u}, x, \xi, t-\tau, \tau)\, d\xi\, d\tau},$$

$$\tilde{u}(t,x) = \frac{\int_{-\infty}^{\infty} \frac{x-\xi}{t}\, Exp\left[-\frac{1}{2\tilde{\nu}}\, F(\tilde{u}_0, x, \xi, t)\right] d\xi + \int_0^t \int_{-\infty}^{\infty} \frac{b\sqrt{t}}{\sqrt{t-\tau}} \frac{x-\xi}{t-\tau}\, G_2(u, \tilde{u}, x, \xi, t-\tau, \tau)\, d\xi\, d\tau}{\int_{-\infty}^{\infty} Exp\left[-\frac{1}{2\tilde{\nu}}\, F(v_0, x, \xi, t)\right] d\xi + b\int_0^t \int_{-\infty}^{\infty} \frac{\sqrt{t}}{\sqrt{t-\tau}}\, G_2(u, \tilde{u}, x, \xi, t-\tau, \tau)\, d\xi\, d\tau}.$$

Corollary 3.2 *Let $u_0(x)$, $\tilde{u}_0(x)$ satisfy the conditions of the theorem. Then for solving the Cauchy problem (1) - (3) we have the formula*

$$u(t,x) = \frac{\int_{-\infty}^{\infty} \frac{x-\xi}{t} \exp\left[-\frac{1}{2\nu}\, F(u_0, x, \xi, t)\right] d\xi}{\int_{-\infty}^{\infty} Exp\left[-\frac{1}{2\nu}\, F(u_0, x, \xi, t)\right] d\xi}$$

$$+\tilde{b}\frac{\int_0^t \int_{-\infty}^{\infty} \sqrt{1+\frac{\tau}{t-\tau}} \left(u(t,x) - \frac{x-\xi}{t-\tau}\right) F_2(u, \tilde{u}, x, \xi, t-\tau, \tau)\, d\xi\, d\tau}{\int_{-\infty}^{\infty} Exp\left[-\frac{1}{2\nu}\, F(u_0, x, \xi, t)\right] d\xi}, \tag{18}$$

$$\tilde{u}(t,x) = \frac{\int_{-\infty}^{\infty} \frac{x-\xi}{t} \exp\left[-\frac{1}{2\tilde{\nu}}\, F(\tilde{u}_0, x, \xi, t)\right] d\xi}{\int_{-\infty}^{\infty} Exp\left[-\frac{1}{2\tilde{\nu}}\, F(\tilde{u}_0, x, \xi, t)\right] d\xi}$$

$$-b\frac{\int_0^t \int_{-\infty}^{\infty} \sqrt{1+\frac{\tau}{t-\tau}} \left(\tilde{u}(t,x) - \frac{x-\xi}{t-\tau}\right) G_2(u, \tilde{u}, x, \xi, t-\tau, \tau)\, d\xi\, d\tau}{\int_{-\infty}^{\infty} Exp\left[-\frac{1}{2\tilde{\nu}}\, F(\tilde{u}_0, x, \xi, t)\right] d\xi}. \tag{19}$$

Remark 3.3 *With disappearance of the friction coefficient b (in the absence of dissipation energy due to friction) solution (18), (19) turns to the famous Cauchy problem for the Burgers equation [7].*

4. The method of weak approximation for the Cauchy problem for the Burgers type system of equations

Let us consider problem (1) - (3) relative to the Cauchy data u_0, $\tilde{u}_0$, assuming that u_0, $\tilde{u}_0 \in C^2(R)$ and

$$\left|\frac{d^n u_0(x)}{dx^n}\right| \le c_n, \quad \left|\frac{d^n \tilde{u}_0(x)}{dx^n}\right| \le \tilde{c}_n, \quad x \in R, \quad n = 0, 1, 2, \tag{20}$$

where c_n, $\tilde{c}_n$ are some non-negative constants.

First, we consider the case of infinitely differentiable Cauchy data. Let us assume that u_0, $\tilde{u}_0 \in C^\infty(R)$ and

$$\left|\frac{d^n u_0(x)}{dx^n}\right| \le c_n, \quad \left|\frac{d^n \tilde{u}_0(x)}{dx^n}\right| \le \tilde{c}_n, \quad x \in R, \quad n = 0, 1, \dots, \tag{21}$$

Following [1],[3], we use a weak approximation of the Cauchy problem (1) - (3) with the help of a series of problems

$$u_t^\tau = 3\,\nu\, u_{xx}^\tau, \quad \tilde{u}_t^\tau = 3\,\tilde{\nu}\,\tilde{u}_{xx}^\tau, \quad n\tau < t \le \left(n + \frac{1}{3}\right)\tau, \tag{22}$$

$$u_t^\tau + 3\,u^\tau\, u_x^\tau = 0, \quad \tilde{u}_t^\tau + 3\,\tilde{u}^\tau\,\tilde{u}_x^\tau = 0, \quad \left(n + \frac{1}{3}\right)\tau < t \le \left(n + \frac{2}{3}\right)\tau, \tag{23}$$

$$u_t^\tau = -3\,\tilde{b}\,(u^\tau - \tilde{u}^\tau), \quad \tilde{u}_t^\tau = 3\,b\,(u^\tau - \tilde{u}^\tau), \quad \left(n + \frac{2}{3}\right)\tau < t \le (n+1)\tau, \tag{24}$$

$$u^\tau(0, x) = u_0(x), \quad \tilde{u}^\tau(0, x) = \tilde{u}_0(x) \tag{25}$$

where $N\,\tau = t^*$, provided that N is integer exceeding 1, $n = 0, 1, ..., N-1$, and the constant t^* satisfies equality (30).

Remark 4.1 *When building the solution of (22) - (25) at the first fractional step we solve the Cauchy problem for the heat equation, and at the second fractional step – the Cauchy problem for the transport equation*

$$v_t + 3\,v\,v_x = 0, \tag{26}$$

and at the third fractional step we solve the Cauchy problem for systems of ordinary differential equations.

It is known that in the case of the Cauchy problem for equation (26) with the initial data

$$v(0, x) = v_0(x) \tag{27}$$

bounded together with their derivatives, the solution may be a gradient catastrophe, that is, there can exist $t_1 > 0$, such that the classical solution v of this problem would exist in the domain $\Gamma_{[0,\,t_1)}$, be bounded in this domain, but the derivative v_x in the neighborhood of a point $(t_1,\, x^0)$ is becoming unbounded: $v_x(t, x) \to \infty$ with $t \to t_1$, $x \to x^0$ [9, 11, 2].

It is easy to show that if

$$\left|\frac{dv_0(x)}{dx}\right| \le c_1, \tag{28}$$

the classical solution of problem (26), (27) is in the domain $\Gamma_{[0,\,t^*]}$, it is bounded and the following estimate holds

$$\mid v_x(t,x) \mid \leq \frac{c_1}{1-3\,c_1\,t}, \quad (t,x) \in \Gamma_{[0,\,t^*]}, \tag{29}$$

where t satisfies the inequality

$$1 - 3\,c_1\,t^* > 0.$$

Let relations (21) be performed, and the constants c_1, $\tilde{c}_1$ and t^* satisfy the conditions

$$1 - c_1\,t^* > 0, \quad 1 - \tilde{c}_1\,t^* > 0. \tag{30}$$

Then the solution of u^τ and $\tilde{u}^\tau$ in the domain $\Gamma_{[0,\,t^*]}$ exists and is bounded together with all its derivatives with respect to the variables t, x.

It is obvious that for any fixed τ, the solution of u^τ and $\tilde{u}^\tau$ of problem (22)-(25) is bounded regardless of the value τ:

$$\mid u^\tau(t,x) \mid \leq c_0, \quad \mid \tilde{u}^\tau(t,x) \mid \leq \tilde{c}_0. \tag{31}$$

Repeating the argument from [1], we can show the boundedness of the derivative solutions u^τ and $\tilde{u}^\tau$ of any order with respect to x :

$$\left| \frac{\partial^k u^\tau(t,x)}{\partial x^k} \right| \leq C_k, \quad \left| \frac{\partial^k \tilde{u}^\tau(t,x)}{\partial x^k} \right| \leq \tilde{C}_k, \quad (t,x) \in \Gamma_{[0,\,t^*]}, \quad k = 0, 1, \ldots, \tag{32}$$

where C_k, $\tilde{C}_k$ are some positive constants such that $C_0 = c_0$, $\tilde{C}_0 = \tilde{c}_0$.

Independent of τ, from expressions (31), (32) and from equations (22) - (24) follow the estimates:

$$\left| \frac{\partial^{k+1} u^\tau(t,x)}{\partial t \partial x^k} \right| \leq s_k, \quad \left| \frac{\partial^{k+1} \tilde{u}^\tau(t,x)}{\partial t \partial x^k} \right| \leq \tilde{s}_k, \quad (t,x) \in \Gamma_{[0,\,t^*]}, \quad k = 0, 1, \ldots,$$

These estimates suggest that u^τ, $\tilde{u}^\tau$ and their derivatives with respect to x of any order are uniformly bounded and equicontinuous in $\Gamma_{[0,\,t^*]}$. By the Arzela Theorem [2], using the diagonal method we can choose subsequences $\{u^{\tau_k}\}$, $\{\tilde{u}^{\tau_k}\}$ sequences $\{u^\tau\}$, $\{\tilde{u}^\tau\}$, converging in $\Gamma_{[0,\,t^*]}$ to the functions u and $\tilde{u}$, respectively, together with all the derivatives with respect to x, uniform in every bounded region of the domain $\Gamma_{[0,\,t^*]}$, whereby the functions u and $\tilde{u}$ have derivatives of any order with respect to x and satisfy the following relations

$$u(0,x) = u_0(x), \qquad \tilde{u}(0,x) = \tilde{u}_0(x),$$

$$\left| \frac{\partial^k u(t,x)}{\partial x^k} \right| \leq C_k, \quad \left| \frac{\partial^k \tilde{u}(t,x)}{\partial x^k} \right| \leq \tilde{C}_k, \quad (t,x) \in \Gamma_{[0,\,t]}, \quad k = 0, 1, \ldots .$$

The uniqueness of the solution is proved in a standard way. Consequently, the sequences of the functions $\{u^\tau\}$, $\{\tilde{u}^\tau\}$ with $\tau \to 0$, uniformly converge in the domain $\Gamma_{[0,\,t]}$ to u and $\tilde{u}$, respectively, together with all their derivatives. The case when u_0, $\tilde{u}_0 \in C^2(R)$ is proved using the average functions [12].

Acknowledgements

The authors are grateful to a reviewer for his valuable remarks, which, in our opinion, appeared to be useful for improving the text of the manuscript.

Journal of Physics: Conference Series **697** (2016) 012024 doi:10.1088/1742-6596/697/1/012024

References

[1] Belov Yu Ya and Demidov G V 1970 In book: *Chislennye metody mekhaniki sploshnoy sredy* (Novosibirsk: Comp. Center USSR Acad. Sci.) **1** 3 (in Russian)

[2] Belov Yu Ya and Kantor S A 1999 *The method of weak approximation* (Krasnoyarsk: Krasnoyarsk state University Press) (in Russian)

[3] Demidov G V and Yanenko N N 1978 *Proc. All-Union Conf. on Partial Diff. Eqs* (Moscow) **1** 100 (in Russian)

[4] Dorovsky V N 1989 *Geologiya i Geofisika.* No. **7** 39

[5] Dorovsky V N and Perepechko Yu V 1992 *J. Appl. Mech. and Techn. Phys.* **33** 94

[6] Dorovsky V N and Perepechko Yu V 1989 *Geologiya i Geofisika.* **9** 56

[7] Hopf E 1950 *Communs Pure and Appl. Math.* **3** 201

[8] Imomnazarov Kh Kh and Mamasoliyev B Zh 2015 In book: *Abst. International Sci. Conf. Modern Methods of Math. Phys. Appl.* (Tashkent) 167 (in Russian)

[9] Kulikovskii A G, Sveshnikov E I and Chugaynova A P 2010 *Mathematical methods for studying discontinuous solutions of nonlinear hyperbolic systems of equations* (Moscow) (in Russian)

[10] Perepechko Yu V, Sorokin K E and Imomnazarov Kh Kh 2014 *Proc. of XVII International conference Modern Problems of Continuum Mechanics (Rostov-on-Don)* 166 (in Russian)

[11] Raputa V F 1975 In book: *Chislennye metody mekhaniki sploshnoy sredy. Novosibirsk: Computing Center of the USSR Academy of Sciences* **6** 93 (in Russian)

[12] Sobolev S L 1991 *Some Applications of Functional Analysis to Mathematical Physics* (3rd edition. American Mathematical Siciety)

[13] Zhabborov N M and Imomnazarov Kh Kh 2012 *Some initial boundary value problems of mechanics of two-velocity media* (Tashkent: National University of Uzbekistan Press) (in Russian)

Spectra of the energy operator of four-electron systems in the triplete state in the Hubbard model

Sadulla M. Tashpulatov

Institute of Nuclear Physics of Academy of Sciences of the Republic of Uzbekistan, Tashkent, Uzbekistan

E-mail: `sadullatashpulatov@yandex.ru, toshpul@inp.uz`

Abstract. We investigate spectral properties of a four-electron system in the Hubbard Model framework in the $\nu-$ dimensional lattice Z^{ν}. We prove that the essential spectrum of the system in a quintet state consists of a single segment and the four-electron bound state or four-electron anti-bound state is absent. We show that the essential spectrum of the system in a triplete states is the union of at most three segments. We also prove that four-electron bound states or a four-electron anti-bound state exists in triplete states. We prove that in the system exists three triplete states and their spectrum are different.

1. Introduction

In the early 1970s, three papers [2, 3, 8], where a simple model of a metal was proposed that has bocome a fundamental model in the theory of strongly correlated electron systems. Note that these papers appeared almost simultaneously and independently. In that model, a single nondegenerate electron band with a local Coulomb interaction was considered. The model Hamiltonian contains only two parameters: the matrix element t of electron hopping from a lattice site to a neighboring site and the parameter U of the one-site Coulomb repulsion of two-electrons. In the secondary quantization representation, the Hamiltonian can be written as follows

$$H = t \sum_{m,\tau,\gamma} a^{+}_{m,\gamma} a_{m+\tau,\gamma} + U \sum_{m} a^{+}_{m,\uparrow} a_{m,\uparrow} a^{+}_{m,\downarrow} a_{m,\downarrow}, \quad (1)$$

where $a^{+}_{m,\gamma}$ and $a_{m,\gamma}$ denote Fermi operators of creation and annihilation of an electron with spin γ on a site m and the summation over τ means summation over the nearest neighbors on the lattice.

The model proposed in [3, 2, 8] was called the Hubbard model after John Hubbard, who made a fundamental contribution to studying the statistical mechanics of that system, although the local form of Coulomb interaction was first introduced for an impurity model in a metal by Anderson [1]. We also recall that the Hubbard model is a particular case of the Shubin-Wonsowsky polaron model [15], which had appeared 30 years before [3, 2, 8]. In the Shubin-Wonsowsky model, along with the on-site Coulomb interaction, the interaction of electrons on neighboring sites is also taken into account.

The Hubbard model is an approximation used in solid state physics to describe the transition between conducting and insulating states. It is the simplest model describing particle interaction on a lattice. Its Hamiltonian contains only two terms: the kinetic term corresponding to the tunneling (hopping) of particles between lattice sites and a term corresponding to the on-site interaction. Particles can be fermions, as in Hubbards original work, and also bosons. The simplicity and sufficiency of Hamiltonian (1) have made the Hubbard model very popular and effective for describing strongly correlated electron systems.

The Hubbard model well describes the behavior of particles in a periodic potential at sufficiently low temperatures such that all particles are in the lower Bloch band and long-range interactions can be neglected. If the interaction between particles at different sites is taken into account, then the model is often called the extended Hubbard model. It was proposed for describing electrons in solids, and it remains especially interesting since then for studying high-temperature superconductivity. Later, the extended Hubbard model also found applications in describing the behavior of ultracold atoms in optical lattices.

In considering electrons in solids, the Hubbard model can be considered a sophisticated version of the model of strongly bound electrons, involving only the electron hopping term in the Hamiltonian. In the case of strong interactions, these two models can give essentially different results. The Hubbard model exactly predicts the existence of so-called Mott insulators, where conductance is absent due to strong repulsion between particles.

The Hubbard model is based on the approximation of strongly coupled electrons. In the strongcoupling approximation, electrons initially occupy orbitals in atoms (lattice sites) and then hop over to other atoms, thus conducting the current. Mathematically, this is represented by the so-called hopping integral. This process can be considered the physical phenomenon underlying the occurrence of electron bands in crystal materials. But the interaction between electrons is not considered in more general band theories. In addition to the hopping integral, which explains the conductance of the material, the Hubbard model contains the so-called on-site repulsion, corresponding to the Coulomb repulsion between electrons. This leads to a competition between the hopping integral, which depends on the mutual position of lattice sites, and the on-site repulsion, which is independent of the atom positions. As a result, the Hubbard model explains the metalinsulator transition in oxides of some transition metals. When such a material is heated, the distance between nearest-neighbor sites increases, the hopping integral decreases, and on-site repulsion becomes dominant.

The Hubbard model is currently one of the most extensively studied multielectron models of metals [9, 10, 11, 17, 4]. But little is known about exact results for the spectrum and wave functions of the crystal described by the Hubbard model, and obtaining the corresponding statements is therefore of great interest.

The spectrum and wave functions of the system of two electrons in a crystal described by the Hubbard Hamiltonian were studied in [9]. It is known that two-electron systems can be in two states, triplet and singlet [9, 10, 11, 17, 4]. It was proved in [9] that the spectrum of the system Hamiltonian H^t in the triplet state is purely continuous and coincides with a segment $[m, M]$, and the operator H^s of the system in the singlet state, in addition to the continuous spectrum $[m, M]$, has a unique antibound state for some values of the quasimomentum. For the antibound state, correlated motion of the electrons is realized under which the contribution of binary states is large. Because the system is closed, the energy must remain constant and large. This prevents the electrons from being separated by long distances. Next, an essential point is that bound states (sometimes called scattering-type states) do not form below the continuous spectrum. This can be easily understood because the interaction is repulsive. We note that a converse situation is realized for $U < 0$: below the continuous spectrum, there is a bound state (antibound states are absent) because the electrons are then attracted to one another.

For the first band, the spectrum is independent of the parameter U of the on-site Coulomb interaction of two electrons and corresponds to the energy of two noninteracting electrons, being exactly equal to the triplet band. The second band is determined by Coulomb interaction to a much greater degree: both the amplitudes and the energy of two electrons depend on U, and the band itself disappears as $U \to 0$ and increases without bound as $U \to \infty$. The second band largely corresponds to a one-particle state, namely, the motion of the doublet, i.e., two-electron bound states.

The spectrum and wave functions of the system of three electrons in a crystal described by the Hubbard Hamiltonian were studied in [16].

Here, we consider the energy operator of four-electron systems in the Hubbard model and describe the structure of the essential and discrete spectra of the system for quintet and triplet states.

The Hamiltonian of the chosen model has the form

$$H = A\sum_{m,\gamma} a^+_{m,\gamma}a_{m,\gamma} + B\sum_{m,\tau,\gamma} a^+_{m,\gamma}a_{m+\tau,\gamma} + U\sum_{m} a^+_{m,\uparrow}a_{m,\uparrow}a^+_{m,\downarrow}a_{m,\downarrow}. \tag{2}$$

Here, A is the electron energy at a lattice site, B is the transfer integral between neighboring sites (we assume that $B > 0$ for convenience), $\tau = \pm e_j, j = 1, 2, ..., \nu$, where e_j are unit mutually orthogonal vectors, which means that summation is taken over the nearest neighbors, U is the parameter of the on-site Coulomb interaction of two electrons, γ is the spin index, $\gamma =\uparrow$ or $\gamma =\downarrow$, $\uparrow$ and $\downarrow$ denote the spin values $\frac{1}{2}$ and $-\frac{1}{2}$, and $a^+_{m,\gamma}$ and $a_{m,\gamma}$ are the respective electron creation and annihilation operators at a site $m \in Z^\nu$.

The energy of the system depends on its total spin S. In the case of a saturated ferromagnetic state ($S = \frac{N_e}{2}$, where N_e is the number of electrons in the system), the solution of the problem is exact and trivial for any admissible number of electrons N_e. In that case, the system is an ideal Fermi gas of electrons with the same direction of the spin projections.

Along with the Hamiltonian, the N_e electron system is characterized by the total spin S,

$$S = S_{max}, S_{max} - 1, ..., S_{min}, S_{max} = \frac{N_e}{2}, S_{min} = 0, \frac{1}{2}.$$

Hamiltonian (2) commutes with all components of the total spin operator $S = (S^+, S^-, S^z)$, and the structure of eigenfunctions and eigenvalues of the system therefore depends on S. The Hamiltonian H acts in the antisymmetric Fock space $\mathcal{H}_{as}$.

2. Quintet state

Let φ_0 be the vacuum vector in the space $\mathcal{H}_{as}$. The quintet state corresponds to the free motion of four electrons over the lattice with the basis functions

$$q^2_{m,n,p,r} = a^+_{m\uparrow}a^+_{n\uparrow}a^+_{p\uparrow}a^+_{r\uparrow}\varphi_0.$$

The subspace $\widetilde{\mathcal{H}}^q_2$, corresponding to the quintet state is the set of all vectors of the form

$$\psi = \sum_{m,n,p,r\in Z^\nu} \widetilde{f}(m,n,p,r)q^2_{m,n,p,r}, \quad \widetilde{f} \in l^{as}_2,$$

where l^{as}_2 is the subspace of antisymmetric functions in the space $l_2((Z^\nu)^4)$.

Theorem 2.1. *The subspace $\mathcal{H}^q_2$ is invariant under the operator H, and the restriction $H^q_2 = H/_{\mathcal{H}^q_2}$ of H to the subspace $\mathcal{H}^q_2$ is a bounded self-adjoint operator. It generates a bounded self-adjoint operator $\overline{H}^q$, acting in the space l^{as}_2 as*

$$(\overline{H}^q f)(p;q;r;t) = 4Af(p;q;r;t)+B\sum_{\tau}(f(p+\tau,q,r,t)+f(p,q+\tau,r,t)+f(p,q,r+\tau,t)+f(p,q,r,t+\tau)). \tag{3}$$

The operator H_2^q acts on a vector $\psi \in \mathcal{H}_2^q$ as

$$H_2^q\psi = \sum_{m,n,p,r} (\overline{H}^q f)(m,n,p,r) q^2_{m,n,p,r}. \tag{4}$$

Proof. We act with the Hamiltonian H on vectors $\psi \in \mathcal{H}_2^q$ using the standard anticommutation relations between electron creation and annihilation operators at lattice sites, $\{a_{m,\gamma}, a^+_{n,\beta}\} = \delta_{m,n}\delta_{\gamma,\beta}$, $\{a_{m,\gamma}, a_{n,\beta}\} = \{a^+_{m,\gamma}, a^+_{n,\beta}\} = \theta$, and also take into account that $a_{m,\gamma}\varphi_0 = \theta$, where θ is the zero element of $\mathcal{H}_2^q$. This yields the statement of the theorem. □

Lemma 2.2. *The spectra of the operators $\overline{H}^q$ and H_2^q coincide.*

The proof follows by using the Weyl criterion [13].

We call $\overline{H}^q$ as the four-electron quintet operator.

We let $\mathcal{F}$ denote the Fourier transform:

$$\mathcal{F} : l_2((Z^\nu)^4) \to L_2((T^\nu)^4) \equiv \mathcal{H}_2^q,$$

where T^ν is the $\nu-$ dimensional torus endowed with the normalized Lebesgue measure $d\lambda$, $\lambda(T^\nu) = 1$.

We set $\widetilde{H}^q = \mathcal{F}\overline{H}^q\mathcal{F}^{-1}$. In the quasi-momentum representation, the operator $\overline{H}^q$ acts in the Hilbert space $L_2^{as}((T^\nu)^4)$ as follows

$$\widetilde{H}^q\widetilde{f}(x;y;z;t) = h(x;y;z;t)\widetilde{f}(x;y;z;t), \widetilde{f} \in L_2((T^\nu)^4),$$

where L_2^{as} is the subspace of antisymmetric functions in $L_2((T^\nu)^4)$ and

$$h(x;y;z;t) = 4A + 2B\sum_{i=1}^{\nu}(\cos x_i + \cos y_i + \cos z_i + \cos t_i). \tag{5}$$

It is obvious that the spectrum of $\widetilde{H}^q$ is purely continuous and coincides with the value set of the function $h(x;y;z;t)$, i.e. with the set

$$Imh(x;y;z) = [m_\nu, M_\nu] = [4A - 8B\nu, 4A + 8B\nu],$$

where

$$m_\nu = min_{x,y,z,t\in T^\nu} h(x;y;z;t), M_\nu = max_{x,y,z,t\in T^\nu} h(x;y;z;t).$$

Therefore, the quintet state spectrum is independent of the Coulomb interaction parameter U and is the set of energies of four noninteracting electrons moving in the crystal. This result is totally natural because the quintet state cannot contain states with two electrons at a site. Hence, in the quintet state, the spectrum of four-electron systems can be evaluated exactly and is purely continuous. The spectral problem that we consider here is a particular case of the problem of finding the spectrum of a system of N noninteracting electrons in a crystal lattice.

By Hunds rule, the minimum-energy state in an $N-$ electron system is the state where all spins are directed upward, i.e., the state $\uparrow\uparrow \ldots \uparrow$. By the Pauli exclusion principle, this state cannot contain states with two electrons at one site. In this case, the spectrum of the system is independent of the Coulomb interaction parameter U and is the band energy of N noninteracting electrons moving in the crystal. The spectrum of the system is then purely continuous.

3. First triplet state

In the system there exist three triplet states. The triplet state corresponds to the basis functions

$$^1t^1_{m,n,p,r} = a^+_{m\uparrow}a^+_{n\uparrow}a^+_{p\uparrow}a^+_{r\downarrow}\varphi_0,\quad ^2t^1_{m,n,p,r} = a^+_{m\uparrow}a^+_{n\uparrow}a^+_{p\downarrow}a^+_{r\uparrow}\varphi_0,\quad ^3t^1_{m,n,p,r} = a^+_{m\uparrow}a^+_{n\downarrow}a^+_{p\uparrow}a^+_{r\uparrow}\varphi_0.$$

We see that there are three such states, and they have different origins.

The subspace $^k\mathcal{H}^t_1, k = 1,2,3$ corresponding to the triplet states in the set of all vectors of the forms $\psi = \sum_{m,n,p,r} f(m,n,p,r)^k t^1_{m,n,p,r}$, $k = 1,2,3$, $f \in l^{as}_2$, where l^{as}_2 is the subspace of antisymmetric functions in the space $l_2((Z^\nu)^4)$.

The restriction $^kH^t_1 = H/_{^k\mathcal{H}^t_1}, k = 1,2,3$ of H to the subspace $^k\mathcal{H}^t_1$ is a bounded self-adjoint operator.

Theorem 3.1. *The subspace $^1\mathcal{H}^t_1$ is invariant under the operator H, and the operator $^1H^t_1$ is a bounded self-adjoint operator. It generates a bounded self-adjoint operator $^1\overline{H}^t_1$ acting in the space l^{as}_2 as*

$$(^1\overline{H}^t_1 f)(m,n,p,r) = 4Af(m,n,p,r) + 2B\sum_\tau [f(m+\tau,n,p,r) + f(m,n+\tau,p,r)+$$

$$+f(m,n,p+\tau,r)+f(m,n,p,r+\tau)]+U[\delta_{m,r}f(m,n,p,r)+\delta_{n,r}f(m,n,p,r)+\delta_{p,r}f(m,n,p,r)], \quad (6)$$

where $\delta_{k,j}$ is the Kronecker symbol. The operator $^1H^t_1$ acts on a vector $\psi \in ^1\mathcal{H}^t_1$ as

$$^1H^t_1\psi = \sum_{m,n,q,r} (^1\overline{H}^t_1 f)(m,n,p,r)^1t^1_{m,n,p,r}. \quad (7)$$

Proof. The proof of the theorem can be obtained from the explicit form of the action of H on vectors $\Psi \in ^1\mathcal{H}^t_1$ using the standard anti-commutation relations between electron creation and annihilation operators. □

We set $^1\widetilde{H}^t_1 = \mathcal{F}\, ^1\overline{H}^t_1 \mathcal{F}^{-1}$. In the quasi-momentum representation, the operator $^1\overline{H}^t_1$ acts in the Hilbert space $L^{as}_2((T^\nu)^4$ as

$$(^1\widetilde{H}^t_1 f)(\lambda,\mu,\gamma,\theta) = h(\lambda,\mu,\gamma,\theta)f(\lambda,\mu,\gamma,\theta) + U[\int_{T^\nu} f(s,\mu,\gamma,\lambda+\theta-s,\theta)ds+$$

$$+\int_{T^\nu} f(\lambda,t,\gamma,\mu+\theta-t)dt + \int_{T^\nu} f(\lambda,\mu,r,\gamma+\theta-r)dr],\quad f \in L^{as}_2((T^\nu)^4), \quad (8)$$

where L^{as}_2 is the subspace of antisymmetric functions in $L_2((T^\nu)^4$, and $h(\lambda,\mu,\gamma,\theta)$ has a form (5).

Using that the function $f(\lambda,\mu,\gamma,\theta)$ is the antisymmetrical function, we verify that the operator $^1\widetilde{H}^t_1$ can be represented in the form

$$^1\widetilde{H}^t_1 = \widetilde{H}_1 \bigotimes I \bigotimes I + I \bigotimes \widetilde{H}_2 \bigotimes I, \quad (9)$$

where I is the unit operator in the two-electron space,

$$(\widetilde{H}_1 f)(\lambda,\theta) = \{2A + 2B\sum_{i=1}^{\nu}[cos\lambda_i + cos\theta_i]\}f(\lambda,\theta) + U\int_{T^\nu} f(s,\lambda+\theta-s)ds,$$

and $(\widetilde{H}_2 f)(\mu,\gamma) = \{2A + 2B\sum_{i=1}^{\nu}[cos\mu_i + cos\gamma_i]\}f(\mu,\gamma)$.

We must therefore investigate the spectrum and bound states of the operators $\widetilde{H}_1$ and $\widetilde{H}_2$, respectively.

Let the total quasimomentum of the two-electron system $\lambda+\theta=\Lambda_1$ be fixed. We let $L_2(\Gamma_\Lambda)$ denote the space of functions that are square integrable on the manifold $\Gamma_{\Lambda_1}=\{(\lambda,\theta):\lambda+\theta=\Lambda_1\}$.

Let $<\mathcal{H}',(.)>$ be a separable Hilbert space, and $<M,\mu>$ be a measurable space with $\sigma-$finite measure μ.

The mapping $f:M\to\mathcal{H}'$ is called weakly measurable, if $(f(m),\eta)_{\mathcal{H}'}$ is a measurable function on $<M,\mu>$ for all $\eta\in\mathcal{H}'$.

By $\mathcal{H}$ we denote a linear space of all weakly measurable mappings $f:M\to\mathcal{H}'$, for which $\int_M ||f(m)||^2_{\mathcal{H}'}d\mu<\infty$ (equally nearly everywhere mappings are identify). The space $\mathcal{H}$ is a Hilbert space with the norm $||f||_{\mathcal{H}'}\equiv(\int_M ||f(m)||^2_{\mathcal{H}'}d\mu)^{\frac{1}{2}}$. This space $\mathcal{H}$ is called direct integral of spaces with identical layers and it is denoted by $\mathcal{H}=\bigoplus\int_M \mathcal{H}'d\mu$.

Let $\mathcal{L}(\mathcal{H}')$ be the algebra of all bounded linear mappings acting in $\mathcal{H}'$. A mapping $A: M\to\mathcal{L}(\mathcal{H}')$ is called measurable, if numerical function $(\varphi,A(.)\psi)$ is measurable in $<M,\mu>$ for all $\varphi,\psi\in\mathcal{H}'$. We denote by $L^\infty(M,d\mu,\mathcal{L}(\mathcal{H}'))$ linear space of all measurable mappings from M into $\mathcal{L}(\mathcal{H}')$, for which $||A||_\infty=esssup||A(m)||_{\mathcal{L}(\mathcal{H}')}<\infty$ (equally nearly everywhere mappings are identify). We say, that $A\in\mathcal{L}(\mathcal{H}')$ is decomposable into direct integral, if exists $A(\cdot)\in L^\infty(M,d\mu,\mathcal{L}(\mathcal{H}'))$ such that

$$(A\psi)(m)=A(m)\psi(m))\ \ a.e.$$

for all $\psi\in\mathcal{H}'$. In this case A is called decomposable and will be written as $A=\bigoplus\int_M A(m)d\mu(m)$.

The following result is a well-known theorem about the description of decomposable operators as direct integral.

Theorem 3.2. *[14]. Let $\mathcal{H}=\bigoplus\int_M\mathcal{H}'d\mu$, where $<M,\mu>$ is separable measurable spaces whit $\sigma-$ finite measure and $\mathcal{H}'$ is separable. Let $\mathcal{A}$ be an algebra of decomposable operators with layers, which multiple to the identity operator. In this case, $A\in\mathcal{L}(\mathcal{H}')$ is decomposable if and only if A is commuted whit every operators from $\mathcal{A}$.*

Since $\widetilde{H}_1U_s=U_s\widetilde{H}_1$ for all $s\in T^\nu$, where $(U_sf)(t)=f(s-t),t,s\in T^\nu$, then, by Theorem 3.2 there exists the decomposition $\mathcal{H}_1=\bigoplus\int_{T^\nu}\mathcal{H}_{1\Lambda_1}d\Lambda_1$ of Hilbert spaces $\mathcal{H}_1$ in the direct integral of spaces $\mathcal{H}_{1\Lambda_1},\Lambda_1\in T^\nu$, such that $\widetilde{H}_1=\bigoplus\int_{T^\nu}\widetilde{H}_{1\Lambda_1}d\Lambda_1$ and $U_t=\bigoplus\int_{T^\nu}e^{it\Lambda_1}I_{\Lambda_1}d\Lambda_1$, where $\widetilde{H}_{1\Lambda_1}$ are operators on $\mathcal{H}_{1\Lambda_1}$ and I_{Λ_1} are identity operator on $\mathcal{H}_{1\Lambda_1}$, and T^ν are $\nu-$ demensional torus.

It is known [12, 14] that the operator $\widetilde{H}_1$ and the space $\mathcal{H}_2\equiv L_2((T^\nu)^2)$ can be decomposed into a direct integral of operators $\widetilde{H}_{1\Lambda_1}$ and spaces $\mathcal{H}_{2\Lambda_1}=L_2(\Gamma_{\Lambda_1})$,

$$\widetilde{H}_1=\bigoplus\int_{T^\nu}\widetilde{H}_{1\Lambda_1}d\Lambda_1,\quad \mathcal{H}_2=\bigoplus\int_{T^\nu}\mathcal{H}_{2\Lambda_1}d\Lambda_1,$$

such that the spaces $\mathcal{H}_{2\Lambda_1}$ are then invariant under the operators $\widetilde{H}_{1\Lambda_1}$ and each operator $\widetilde{H}_{1\Lambda_1}$ acts in $\mathcal{H}_{2\Lambda_1}$ as $(\widetilde{H}_{1\Lambda_1}f_{\Lambda_1})(\lambda)=\{2A+4B\sum_{i=1}^{\nu}\cos\frac{\Lambda_1^i}{2}\cos(\frac{\Lambda_1^i}{2}-\lambda_i)\}f_{\Lambda_1}(\lambda)+U\int_{T^\nu}f_{\Lambda_1}(s)ds$, where $f_{\Lambda_1}(\lambda)=f(\lambda,\Lambda_1-\lambda)$.

It is known that the continuous spectrum of $\widetilde{H}_{1\Lambda_1}$ is independent of U and coincides with the segment $[m_{\Lambda_1},M_{\Lambda_1}]$, where $m_{\Lambda_1}=inf_{x\in T^\nu}h_{\Lambda_1}(x),M_{\Lambda_1}=sup_{x\in T^\nu}h_{\Lambda_1}(x)$, here $h_{\Lambda_1}(x)=2A+4B\sum_{i=1}^{\nu}\cos\frac{\Lambda_1^i}{2}\cos(\frac{\Lambda_1^i}{2}-x_i)$.

We set $\Delta^\nu_{\Lambda_1}(z)=1+U\int_{T^\nu}\frac{ds_1ds_2...ds_\nu}{2A+4B\sum_{i=1}^{\nu}\cos\frac{\Lambda_1^i}{2}\cos(\frac{\Lambda_1^i}{2}-s_i)-z}$.

Lemma 3.3. *The number $z = z_0$ not belonging to the continuous spectrum of the operator $\widetilde{H}_1$ is an eigenvalue of that operator if and only if it is a zero of the function $\Delta^{\nu}_{\Lambda_1}(z)$, i.e., $\Delta^{\nu}_{\Lambda_1}(z_0) = 0$.*

Proof. If z_0 is an eigenvalue of $\widetilde{H}_1$, then

$$\{2A + 4B\sum_{i=1}^{\nu}\cos\frac{\Lambda_1^i}{2}\cos(\frac{\Lambda_1^i}{2} - x_i)\}\widetilde{f}_{\Lambda_1}(x) + U\int_{T^\nu}\widetilde{f}_{\Lambda_1}(s)ds_1ds_2...ds_\nu = z_0\widetilde{f}_{\Lambda_1}(x),$$

whence $\Delta^{\nu}_{\Lambda_1}(z_0) = 0$.

Now let $\Delta^{\nu}_{\Lambda_1}(z_0) = 0$. Then the homogeneous equation

$$\widetilde{f}_{\Lambda_1}(x) + U\int_{T^\nu}\frac{\widetilde{f}_{\Lambda_1}(s)ds_1ds_2...ds_\nu}{2A + 4B\sum_{i=1}^{\nu}\cos\frac{\Lambda_1^i}{2}\cos(\frac{\Lambda_1^i}{2} - s_i) - z} = 0$$

has a nontrivial solution. It then follows that the number $z = z_0$ is an eigenvalue of $\widetilde{H}_1$. □

It is clear that the continuous spectrum of $\widetilde{H}_1$ coincides with the segment

$$[m^{\nu}_{\Lambda_1}, M^{\nu}_{\Lambda_1}] = [2A - 4B\sum_{i=1}^{\nu} cos\frac{\Lambda_1^i}{2}, 2A + 4B\sum_{i=1}^{\nu} cos\frac{\Lambda_1^i}{2}].$$

We consider the one-dimensional case.

Theorem 3.4. *At $\nu = 1$ and $U > 0$ and for all values of the parameter of the Hamiltonian, the operator $\widetilde{H}_1$ has a unique eigenvalue $z = 2A + \sqrt{U^2 + 16B^2cos^2\frac{\Lambda_1}{2}}$ that is above the continuous spectrum of $\widetilde{H}_1$, i.e., $z > M^1_{\Lambda_1}$.*

Proof. In the one-dimensional case, the function $\Delta^1_{\Lambda_1}(z)$ increases monotonically outside the continuous spectrum domain of the singlet-state operator $\widetilde{H}_1$. For $z < m^1_{\Lambda_1}$, the function $\Delta^1_{\Lambda_1}(z)$ increases from 1 to $+\infty$, and one has

$$\Delta^1_{\Lambda_1}(z) \to 1 \text{ as } z \to -\infty$$

$$\Delta^1_{\Lambda_1}(z) \to +\infty \text{ as } z \to m^1_{\Lambda_1} - 0.$$

Therefore, $\Delta^1_{\Lambda_1}(z)$ cannot vanish below the value $m^1_{\Lambda_1}$. For $z > M^1_{\Lambda_1}$, $\Delta^1_{\Lambda_1}(z)$ increases from $-\infty$ to 1,

$$\Delta^1_{\Lambda_1}(z) \to -\infty \text{ as } z \to M^1_{\Lambda_1} + 0,$$

$$\Delta^1_{\Lambda_1}(z) \to +1 \text{ as } z \to +\infty.$$

Therefore, above the value $M^1_{\Lambda_1}$, $\Delta^1_{\Lambda_1}(z)$ vanishes at a single point $z = 2A + \sqrt{U^2 + 16B^2\cos^2\frac{\Lambda}{2}}$. □

We now consider the two-dimensional case.

Theorem 3.5. *At $\nu = 2$ and $U > 0$ and for all values of the parameter of the Hamiltonian, the operator $\widetilde{H}_1$ has a unique eigenvalue $\widetilde{z}$ that is above the continuous spectrum of $\widetilde{H}_1$, i.e., $z > M^2_{\Lambda_1}$.*

Proof. In the two-dimensional case, the function $\Delta^2_{\Lambda_1}(z)$ increases monotonically outside the continuous spectrum domain of $\widetilde{H}_1$. For $z < m^2_{\Lambda_1}$, the function $\Delta^2_{\Lambda_1}(z)$ increases from 1 to $+\infty$,

$\Delta^2_{\Lambda_1}(z) \to 1$ as $z \to -\infty$

$\Delta^2_{\Lambda_1} \to +\infty$ as $z \to m^2_{\Lambda_1} - 0$.

Therefore, $\Delta^2_{\Lambda_1}(z)$ cannot vanish below the value $m^2_{\Lambda_1}$. For $z > M^2_{\Lambda_1}$, $\Delta^2_{\Lambda_1}(z)$ increases from $-\infty$ to 1,

$\Delta^2_{\Lambda_1}(z) \to -\infty$ as $z \to M^2_{\Lambda_1} = 0$,

$\Delta^2_{\Lambda_1}(z) \to 1$ as $z \to +\infty$.

Therefore, above $M^2_{\Lambda_1}$, the function $\Delta^2_{\Lambda_1}(z)$ has a single zero at the point $\widetilde{z} > M^2_{\Lambda_1}$. □

In the three-dimensional case, the operator $\widetilde{H}_1$ for some values of the parameters in the Hamiltonian has a unique eigenvalue $\widetilde{\widetilde{z}}$, such that $\widetilde{\widetilde{z}} > M^3_{\Lambda_1}$.

We consider the Watson integral [18]

$$W = \frac{1}{\pi^3}\int_0^\pi\int_0^\pi\int_0^\pi \frac{3dxdydz}{3-\cos x-\cos y-\cos z} = 1,516.$$

Because the measure ν is normalized,

$$J = \int_{-\pi}^\pi\int_{-\pi}^\pi\int_{-\pi}^\pi \frac{dxdydz}{3-\cos x-\cos y-\cos z} =$$

$$= \int_{-\pi}^\pi\int_{-\pi}^\pi\int_{-\pi}^\pi \frac{dxdydz}{3+\cos x+\cos y+\cos z} = \frac{W}{3}.$$

Theorem 3.6. *At $\nu = 3$ and $U > 0$ and the total quasimomentum Λ_1 of the system have the form*

$$\Lambda_1 = (\Lambda_1^1; \Lambda_1^2; \Lambda_1^3) = (\Lambda_1^0; \Lambda_1^0; \Lambda_1^0).$$

Then the operator $\widetilde{H}_1$ has a unique eigenvalue $\widetilde{\widetilde{z}}$ if

$$U > \frac{12B\cos\frac{\Lambda_1^0}{2}}{W}.$$

Otherwise, the operator $\widetilde{H}_1$ has no eigenvalue.

Proof. In the three-dimensional case,

$\Delta^3_{\Lambda_1}(z) \to 1$ as $z \to -\infty$

$\Delta^3_{\Lambda_1}(z) \to Q > 1$ as $z \to m^3_{\Lambda_1} - 0$.

Hence, the function $\Delta^3_{\Lambda_1}(z)$ cannot vanish below the continuous spectrum. Above the continuous spectrum,

$\Delta^3_{\Lambda_1}(z) \to 1 - \frac{UW}{12Bcos(\Lambda_0/2}$ as $z \to M^3_{\Lambda_1} + 0$,

$\Delta^3_{\Lambda_1}(z) \to 1$ as $z \to +\infty$.

Inspecting the equation $\Delta^3_{\Lambda_1}(z) = 0$ above the continuous spectrum domain of the operator $\widetilde{H}_1$, we obtain the statement of the theorem. □

We note that the converse situation is realized for $U < 0$.

We must therefore investigate the spectrum of the operator $\widetilde{H}_2$.

$(\widetilde{H}_2 f)(\mu,\gamma) = \{2A + 2B\sum_{i=1}^{\nu}[cos\mu_i + cos\gamma_i]\}f(\mu,\gamma)$.

Now let $\Lambda_2 = \mu + \gamma$. Then $(\widetilde{H}_{2\Lambda_2} f_{\Lambda_2})(\mu) = \{2A + 4B\sum_{i=1}^{\nu}[cos\frac{\Lambda_2}{2}cos(\frac{\Lambda_2}{2} - \mu)]\}f_{\Lambda_2}(\mu)$.

It is obvious that the spectrum of $\widetilde{H}_{2\Lambda_2}$ is purely continuous and coincides with the value set of the function $h_{\Lambda_2}(\mu) = 2A + 4B\sum_{i=1}^{\nu}[cos\frac{\Lambda_2^i}{2}cos(\frac{\Lambda_2^i}{2} - \mu^i)]$, i.e., with the set $[2A - 4B\sum_{i=1}^{\nu} cos\frac{\Lambda_2^i}{2}, 2A + 4B\sum_{i=1}^{\nu} cos\frac{\Lambda_2^i}{2}]$.

Now, using the obtained results and representation (9), we describe the structure of the essential spectrum and the discrete spectrum of the operator ${}^1\widetilde{H}_t^1$.

The spectrum of the operator $A\bigotimes I + I\bigotimes B$, where A and B are densely defined bounded linear operators, was studied in [5],[6],[7]. Explicit formulas were given there that express the essential spectrum $\sigma_{ess}(A\bigotimes I + I\bigotimes B)$ and the discrete spectrum $\sigma_{disc}(A\bigotimes I + I\bigotimes B)$ of $A\bigotimes I + I\bigotimes B$ in terms of the spectrum $\sigma(A)$ and the discrete spectrum $\sigma_{disc}(A)$ of A and in terms of the spectrum $\sigma(B)$ and the discrete spectrum $\sigma_{disc}(B)$ of B :

$$\sigma_{disc}(A\bigotimes I + I\bigotimes B) = \{\sigma(A)\backslash\sigma_{ess}(A) + \sigma(B)\backslash\sigma_{ess}(B)\}\backslash\{(\sigma_{ess}(A)+$$

$$+\sigma(B)) \cup (\sigma(A) + \sigma_{ess}(B))\},$$

$$\sigma_{ess}(A\bigotimes I + I\bigotimes B) = (\sigma_{ess}(A) + \sigma(B)) \cup (\sigma(A) + \sigma_{ess}(B)).$$

It is clear that $\sigma(A\bigotimes I + I\bigotimes B) = \{\lambda + \mu : \lambda \in \sigma(A), \mu \in \sigma(B)\}$.

Theorem 3.7. *At $\nu = 1$ and $U > 0$ the essential spectrum of the system first triplet-state operator ${}^1\widetilde{H}_t^1$ is exactly the union of two segments: $\sigma_{ess}({}^1\widetilde{H}_t^1) = [4A - 4B\cos\frac{\Lambda_1}{2} - 4B\cos\frac{\Lambda_2}{2}, 4A + 4B\cos\frac{\Lambda_1}{2} + 4B\cos\frac{\Lambda_2}{2}] \cup [4A + \sqrt{U^2 + 16B^2cos^2\frac{\Lambda_1}{2}} - 4B\cos\frac{\Lambda_2}{2}, 4A + \sqrt{U^2 + 16B^2cos^2\frac{\Lambda_1}{2}} + 4B\cos\frac{\Lambda_2}{2}$. The discrete spectrum of ${}^1\widetilde{H}_t^1$ is empty.*

Proof. It follows from representation (9) that

$$\sigma({}^1\widetilde{H}_t^1) = \{\lambda + \mu : \lambda \in \sigma(\widetilde{H}_1), \mu \in \sigma(\widetilde{H}_2)\},$$

and in the one-dimensional case, the continuous spectrum of $\widetilde{H}_1$ is

$$\sigma_{cont}(\widetilde{H}_1) = [2A - 4B\cos\frac{\Lambda_1}{2}, 2A + 4B\cos\frac{\Lambda_1}{2}],$$

and the discrete spectrum of $\widetilde{H}_1$ consists of a single point $z = 2A + \sqrt{U^2 + 16B^2cos^2\frac{\Lambda_1}{2}}$.

The continuous spectrum of the operator $\widetilde{H}_2$ is the segment $[2A - 4B\cos\frac{\Lambda_1}{2}, 2A + 4B\cos\frac{\Lambda_1}{2}]$, and the discrete spectrum of $\widetilde{H}_2$ is empty. Therefore, the essential spectrum of the system first triplet-state operator ${}^1\widetilde{H}_t^1$ is the union of two segments, and the first triplet-state operator ${}^1\widetilde{H}_t^1$ has no eigenvalue. □

The following theorem is proved totally similarly to Theorem 3.6.

Theorem 3.8. *At $\nu = 2$ and $U > 0$ the essential spectrum of the system first triplet-state operator ${}^1\widetilde{H}_t^1$ is exactly the union of two segments: $\sigma_{ess}({}^1\widetilde{H}_t^1) = [4A - 4B\sum_{i=1}^{2}\cos\frac{\Lambda_1^i}{2} - 4B\sum_{i=1}^{2}\cos\frac{\Lambda_2^i}{2}, 4A + 4B\sum_{i=1}^{2}\cos\frac{\Lambda_1^i}{2} + 4B\sum_{i=1}^{2}\cos\frac{\Lambda_2^i}{2}] \cup [2A + \widetilde{z} - 4B\sum_{i=1}^{2}\cos\frac{\Lambda_2^i}{2}, 2A + \widetilde{z} + 4B\sum_{i=1}^{2}\cos\frac{\Lambda_2^i}{2}$. The discrete spectrum of ${}^1\widetilde{H}_t^1$ is empty.*

Let $\nu = 3$ and $\Lambda_1 = (\Lambda_1^0, \Lambda_1^0, \Lambda_1^0)$.

Theorem 3.9. *The following statements hold:*
1. Let

$$U > \frac{12B cos\frac{\Lambda_1^0}{2}}{W}.$$

Then the essential spectrum of the system first triplet-state operator ${}^1\widetilde{H}_t^1$ *is the union of two segments* $\sigma_{ess}({}^1\widetilde{H}_t^1) = [4A - 12B\cos\frac{\Lambda_1^0}{2} - 4B\sum_{i=1}^3 \cos\frac{\Lambda_2^i}{2}, 4A + 12B\cos\frac{\Lambda_1^0}{2} + 4B\sum_{i=1}^3 \cos\frac{\Lambda_2^i}{2}] \cup [\widetilde{\widetilde{z}} + 2A - 4B\sum_{i=1}^3 \cos\frac{\Lambda_2^i}{2}, \widetilde{\widetilde{z}} + 2A + 4B\sum_{i=1}^3 \cos\frac{\Lambda_2^i}{2}]$. *The discrete spectrum of* ${}^1\widetilde{H}_t^1$ *is empty. Here and hereafter,* $\widetilde{\widetilde{z}}$ *is an eigenvalue of* $\widetilde{H}_1$.
2. Let

$$U \leq \frac{12B cos\frac{\Lambda_1^0}{2}}{W}.$$

Then the essential spectrum of the system first triplet-state operator ${}^1\widetilde{H}_t^1$ *is the segment* $\sigma_{ess}({}^1\widetilde{H}_t^1) = [4A - 12B\cos\frac{\Lambda_1^0}{2} - 4B\sum_{i=1}^3 \cos\frac{\Lambda_2^i}{2}, 4A + 12B\cos\frac{\Lambda_1^0}{2} + 4B\sum_{i=1}^3 \cos\frac{\Lambda_2^i}{2}]$, *and the discrete spectrum is empty.*

Proof. The proof uses representation (9) and the results for the spectra of $\widetilde{H}_1$ and $\widetilde{H}_2$ in the case $\nu = 3$. □

4. Second triplet state

Theorem 4.1. *The subspace* ${}^2\mathcal{H}_1^t$ *is invariant under the operator* H, *and the operator* ${}^2H_1^t$ *is a bounded self-adjoint operator. It generates a bounded self-adjoint operator* ${}^2\overline{H}_1^t$ *acting in the space* l_2^{as} *as*

$$({}^2\overline{H}_1^t f)(m,n,p,r) = 4Af(m,n,p,r) + 2B\sum_\tau [f(m+\tau,n,p,r) + f(m,n+\tau,p,r) + f(m,n,p+\tau,r) +$$

$$+ f(m,n,p,r+\tau)] + U[\delta_{m,p}f(m,n,p,r) + \delta_{n,p}f(m,n,p,r) + \delta_{p,r}f(m,n,p,r)], \qquad (10)$$

where $\delta_{k,j}$ *is the Kronecker symbol. The operator* ${}^2H_1^t$ *acts on a vector* $\psi \in {}^2\mathcal{H}_1^t$ *as*

$${}^2H_1^t\psi = \sum_{m,n,q,r} ({}^2\overline{H}_1^t f)(m,n,p,r)^2 t_{m,n,p,r}^1. \qquad (11)$$

Proof. The proof of the theorem follows from the explicit form of the action of the Hamiltonian H on vectors $\psi \in^2 \mathcal{H}_1^t$ using standard anticommutation relations for electron creation and annihilation operators. □

We set ${}^2\widetilde{H}_1^t = \mathcal{F}\,{}^2\overline{H}_1^t\mathcal{F}^{-1}$. In the quasimomentum representation, the operator ${}^2\overline{H}_1^t$ acts in the Hilbert space $L_2^{as}((T^\nu)^4$ as

$$({}^2\widetilde{H}_1^t f)(\lambda,\mu,\gamma,\theta) = h(\lambda,\mu,\gamma,\theta)f(\lambda,\mu,\gamma,\theta) + U[\int_{T^\nu} f(s,\mu,\lambda+\gamma-s,\theta)ds +$$

$$+ \int_{T^\nu} f(\lambda,t,\mu+\gamma-t,\theta)dt + \int_{T^\nu} f(\lambda,\mu,r,\gamma+\theta-r)dr], \quad f \in L_2^{as}((T^\nu)^4), \qquad (12)$$

where L_2^{as} is the subspace of antisymmetric functions in $L_2((T^\nu)^4$, and $h(\lambda,\mu,\gamma,\theta)$ has form (5).

Taking into account that the function $f(\lambda,\mu,\gamma,\theta)$ is antisymmetric, we can rewrite formula (12) as

$$({}^2\widetilde{H}_1^t f)(\lambda,\mu,\gamma,\theta) = \{-2A - 2B\sum_{i=1}^{\nu}[\cos\lambda_i + \cos\gamma_i]\}f(\lambda,\gamma,\mu,\theta) -$$

$$-U\int_{T^\nu} f(s,\lambda+\gamma-s,\mu,\theta)ds + U\int_{T^\nu} f(\lambda,s,\mu+\gamma-s,\theta)ds +$$

$$+\{-2A - 2B\sum_{i=1}^{\nu}[\cos\lambda_i + \cos\gamma_i]\}f(\lambda,\gamma,\mu,\theta) + U\int_{T^\nu} f(\lambda,\mu,s,\gamma+\theta-s)ds. \quad (13)$$

We verify that the operator ${}^2\widetilde{H}_1^t$ can be represented in the form

$$ {}^2\widetilde{H}_1^t = \widetilde{H}_3 \bigotimes I \bigotimes I + I \bigotimes \widetilde{H}_4 \bigotimes I + I \bigotimes I \bigotimes \widetilde{H}_5, \quad (14)$$

where $(\widetilde{H}_3 f)(\lambda,\gamma) = \{-2A - 2B\sum_{i=1}^{\nu}[\cos\lambda_i + \cos\gamma_i]\}f(\lambda,\gamma) - U\int_{T^\nu} f(s,\lambda+\gamma-s)ds$,
$(\widetilde{H}_4 f)(\mu,\gamma) = U\int_{T^\nu} f(s,\mu+\gamma-s)ds$, $(\widetilde{H}_5 f)(\gamma,\theta) = \{-2A-2B\sum_{i=1}^{\nu}[\cos\mu_i + \cos\theta_i]\}f(\mu,\theta) + U\int_{T^\nu} f(s,\gamma+\theta-s)ds$.

We must therefore investigate the spectrum and bound states of the operators $\widetilde{H}_3$, and $\widetilde{H}_4$, and $\widetilde{H}_5$.

Let the total quasimomentum of the two-electron system $\lambda + \gamma = \Lambda_3$ be fixed.

We let $L_2(\Gamma_{\Lambda_3})$ denote the space of functions that are square integrable on the manifold $\Gamma_{\Lambda_3} = \{(\lambda,\gamma) : \lambda + \gamma = \Lambda_3\}$.

That the operator $\widetilde{H}_3$ and the space $\mathcal{H}_2 \equiv L_2((T^\nu)^2)$ can be decomposed into a direct integral of operators $\widetilde{H}_{3\Lambda_3}$ and spaces $\mathcal{H}_{2\Lambda_3} = L_2(\Gamma_{\Lambda_3})$,

$$\widetilde{H}_3 = \bigoplus\int_{T^\nu} \widetilde{H}_{3\Lambda_3} d\Lambda_3, \quad \mathcal{H}_2 = \bigoplus\int_{T^\nu} \mathcal{H}_{2\Lambda_3} d\Lambda_3,$$

such that the spaces $\mathcal{H}_{2\Lambda_3}$ are then invariant under the operators $\widetilde{H}_{3\Lambda_3}$ and each operator $\widetilde{H}_{3\Lambda_3}$ acts in $\mathcal{H}_{2\Lambda_3}$ as $(\widetilde{H}_{3\Lambda_3} f_{\Lambda_3})(\lambda) = \{-2A - 4B\sum_{i=1}^{\nu}\cos\frac{\Lambda_3^i}{2}\cos(\frac{\Lambda_3^i}{2} - \lambda_i)\}f_{\Lambda_3}(\lambda) - U\int_{T^\nu} f_{\Lambda_3}(s)ds$, where $f_{\Lambda_3}(\lambda) = f(\lambda, \Lambda_3 - \lambda)$.

It is known that the continuous spectrum of $\widetilde{H}_{3\Lambda_3}$ is independent of U and coincides with the segment $[m_{\Lambda_3}, M_{\Lambda_3}]$, where $m_{\Lambda_3} = inf_{x\in T^\nu} h_{\Lambda_3}(x)$, $M_{\Lambda_3} = sup_{x\in T^\nu} h_{\Lambda_3}(x)$, here $h_{\Lambda_3}(x) = -2A - 4B\sum_{i=1}^{\nu}\cos\frac{\Lambda_3^i}{2}\cos(\frac{\Lambda_3^i}{2} - x_i)$.

We set $\Delta_{\Lambda_3}^{\nu}(z) = 1 - U\int_{T^\nu}\frac{ds_1 ds_2 \ldots ds_\nu}{-2A-4B\sum_{i=1}^{\nu}\cos\frac{\Lambda_3^i}{2}\cos(\frac{\Lambda_3^i}{2} - s_i) - z}$.

Lemma 4.2. *The number $z = z_0$ not belonging to the continuous spectrum of the operator $\widetilde{H}_3$ is an eigenvalue of that operator if and only if it is a zero of the function $\Delta_{\Lambda_3}^{\nu}(z)$, i.e., $\Delta_{\Lambda_3}^{\nu}(z_0) = 0$.*

Proof. The proof of the statements in the lemma 4.2 is similar to the proof of lemma 3.2. □

It is clear that the continuous spectrum of $\widetilde{H}_3$ coincides with the segment

$$[m_{\Lambda_3}^{\nu}, M_{\Lambda_3}^{\nu}] = [-2A - 4B\sum_{i=1}^{\nu} cos\frac{\Lambda_3^i}{2}, -2A + 4B\sum_{i=1}^{\nu} cos\frac{\Lambda_3^i}{2}].$$

We consider the one-dimensional case.

Theorem 4.3. *At $\nu = 1$ and $U > 0$ and for all values of the parameter of the Hamiltonian, the operator $\widetilde{H}_3$ has a unique eigenvalue $z = -2A - \sqrt{U^2 + 16B^2 cos^2\frac{\Lambda_3}{2}}$ that is below the continuous spectrum of $\widetilde{H}_3$, i.e., $z < m_{\Lambda_3}^1$.*

Proof. In the one-dimensional case, the function $\Delta^1_{\Lambda_3}(z)$ decreases monotonically outside the continuous spectrum domain of the operator $\widetilde{H}_3$. For $z < m^1_{\Lambda_3}$, the function $\Delta^1_{\Lambda_3}(z)$ decreases from 1 to $-\infty$,

$\Delta^1_{\Lambda_3}(z) \to 1$ as $z \to -\infty$

$\Delta^1_{\Lambda_3}(z) \to -\infty$ as $z \to m^1_{\Lambda_3} - 0.$

Therefore, below the value $m^1_{\Lambda_3}$, $\Delta^1_{\Lambda_3}(z)$ vanishes at a single point $z = -2A - \sqrt{U^2 + 16B^2 cos^2 \frac{\Lambda_3}{2}}$. For $z > M^1_{\Lambda_3}$, $\Delta^1_{\Lambda_3}(z)$ decreases from $+\infty$ to 1,

$\Delta^1_{\Lambda_3}(z) \to +\infty$ as $z \to M^1_{\Lambda_3} + 0,$

$\Delta^1_{\Lambda_3}(z) \to +1$ as $z \to +\infty.$

Therefore, above the value $M^1_{\Lambda_3}$, $\Delta^1_{\Lambda_3}(z)$ cannot vanish. □

We now consider the two-dimensional case.

Theorem 4.4. *At $\nu = 2$ and $U > 0$ and for all values of the parameter of the Hamiltonian, the operator $\widetilde{H}_3$ has a unique eigenvalue $\widetilde{z}$ that is below the continuous spectrum of $\widetilde{H}_3$, i.e., $z < m^2_{\Lambda_3}$.*

Proof. In the two-dimensional case, the function $\Delta^2_{\Lambda_3}(z)$ decreases monotonically outside the continuous spectrum domain of $\widetilde{H}_3$. For $z < m^2_{\Lambda_3}$, the function $\Delta^2_{\Lambda_1}(z)$ decreases from 1 to $-\infty$,

$\Delta^2_{\Lambda_3}(z) \to 1$ as $z \to -\infty$

$\Delta^2_{\Lambda_3} \to -\infty$ as $z \to m^2_{\Lambda_3} - 0.$

Therefore, below $m^2_{\Lambda_3}$, the function $\Delta^2_{\Lambda_3}(z)$ has a single zero at the point $\widetilde{z} < m^2_{\Lambda_3}$. For $z > M^2_{\Lambda_3}$, $\Delta^2_{\Lambda_3}(z)$ decreases from $+\infty$ to 1,

$\Delta^2_{\Lambda_3}(z) \to +\infty$ as $z \to M^2_{\Lambda_3} = 0,$

$\Delta^2_{\Lambda_3}(z) \to 1$ as $z \to +\infty.$

Therefore, above $M^2_{\Lambda_3}$, the function $\Delta^2_{\Lambda_3}(z)$ cannot vanish. □

In the three-dimensional case, the operator $\widetilde{H}_3$ for some values of the parameters in the Hamiltonian has a unique eigenvalue $\widetilde{\widetilde{z}}$, such that $\widetilde{\widetilde{z}} < m^3_{\Lambda_3}$.

Theorem 4.5. *At $\nu = 3$ and $U > 0$ and the total quasimomentum Λ_3 of the system have the form*

$$\Lambda_3 = (\Lambda_3^1; \Lambda_3^2; \Lambda_3^3) = (\Lambda_3^0; \Lambda_3^0; \Lambda_3^0).$$

Then the operator $\widetilde{H}_3$ has a unique eigenvalue $\widetilde{\widetilde{z}}$ if

$$U > \frac{12B \cos \frac{\Lambda_3^0}{2}}{W}.$$

Otherwise, the operator $\widetilde{H}_3$ has no eigenvalue.

Proof. In the three-dimensional case,

$\Delta^3_{\Lambda_3}(z) \to 1$ as $z \to -\infty$

$\Delta^3_{\Lambda_3}(z) \to 1 - \frac{UW}{12Bcos(\Lambda_3^0/2)}$ as $z \to m^3_{\Lambda_3} - 0.$

Hence, the function $\Delta^3_{\Lambda_3}(z)$ has a single zero below the continuous spectrum, if $U > \frac{12B \cos \frac{\Lambda_3^0}{2}}{W}$. Above the continuous spectrum,

$\Delta^3_{\Lambda_3}(z) \to +\infty$ as $z \to M^3_{\Lambda_3} + 0,$

$\Delta^3_{\Lambda_3}(z) \to 1 + \frac{UW}{12Bcos(\Lambda_3^0/2}$ as $z \to +\infty.$

Inspecting the equation $\Delta^3_{\Lambda_3}(z) = 0$ above the continuous spectrum domain of the operator $\widetilde{H}_3$, we obtain the statement of the theorem. □

We note that the converse situation is realized for $U < 0$.

The spectrum of operator $(\widetilde{H}_4 f)(\mu,\gamma) = U\int_{T^\nu} f(s,\mu+\gamma-s)ds$ is purely discrete and consists is a single point $z = U$, for arbitrary value ν.

Now we investigated the spectrum of the operator
$(\widetilde{H}_5 f)(\gamma,\theta) = \{-2A - 2B\sum_{i=1}^{\nu}[\cos\mu_i + \cos\theta_i]\}f(\mu,\theta) + U\int_{T^\nu} f(s,\gamma+\theta-s)ds$.

We let $\Lambda_4 = \mu+\theta$, and $\Lambda_5 = \gamma+\theta$.

It is known that the continuous spectrum of $\widetilde{H}_5$ is independent of U and coincides with the segment $[m^\nu_{\Lambda_4}, M^\nu_{\Lambda_4}]$, where $m^\nu_{\Lambda_4} = -2A - 4B\sum_{i=1}^{\nu} cos\frac{\Lambda_4^i}{2}$, $M^\nu_{\Lambda_4} = -2A + 4B\sum_{i=1}^{\nu} cos\frac{\Lambda_4^i}{2}$.

We let $L_2(\Gamma_{\Lambda_4})$ denote the space of functions that are square integrable on the manifold $\Gamma_{\Lambda_4} = \{(\mu,\theta) : \mu+\theta = \Lambda_4\}$.

That the operator $\widetilde{H}_5$ and the space $\mathcal{H}_2 \equiv L_2((T^\nu)^2)$ can be decomposed into a direct integral of operators $\widetilde{H}_{5\Lambda_4}$ and spaces $\mathcal{H}_{2\Lambda_4} = L_2(\Gamma_{\Lambda_4})$,

$$\widetilde{H}_5 = \bigoplus\int_{T^\nu} \widetilde{H}_{5\Lambda_4} d\Lambda_4, \quad \mathcal{H}_2 = \bigoplus\int_{T^\nu} \mathcal{H}_{2\Lambda_4} d\Lambda_4,$$

such that the spaces $\mathcal{H}_{2\Lambda_4}$ are then invariant under the operators $\widetilde{H}_{5\Lambda_4}$ and each operator $\widetilde{H}_{5\Lambda_4}$ acts in $\mathcal{H}_{2\Lambda_4}$ as $(\widetilde{H}_{5\Lambda_4} f_{\Lambda_4})(\lambda) = \{-2A - 4B\sum_{i=1}^{\nu}\cos\frac{\Lambda_4^i}{2}\cos(\frac{\Lambda_4^i}{2} - \lambda_i)\}f_{\Lambda_4}(\lambda) + U\int_{T^\nu} f_{\Lambda_4}(s)ds$, where $f_{\Lambda_4}(\lambda) = f(\lambda, \Lambda_4 - \lambda)$.

We set $\Delta^\nu_{\Lambda_5}(z) = 1 + U\int_{T^\nu} \frac{ds_1 ds_2 \ldots ds_\nu}{-2A - 4B\sum_{i=1}^{\nu}\cos\frac{\Lambda_5^i}{2}\cos(\frac{\Lambda_5^i}{2} - s_i) - z}$.

Lemma 4.6. *The number $z = z_0$ not belonging to the continuous spectrum of the operator $\widetilde{H}_4$ is an eigenvalue of that operator if and only if it is a zero of the function $\Delta^\nu_{\Lambda_5}(z)$, i.e., $\Delta^\nu_{\Lambda_5}(z_0) = 0$.*

Proof. The proof of the statements in the lemma 4.6 is similar to the proof of Lemma 3.2. □

We consider the one-dimensional case.

Theorem 4.7. *At $\nu = 1$ and $U > 0$ and for all values of the parameter of the Hamiltonian, the operator $\widetilde{H}_5$ has a unique eigenvalue $z = -2A + \sqrt{U^2 + 16B^2 cos^2\frac{\Lambda_5}{2}}$ that is above the continuous spectrum of $\widetilde{H}_5$, i.e., $z > M^1_{\Lambda_4}$.*

Proof. In the one-dimensional case, the function $\Delta^1_{\Lambda_5}(z)$ increases monotonically outside the continuous spectrum domain of the operator $\widetilde{H}_5$. For $z < m^1_{\Lambda_4}$, the function $\Delta^1_{\Lambda_5}(z)$ increases from 1 to $+\infty$,

$\Delta^1_{\Lambda_5}(z) \to 1$ as $z \to -\infty$
$\Delta^1_{\Lambda_5}(z) \to +\infty$ as $z \to m^1_{\Lambda_4} - 0$.

Therefore, below the value $m^1_{\Lambda_4}$, the function $\Delta^1_{\Lambda_5}(z)$ cannot vanish. For $z > M^1_{\Lambda_4}$, the function $\Delta^1_{\Lambda_5}(z)$ increases from $-\infty$ to 1,

$\Delta^1_{\Lambda_5}(z) \to -\infty$ as $z \to M^1_{\Lambda_4} + 0$,
$\Delta^1_{\Lambda_5}(z) \to +1$ as $z \to +\infty$.

Therefore, above the value $M^1_{\Lambda_4}$, the function $\Delta^1_{\Lambda_5}(z)$ has a single zero at the point $z = -2A + \sqrt{U^2 + 16B^2 cos^2\frac{\Lambda_5}{2}}$. □

We now consider the two-dimensional case.

Theorem 4.8. *At $\nu = 2$ and $U > 0$ and for all values of the parameter of the Hamiltonian, the operator $\widetilde{H}_5$ has a unique eigenvalue z' that is above the continuous spectrum of $\widetilde{H}_5$, i.e., $z' > M^2_{\Lambda_4}$.*

Proof. In the two-dimensional case, the function $\Delta^2_{\Lambda_5}(z)$ increases monotonically outside the continuous spectrum domain of $\widetilde{H}_5$. For $z < m^2_{\Lambda_4}$, the function $\Delta^2_{\Lambda_5}(z)$ increases from 1 to $+\infty$,

$\Delta^2_{\Lambda_5}(z) \to 1$ as $z \to -\infty$

$\Delta^2_{\Lambda_5} \to +\infty$ as $z \to m^2_{\Lambda_4} - 0$.

Therefore, below $m^2_{\Lambda_4}$, the function $\Delta^2_{\Lambda_5}(z)$ cannot vanishes. For $z > M^2_{\Lambda_4}$, the function $\Delta^2_{\Lambda_5}(z)$ increases from $-\infty$ to 1,

$\Delta^2_{\Lambda_5}(z) \to -\infty$ as $z \to M^2_{\Lambda_4} + 0$,

$\Delta^2_{\Lambda_5}(z) \to 1$ as $z \to +\infty$.

Therefore, above $M^2_{\Lambda_4}$, the function $\Delta^2_{\Lambda_5}(z)$ has a single zero at the point $\widetilde{z} > M^2_{\Lambda_4}$. □

In the three-dimensional case, the operator $\widetilde{H}_5$ for some values of the parameters in the Hamiltonian has a unique eigenvalue $\widetilde{\widetilde{z}}$, such that $\widetilde{\widetilde{z}} > M^3_{\Lambda_4}$.

Theorem 4.9. *At $\nu = 3$ and $U > 0$ and the total quasimomentum Λ_5 of the system have the form*

$$\Lambda_5 = (\Lambda^1_5; \Lambda^2_5; \Lambda^3_5) = (\Lambda^0_5; \Lambda^0_3; \Lambda^0_3).$$

Then the operator $\widetilde{H}_5$ has a unique eigenvalue z'' if

$$U > \frac{12B\cos\frac{\Lambda^0_5}{2}}{W}.$$

Otherwise, the operator $\widetilde{H}_5$ has no eigenvalue.

Proof. In the three-dimensional case,

$\Delta^3_{\Lambda_5}(z) \to 1$ as $z \to -\infty$

$\Delta^3_{\Lambda_5}(z) \to 1 + \frac{UW}{12Bcos(\Lambda^0_5/2)}$ as $z \to m^3_{\Lambda_4} - 0$.

Hence, the function $\Delta^3_{\Lambda_5}(z)$ cannot vanishes below the continuous spectrum. Above the continuous spectrum,

$\Delta^3_{\Lambda_5}(z) \to 1 - \frac{UW}{12Bcos(\Lambda^0_5/2)}$ as $z \to M^3_{\Lambda_4} + 0$,

$\Delta^3_{\Lambda_5}(z) \to 1$ as $z \to +\infty$.

Inspecting the equation $\Delta^3_{\Lambda_5}(z) = 0$ above the continuous spectrum domain of the operator $\widetilde{H}_5$, we obtain the statement of the theorem. □

Now, using the obtained results and representation (9), we describe the structure of the essential spectrum and the discrete spectrum of the operator ${}^2\widetilde{H}^1_t$.

Theorem 4.10. *At $\nu = 1$ and $U > 0$ the essential spectrum of the system second triplet-state operator ${}^2\widetilde{H}^1_t$ is exactly the union of three segments: $\sigma_{ess}({}^2\widetilde{H}^1_t) = [U - 4A - 4B\cos\frac{\Lambda_3}{2} - 4B\cos\frac{\Lambda_4}{2}, U - 4A + 4B\cos\frac{\Lambda_3}{2} + 4B\cos\frac{\Lambda_4}{2}] \cup [U - 4A - 4B\cos\frac{\Lambda_4}{2} - \sqrt{U^2 + 16B^2cos^2\frac{\Lambda_3}{2}}, U - 4A + 4B\cos\frac{\Lambda_4}{2} - \sqrt{U^2 + 16B^2cos^2\frac{\Lambda_3}{2}}] \bigcup [U - 4A - 4Bcos^2\frac{\Lambda_3}{2} + \sqrt{U^2 + 16B^2cos^2\frac{\Lambda_5}{2}}, U - 4A + 4Bcos^2\frac{\Lambda_3}{2} + \sqrt{U^2 + 16B^2cos^2\frac{\Lambda_5}{2}}]$. The discrete spectrum of the operator ${}^2\widetilde{H}^1_t$ is consists no more than one point.*

Proof. It follows from representation (14) that

$$\sigma({}^2\widetilde{H}^1_t) = \{\lambda + \mu + \theta : \lambda \in \sigma(\widetilde{H}_3), \mu \in \sigma(\widetilde{H}_4), \theta \in \sigma(\widetilde{H}_5)\},$$

and in the one-dimensional case, the continuous spectrum of $\widetilde{H}_3$ is

$$\sigma_{cont}(\widetilde{H}_5) = [-2A - 4B\cos\frac{\Lambda_3}{2}, -2A + 4B\cos\frac{\Lambda_3}{2}],$$

and the discrete spectrum of $\widetilde{H}_3$ consists of a single point $z_1 = -2A - \sqrt{U^2 + 16B^2 cos^2\frac{\Lambda_3}{2}}$.

The spectra of the operator $\widetilde{H}_4$ is a purely discrete and consists is a single point $z = U$, for arbitrary value ν.

The continuous spectrum of the operator $\widetilde{H}_5$ consists of the segment $[-2A - 4B\cos\frac{\Lambda_4}{2}, -2A + 4B\cos\frac{\Lambda_4}{2}]$, and the discrete spectrum of $\widetilde{H}_5$ consists of a single point $z_2 = -2A + \sqrt{U^2 + 16B^2 cos^2\frac{\Lambda_5}{2}}$. Therefore, the essential spectrum of the system second triplet-state operator ${}^2\widetilde{H}_t^1$ is the union of three segments, and the second triplet-state operator ${}^2\widetilde{H}_t^1$ has a single eigenvalue $z = z_1 + z_2 + U$. If these eigenvalue lie the outside the essential spectrum of operator ${}^2\widetilde{H}_t^1$, then these eigenvalue is belonging the discrete spectrum of the operator ${}^2\widetilde{H}_t^1$, otherwise, the discrete spectrum of operator ${}^2\widetilde{H}_t^1$, is empty. □

The following theorem is proved totally similarly to Theorem 4.10.

Theorem 4.11. *At $\nu = 2$ and $U > 0$ the essential spectrum of the system second triplet-state operator ${}^2\widetilde{H}_t^1$ is exactly the union of three segments: $\sigma_{ess}({}^2\widetilde{H}_t^1) = [U - 4A - 4B\sum_{i=1}^2 \cos\frac{\Lambda_3^i}{2} - 4B\sum_{i=1}^2 \cos\frac{\Lambda_4^i}{2}, U - 4A + 4B\sum_{i=1}^2 \cos\frac{\Lambda_3^i}{2} + 4B\sum_{i=1}^2 \cos\frac{\Lambda_4^i}{2}] \cup [U - 2A + \widetilde{z} - 4B\sum_{i=1}^2 \cos\frac{\Lambda_3^i}{2}, U - 2A + \widetilde{z} + 4B\sum_{i=1}^2 \cos\frac{\Lambda_3^i}{2}] \cup [U - 2A + z' - 4B\sum_{i=1}^2 \cos\frac{\Lambda_3^i}{2}, U - 2A + z' + 4B\sum_{i=1}^2 \cos\frac{\Lambda_3^i}{2}]$. The discrete spectrum of the operator ${}^2\widetilde{H}_t^1$ is consists no more than one point.*

Let $\nu = 3$ and $\Lambda_3 = (\Lambda_3^0, \Lambda_3^0, \Lambda_3^0)$, and $\Lambda_4 = (\Lambda_4^0, \Lambda_4^0, \Lambda_4^0)$.

Theorem 4.12. *The following statements hold:*

1. Let

$$U > \frac{12Bcos\frac{\Lambda_3^0}{2}}{W}, cos\frac{\Lambda_3^0}{2} \geq \cos\frac{\Lambda_5^0}{2}.$$

Then the essential spectrum of the system second triplet-state operator ${}^2\widetilde{H}_t^1$ is the union of three segments $\sigma_{ess}({}^2\widetilde{H}_t^1) = [U - 4A - 12B\cos\frac{\Lambda_3^0}{2} - 12B\cos\frac{\Lambda_4^0}{2}, U - 4A + 12B\cos\frac{\Lambda_3^0}{2} + 12B\cos\frac{\Lambda_4^0}{2}] \cup [\widetilde{\widetilde{z}} + U - 2A - 12B\cos\frac{\Lambda_4^0}{2}, \widetilde{\widetilde{z}} + U - 2A + 12B\cos\frac{\Lambda_4^0}{2}] \bigcup [z'' + U - 2A - 12B\cos\frac{\Lambda_3^0}{2}, z'' + U - 2A + 12B\cos\frac{\Lambda_3^0}{2}]$. The discrete spectrum of the operator ${}^2\widetilde{H}_t^1$ is consists no more than one point. Here and hereafter, $\widetilde{\widetilde{z}}$ is an eigenvalue of the operator $\widetilde{H}_3$, and z'' is an eigenvalue of the operator $\widetilde{H}_5$.

2. Let

$$\frac{12Bcos\frac{\Lambda_3^0}{2}}{W} < U \leq \frac{12Bcos\frac{\Lambda_5^0}{2}}{W}.$$

Then the essential spectrum of the system second triplet-state operator ${}^2\widetilde{H}_t^1$ is the union of two segment $\sigma_{ess}({}^2\widetilde{H}_t^1) = [U - 4A - 12B\cos\frac{\Lambda_3^0}{2} - 12B\cos\frac{\Lambda_4^0}{2}, U - 4A + 12B\cos\frac{\Lambda_3^0}{2} + 12B\cos\frac{\Lambda_4^0}{2}] \bigcup [\widetilde{\widetilde{z}} + U - 2A - 12B\cos\frac{\Lambda_4^0}{2}, \widetilde{\widetilde{z}} + U - 2A + 12B\cos\frac{\Lambda_4^0}{2}]$, and the discrete spectrum of the system second triplet-state operator ${}^2\widetilde{H}_t^1$ is empty.

3. Let

$$\frac{12Bcos\frac{\Lambda_5^0}{2}}{W} < U \leq \frac{12Bcos\frac{\Lambda_3^0}{2}}{W}.$$

Then the essential spectrum of the system second triplet-state operator ${}^2\widetilde{H}_t^1$ is the union of two segment $\sigma_{ess}({}^2\widetilde{H}_t^1) = [U - 4A - 12B\cos\frac{\Lambda_3^0}{2} - 12B\cos\frac{\Lambda_4^0}{2}, U - 4A + 12B\cos\frac{\Lambda_3^0}{2} + 12B\cos\frac{\Lambda_4^0}{2}] \bigcup [z'' +$

$U-2A-12B\cos\frac{\Lambda_3^0}{2}, z''+U-2A+12B\cos\frac{\Lambda_3^0}{2}]$, *and the discrete spectrum of the system second triplet-state operator* ${}^2\widetilde{H}_t^1$ *is empty.*

4. Let

$$U \leq \frac{12Bcos\frac{\Lambda_3^0}{2}}{W}, cos\frac{\Lambda_3^0}{2} \leq cos\frac{\Lambda_5^0}{2}.$$

Then the essential spectrum of the system second triplet-state operator ${}^2\widetilde{H}_t^1$ *is consists of single segment* $\sigma_{ess}({}^2\widetilde{H}_t^1) = [U-4A-12B\cos\frac{\Lambda_3^0}{2}-12B\cos\frac{\Lambda_4^0}{2}, U-4A+12B\cos\frac{\Lambda_3^0}{2}+12B\cos\frac{\Lambda_4^0}{2}]$, *and the discrete spectrum of the system second triplet-state operator* ${}^2\widetilde{H}_t^1$ *is empty.*

Proof. The proof uses representation (14) and the results for the spectra of $\widetilde{H}_3$, $\widetilde{H}_4$, and $\widetilde{H}_5$ in the case $\nu = 3$. □

We have considered the case $\cos\frac{\Lambda_3^0}{2} > 0$, and $\cos\frac{\Lambda_5^0}{2} > 0$, The case $\cos\frac{\Lambda_3^0}{2} < 0$, and $\cos\frac{\Lambda_5^0}{2} < 0$, and other case is investigated similarly. Hence, in second triplet state, the four-electron system with $U > 0$ or $U < 0$ respectively has at most one antibound state or at most one bound state. The essential spectrum of the system is the union of at most three segments.

5. Threed triplet state

Theorem 5.1. *The subspace* ${}^3\mathcal{H}_1^t$ *is invariant under the operator* H, *and the operator* ${}^3H_1^t$ *is a bounded self-adjoint operator. It generates a bounded self-adjoint operator* ${}^3\overline{H}_1^t$ *acting in the space* l_2^{as} *as*

$$({}^3\overline{H}_1^t f)(m,n,p,r) = 4Af(m,n,p,r) + 2B\sum_\tau[f(m+\tau,n,p,r)+f(m,n+\tau,p,r)+f(m,n,p+\tau,r)+$$

$$+f(m,n,p,r+\tau)] + U[\delta_{m,n}f(m,n,p,r)+\delta_{n,p}f(m,n,p,r)+\delta_{n,r}f(m,n,p,r)], \tag{15}$$

where $\delta_{k,j}$ *is the Kronecker symbol. The operator* ${}^3H_1^t$ *acts on a vector* $\psi \in^3 \mathcal{H}_1^t$ *as*

$${}^3H_1^t\psi = \sum_{m,n,q,r}({}^3\overline{H}_1^t f)(m,n,p,r)^3 t_{m,n,p,r}^1. \tag{16}$$

Proof. The proof of the theorem follows from the explicit form of the action of the Hamiltonian H on vectors $\psi \in^3 \mathcal{H}_1^t$ using standard anticommutation relations for electron creation and annihilation operators. □

We set ${}^3\widetilde{H}_1^t = \mathcal{F}\,{}^3\overline{H}_1^t\mathcal{F}^{-1}$. In the quasimomentum representation, the operator ${}^3\overline{H}_1^t$ acts in the Hilbert space $L_2^{as}((T^\nu)^4$ as

$$({}^3\widetilde{H}_1^t f)(\lambda,\mu,\gamma,\theta) = h(\lambda,\mu,\gamma,\theta)f(\lambda,\mu,\gamma,\theta) + U[\int_{T^\nu} f(s,\lambda+\mu-s,\gamma,\theta)ds+$$

$$+\int_{T^\nu} f(\lambda,t,\mu+\gamma-t,\theta)dt + \int_{T^\nu} f(\lambda,t,\gamma,\mu,+\theta-t)dt],\ f\in L_2^{as}((T^\nu)^4), \tag{17}$$

where L_2^{as} is the subspace of antisymmetric functions in $L_2((T^\nu)^4$, and $h(\lambda,\mu,\gamma,\theta)$ has form (5).

Taking into account that the function $f(\lambda,\mu,\gamma,\theta)$ is antisymmetric, we can rewrite formula (17) as

$$({}^3\widetilde{H}_1^t f)(\lambda,\mu,\gamma,\theta) = \{2A+2B\sum_{i=1}^{\nu}[\cos\lambda_i+\cos\mu_i]\}f(\lambda,\mu,\gamma,\theta)+$$

$$+U\int_{T^\nu} f(s,\lambda+\mu-s,\gamma,\theta)ds+U\int_{T^\nu} f(\lambda,s,\mu+\gamma-s,\theta)ds+$$

$$+\{2A+2B\sum_{i=1}^{\nu}[\cos\gamma_i+\cos\theta_i]\}f(\lambda,\mu,\gamma,\theta)-U\int_{T^\nu} f(\lambda,\mu,s,\mu+\theta-s)ds. \quad (18)$$

We verify that the operator ${}^3\widetilde{H}_1^t$ can be represented in the form

$${}^3\widetilde{H}_1^t=\widetilde{H}_6\bigotimes I\bigotimes I+I\bigotimes\widetilde{H}_7\bigotimes I+I\bigotimes I\bigotimes\widetilde{H}_8, \quad (19)$$

where $(\widetilde{H}_6 f)(\lambda,\mu)=\{2A+2B\sum_{i=1}^{\nu}[\cos\lambda_i+\cos\mu_i]\}f(\lambda,\mu)+U\int_{T^\nu} f(s,\lambda+\mu-s)ds$, $(\widetilde{H}_7 f)(\mu,\gamma)=U\int_{T^\nu} f(s,\mu+\gamma-s)ds$, $(\widetilde{H}_8 f)(\gamma,\theta)=\{2A+2B\sum_{i=1}^{\nu}[\cos\gamma_i+\cos\theta_i]\}f(\gamma,\theta)-U\int_{T^\nu} f(s,\mu+\theta-s)ds$.

We must therefore investigate the spectrum and bound states of the operators $\widetilde{H}_6$, and $\widetilde{H}_7$, and $\widetilde{H}_8$.

Let the total quasimomentum of the two-electron system $\lambda+\mu=\Lambda_6$ be fixed.

By $L_2(\Gamma_{\Lambda_6})$ we denote the space of functions that are square integrable on the manifold $\Gamma_{\Lambda_6}=\{(\lambda,\mu):\lambda+\mu=\Lambda_6\}$.

That the operator $\widetilde{H}_6$ and the space $\mathcal{H}_2\equiv L_2((T^\nu)^2)$ can be decomposed into a direct integral of operators $\widetilde{H}_{6\Lambda_6}$ and spaces $\mathcal{H}_{2\Lambda_6}=L_2(\Gamma_{\Lambda_6})$,

$$\widetilde{H}_6=\bigoplus\int_{T^\nu}\widetilde{H}_{6\Lambda_6}d\Lambda_6,\quad \mathcal{H}_2=\bigoplus\int_{T^\nu}\mathcal{H}_{2\Lambda_6}d\Lambda_6,$$

such that the spaces $\mathcal{H}_{2\Lambda_6}$ are then invariant under the operators $\widetilde{H}_{6\Lambda_6}$ and each operator $\widetilde{H}_{6\Lambda_6}$ acts in $\mathcal{H}_{2\Lambda_6}$ as $(\widetilde{H}_{6\Lambda_6} f_{\Lambda_6})(\lambda)=\{2A+4B\sum_{i=1}^{\nu}\cos\frac{\Lambda_6^i}{2}\cos(\frac{\Lambda_6^i}{2}-\lambda_i)\}f_{\Lambda_6}(\lambda)+U\int_{T^\nu} f_{\Lambda_6}(s)ds$, where $f_{\Lambda_6}(\lambda)=f(\lambda,\Lambda_6-\lambda)$.

It is known that the continuous spectrum of $\widetilde{H}_{6\Lambda_6}$ is independent of U and coincides with the segment $[m_{\Lambda_6},M_{\Lambda_6}]$, where $m_{\Lambda_6}=inf_{x\in T^\nu}h_{\Lambda_6}(x), M_{\Lambda_6}=sup_{x\in T^\nu}h_{\Lambda_6}(x)$, here $h_{\Lambda_6}(x)=2A+4B\sum_{i=1}^{\nu}\cos\frac{\Lambda_6^i}{2}\cos(\frac{\Lambda_6^i}{2}-x_i)$.

We set $\Delta_{\Lambda_6}^{\nu}(z)=1+U\int_{T^\nu}\frac{ds_1ds_2...ds_\nu}{2A+4B\sum_{i=1}^{\nu}\cos\frac{\Lambda_6^i}{2}\cos(\frac{\Lambda_6^i}{2}-s_i)-z}$.

Lemma 5.2. *The number $z=z_0$ not belonging to the continuous spectrum of the operator $\widetilde{H}_6$ is an eigenvalue of that operator if and only if it is a zero of the function $\Delta_{\Lambda_6}^{\nu}(z)$, i.e., $\Delta_{\Lambda_6}^{\nu}(z_0)=0$.*

Proof. The proof of the statements in the lemma 5.2 is similar to the proof of lemma 3.2. □

It is clear that the continuous spectrum of $\widetilde{H}_6$ coincides with the segment

$$[m_{\Lambda_6}^{\nu},M_{\Lambda_6}^{\nu}]=[2A-4B\sum_{i=1}^{\nu}cos\frac{\Lambda_6^i}{2},2A+4B\sum_{i=1}^{\nu}cos\frac{\Lambda_6^i}{2}].$$

We consider the one-dimensional case.

Theorem 5.3. *At $\nu=1$ and $U>0$ and for all values of the parameter of the Hamiltonian, the operator $\widetilde{H}_6$ has a unique eigenvalue $z=2A+\sqrt{U^2+16B^2cos^2\frac{\Lambda_6}{2}}$ that is above the continuous spectrum of $\widetilde{H}_6$, i.e., $z>M_{\Lambda_6}^1$.*

Proof. In the one-dimensional case, the function $\Delta^1_{\Lambda_6}(z)$ increases monotonically outside the continuous spectrum domain of the operator $\widetilde{H}_6$. For $z < m^1_{\Lambda_6}$, the function $\Delta^1_{\Lambda_6}(z)$ increases from 1 to $+\infty$,

$\Delta^1_{\Lambda_6}(z) \to 1$ as $z \to -\infty$
$\Delta^1_{\Lambda_6}(z) \to +\infty$ as $z \to m^1_{\Lambda_6} - 0$.

Therefore, below the value $m^1_{\Lambda_6}$, $\Delta^1_{\Lambda_6}(z)$ cannot vanish. For $z > M^1_{\Lambda_6}$, the function $\Delta^1_{\Lambda_6}(z)$ increases from $-\infty$ to 1,

$\Delta^1_{\Lambda_6}(z) \to -\infty$ as $z \to M^1_{\Lambda_6} + 0$,
$\Delta^1_{\Lambda_6}(z) \to +1$ as $z \to +\infty$.

Therefore, above the value $M^1_{\Lambda_6}$, the function $\Delta^1_{\Lambda_6}(z)$ has a single zero at the point $z = 2A + \sqrt{U^2 + 16B^2 cos^2 \frac{\Lambda_6}{2}}$. □

We now consider the two-dimensional case. at a single point

Theorem 5.4. *At $\nu = 2$ and $U > 0$ and for all values of the parameter of the Hamiltonian, the operator $\widetilde{H}_6$ has a unique eigenvalue $\widetilde{z}$ that is above the continuous spectrum of $\widetilde{H}_6$, i.e., $z > M^2_{\Lambda_6}$.*

Proof. In the two-dimensional case, the function $\Delta^2_{\Lambda_6}(z)$ increases monotonically outside the continuous spectrum domain of $\widetilde{H}_6$. For $z < m^2_{\Lambda_6}$, the function $\Delta^2_{\Lambda_6}(z)$ increases from 1 to $+\infty$,

$\Delta^2_{\Lambda_6}(z) \to 1$ as $z \to +\infty$
$\Delta^2_{\Lambda_6} \to -\infty$ as $z \to m^2_{\Lambda_6} - 0$.

Therefore, below $m^2_{\Lambda_6}$, the function $\Delta^2_{\Lambda_6}(z)$ cannot vanish. For $z > M^2_{\Lambda_6}$, the function $\Delta^2_{\Lambda_6}(z)$ increases from $-\infty$ to 1,

$\Delta^2_{\Lambda_6}(z) \to -\infty$ as $z \to M^2_{\Lambda_6} + 0$,
$\Delta^2_{\Lambda_6}(z) \to 1$ as $z \to +\infty$.

Therefore, above $M^2_{\Lambda_6}$, the function $\Delta^2_{\Lambda_6}(z)$ has a single zero at the point $\widetilde{z} > M^2_{\Lambda_6}$. □

In the three-dimensional case, the operator $\widetilde{H}_6$ for some values of the parameters in the Hamiltonian has a unique eigenvalue $\widetilde{\widetilde{z}}$, such that $\widetilde{\widetilde{z}} > M^3_{\Lambda_6}$.

Theorem 5.5. *At $\nu = 3$ and $U > 0$ and the total quasimomentum Λ_6 of the system have the form*

$$\Lambda_6 = (\Lambda_6^1; \Lambda_6^2; \Lambda_6^3) = (\Lambda_6^0; \Lambda_6^0; \Lambda_6^0).$$

Then the operator $\widetilde{H}_6$ has a unique eigenvalue $\widetilde{\widetilde{z}}$ if

$$U > \frac{12B \cos \frac{\Lambda_6^0}{2}}{W}.$$

Otherwise, the operator $\widetilde{H}_6$ has no eigenvalue.

Proof. In the three-dimensional case,

$\Delta^3_{\Lambda_6}(z) \to 1$ as $z \to -\infty$
$\Delta^3_{\Lambda_6}(z) \to 1 + \frac{UW}{12Bcos(\Lambda_6^0/2)}$ as $z \to m^3_{\Lambda_6} - 0$.

Hence, the function $\Delta^3_{\Lambda_6}(z)$ cannot vanish below the continuous spectrum. Above the continuous spectrum,

$\Delta^3_{\Lambda_6}(z) \to 1 - \frac{UW}{12Bcos(\Lambda_6^0/2}$ as $z \to M^3_{\Lambda_6} + 0$,
$\Delta^3_{\Lambda_6}(z) \to 1$ as $z \to +\infty$.

Inspecting the equation $\Delta^3_{\Lambda_6}(z) = 0$ above the continuous spectrum domain of the operator $\widetilde{H}_6$, we obtain the statement of the theorem. □

We note that the converse situation is realized for $U < 0$.

The spectrum of operator $(\widetilde{H}_7 f)(\mu,\gamma) = U\int_{T^\nu} f(s,\mu+\gamma-s)ds$ is purely discrete and consists is a single point $z = U$, for arbitrary value ν.

Now we investigated the spectrum of the operator
$(\widetilde{H}_8 f)(\gamma,\theta) = \{2A + 2B\sum_{i=1}^{\nu}[\cos\gamma_i + \cos\theta_i]\}f(\gamma,\theta) - U\int_{T^\nu} f(s,\mu+\theta-s)ds$.

We let $\Lambda_4 = \mu+\theta$, and $\Lambda_5 = \gamma+\theta$.

It is known that the continuous spectrum of $\widetilde{H}_8$ is independent of U and coincides with the segment $[m^\nu_{\Lambda_5}, M^\nu_{\Lambda_5}]$, where $m^\nu_{\Lambda_5} = 2A - 4B\sum_{i=1}^{\nu} cos\frac{\Lambda_5^i}{2}$, $M^\nu_{\Lambda_5} = 2A + 4B\sum_{i=1}^{\nu} cos\frac{\Lambda_5^i}{2}$.

We let $L_2(\Gamma_{\Lambda_5})$ denote the space of functions that are square integrable on the manifold $\Gamma_{\Lambda_5} = \{(\gamma,\theta) : \gamma+\theta = \Lambda_5\}$.

That the operator $\widetilde{H}_8$ and the space $\mathcal{H}_2 \equiv L_2((T^\nu)^2)$ can be decomposed into a direct integral of operators $\widetilde{H}_{8\Lambda_5}$ and spaces $\mathcal{H}_{2\Lambda_5} = L_2(\Gamma_{\Lambda_5})$,

$$\widetilde{H}_8 = \bigoplus\int_{T^\nu} \widetilde{H}_{8\Lambda_5} d\Lambda_5, \quad \mathcal{H}_2 = \bigoplus\int_{T^\nu} \mathcal{H}_{2\Lambda_5} d\Lambda_5,$$

such that the spaces $\mathcal{H}_{2\Lambda_5}$ are then invariant under the operators $\widetilde{H}_{8\Lambda_5}$ and each operator $\widetilde{H}_{8\Lambda_5}$ acts in $\mathcal{H}_{2\Lambda_5}$ as $(\widetilde{H}_{8\Lambda_5} f_{\Lambda_5})(\lambda) = \{2A + 4B\sum_{i=1}^{\nu}\cos\frac{\Lambda_5^i}{2}\cos(\frac{\Lambda_5^i}{2} - \lambda_i)\}f_{\Lambda_5}(\lambda) - U\int_{T^\nu} f_{\Lambda_4}(s)ds$, where $f_{\Lambda_5}(\lambda) = f(\lambda, \Lambda_5 - \lambda)$.

We set $\Delta^\nu_{\Lambda_5}(z) = 1 - U\int_{T^\nu}\frac{ds_1 ds_2 \ldots ds_\nu}{2A+4B\sum_{i=1}^{\nu}\cos\frac{\Lambda_5^i}{2}\cos(\frac{\Lambda_5^i}{2}-s_i)-z}$.

Lemma 5.6. *The number $z = z_0$ not belonging to the continuous spectrum of the operator $\widetilde{H}_8$ is an eigenvalue of that operator if and only if it is a zero of the function $\Delta^\nu_{\Lambda_5}(z)$, i.e., $\Delta^\nu_{\Lambda_5}(z_0) = 0$.*

Proof. The proof of the statements in the lemma 5.6 is similar to the proof of lemma 3.2. □

We consider the one-dimensional case.

Theorem 5.7. *At $\nu = 1$ and $U > 0$ and for all values of the parameter of the Hamiltonian, the operator $\widetilde{H}_8$ has a unique eigenvalue $z = 2A - \sqrt{U^2 + 16B^2 cos^2\frac{\Lambda_5}{2}}$ that is below the continuous spectrum of $\widetilde{H}_8$, i.e., $z < m^1_{\Lambda_5}$.*

Proof. In the one-dimensional case, the function $\Delta^1_{\Lambda_5}(z)$ decreases monotonically outside the continuous spectrum domain of the operator $\widetilde{H}_5$. For $z < m^1_{\Lambda_5}$, the function $\Delta^1_{\Lambda_5}(z)$ decreases from 1 to $-\infty$,

$\Delta^1_{\Lambda_5}(z) \to 1$ as $z \to -\infty$
$\Delta^1_{\Lambda_5}(z) \to -\infty$ as $z \to m^1_{\Lambda_5} - 0$.

Therefore, below the value $m^1_{\Lambda_5}$, the function $\Delta^1_{\Lambda_5}(z)$ has a single zero at the point $z = 2A - \sqrt{U^2 + 16B^2 cos^2\frac{\Lambda_5}{2}}$. For $z > M^1_{\Lambda_5}$, the function $\Delta^1_{\Lambda_5}(z)$ decreases from $+\infty$ to 1,

$\Delta^1_{\Lambda_5}(z) \to +\infty$ as $z \to M^1_{\Lambda_5} + 0$,
$\Delta^1_{\Lambda_5}(z) \to +1$ as $z \to +\infty$.

Therefore, above the value $M^1_{\Lambda_5}$, the function $\Delta^1_{\Lambda_5}(z)$ cannot vanish. □

We now consider the two-dimensional case.

Theorem 5.8. *At $\nu = 2$ and $U > 0$ and for all values of the parameter of the Hamiltonian, the operator $\widetilde{H}_8$ has a unique eigenvalue z' that is below the continuous spectrum of $\widetilde{H}_8$, i.e., $z' < m^2_{\Lambda_5}$.*

Proof. In the two-dimensional case, the function $\Delta^2_{\Lambda_5}(z)$ decreases monotonically outside the continuous spectrum domain of $\widetilde{H}_8$. For $z < m^2_{\Lambda_5}$, the function $\Delta^2_{\Lambda_5}(z)$ decreases from 1 to $-\infty$,

$\Delta^2_{\Lambda_5}(z) \to 1$ as $z \to -\infty$

$\Delta^2_{\Lambda_5} \to -\infty$ as $z \to m^2_{\Lambda_5} - 0.$

Therefore, below $m^2_{\Lambda_5}$, the function $\Delta^2_{\Lambda_5}(z)$ has a single zero at the point $z' < m^2_{\Lambda_5}$. For $z > M^2_{\Lambda_5}$, the function $\Delta^2_{\Lambda_5}(z)$ decreases from $+\infty$ to 1,

$\Delta^2_{\Lambda_5}(z) \to +\infty$ as $z \to M^2_{\Lambda_5} + 0,$

$\Delta^2_{\Lambda_5}(z) \to 1$ as $z \to +\infty.$

Therefore, above $M^2_{\Lambda_5}$, the function $\Delta^2_{\Lambda_5}(z)$ cannot vanish. □

In the three-dimensional case, the operator $\widetilde{H}_8$ for some values of the parameters in the Hamiltonian has a unique eigenvalue z'', such that $z'' < m^3_{\Lambda_5}$.

Theorem 5.9. *At $\nu = 3$ and $U > 0$ and the total quasimomentum Λ_5 of the system have the form*

$$\Lambda_5 = (\Lambda_5^1; \Lambda_5^2; \Lambda_5^3) = (\Lambda_5^0; \Lambda_3^0; \Lambda_3^0).$$

Then the operator $\widetilde{H}_8$ has a unique eigenvalue z'' if

$$U > \frac{12B\cos\frac{\Lambda_5^0}{2}}{W}.$$

Otherwise, the operator $\widetilde{H}_8$ has no eigenvalue.

Proof. In the three-dimensional case,

$\Delta^3_{\Lambda_5}(z) \to 1$ as $z \to -\infty$

$\Delta^3_{\Lambda_5}(z) \to 1 - \frac{UW}{12Bcos(\Lambda_5^0/2)}$ as $z \to m^3_{\Lambda_5} - 0.$

Hence, the function $\Delta^3_{\Lambda_5}(z)$ has a single zero at the point z'' below the continuous spectrum. Above the continuous spectrum,

$\Delta^3_{\Lambda_5}(z) \to 1 + \frac{UW}{12Bcos(\Lambda_5^0/2)}$ as $z \to M^3_{\Lambda_5} + 0,$

$\Delta^3_{\Lambda_5}(z) \to 1$ as $z \to +\infty.$

Inspecting the equation $\Delta^3_{\Lambda_5}(z) = 0$ above the continuous spectrum domain of the operator $\widetilde{H}_8$, we obtain the statement of the theorem. □

Now, using the obtained results and representation (19), we describe the structure of the essential spectrum and the discrete spectrum of the operator ${}^3\widetilde{H}_t^1$.

Theorem 5.10. *At $\nu = 1$ and $U > 0$ the essential spectrum of the system third triplet-state operator ${}^3\widetilde{H}_t^1$ is exactly the union of three segments: $\sigma_{ess}({}^3\widetilde{H}_t^1) = [U + 4A - 4B\cos\frac{\Lambda_5}{2} - 4B\cos\frac{\Lambda_6}{2}, U + 4A + 4B\cos\frac{\Lambda_5}{2} + 4B\cos\frac{\Lambda_6}{2}] \cup [U + 4A - 4B\cos\frac{\Lambda_6}{2} - \sqrt{U^2 + 16B^2cos^2\frac{\Lambda_5}{2}}, U + 4A + 4B\cos\frac{\Lambda_6}{2} - \sqrt{U^2 + 16B^2cos^2\frac{\Lambda_5}{2}}] \bigcup [U + 4A - 4Bcos^2\frac{\Lambda_5}{2} + \sqrt{U^2 + 16B^2cos^2\frac{\Lambda_6}{2}}, U + 4A + 4Bcos^2\frac{\Lambda_5}{2} + \sqrt{U^2 + 16B^2cos^2\frac{\Lambda_6}{2}}]$. The discrete spectrum of the operator ${}^3\widetilde{H}_t^1$ is consists no more than one point.*

Proof. It follows from representation (19) that

$$\sigma({}^3\widetilde{H}_t^1) = \{\lambda + \mu + \theta : \lambda \in \sigma(\widetilde{H}_6), \mu \in \sigma(\widetilde{H}_7), \theta \in \sigma(\widetilde{H}_8)\},$$

and in the one-dimensional case, the continuous spectrum of $\widetilde{H}_6$ is

$$\sigma_{cont}(\widetilde{H}_6) = [2A - 4B\cos\frac{\Lambda_6}{2}, 2A + 4B\cos\frac{\Lambda_6}{2}],$$

and the discrete spectrum of $\widetilde{H}_6$ consists of a single point $z_1 = 2A + \sqrt{U^2 + 16B^2cos^2\frac{\Lambda_6}{2}}$.

The spectra of the operator $\widetilde{H}_7$ is a purely discrete and consists is a single point $z = U$, for arbitrary value ν.

The continuous spectrum of the operator $\widetilde{H}_8$ consists of the segment $[2A - 4B\cos\frac{\Lambda_5}{2}, 2A + 4B\cos\frac{\Lambda_5}{2}]$, and the discrete spectrum of $\widetilde{H}_8$ consists of a single point $z_2 = 2A - \sqrt{U^2 + 16B^2cos^2\frac{\Lambda_5}{2}}$. Therefore, the essential spectrum of the system third triplet-state operator ${}^3\widetilde{H}_t^1$ is the union of three segments, and the third triplet-state operator ${}^3\widetilde{H}_t^1$ has a single eigenvalue $z = z_1 + z_2 + U$. If these eigenvalue lie the outside the essential spectrum of operator ${}^3\widetilde{H}_t^1$, then these eigenvalue is belonging the discrete spectrum of the operator ${}^3\widetilde{H}_t^1$, otherwise, the discrete spectrum of operator ${}^3\widetilde{H}_t^1$, is empty. □

The following theorem is proved totally similarly to Theorem 5.10.

Theorem 5.11. *At $\nu = 2$ and $U > 0$ the essential spectrum of the system third triplet-state operator ${}^3\widetilde{H}_t^1$ is exactly the union of three segments:* $\sigma_{ess}({}^3\widetilde{H}_t^1) = [U + 4A - 4B\sum_{i=1}^2 \cos\frac{\Lambda_6^i}{2} - 4B\sum_{i=1}^2 \cos\frac{\Lambda_5^i}{2}, U + 4A + 4B\sum_{i=1}^2 \cos\frac{\Lambda_6^i}{2} + 4B\sum_{i=1}^2 \cos\frac{\Lambda_5^i}{2}] \cup [U + 2A + \widetilde{z} - 4B\sum_{i=1}^2 \cos\frac{\Lambda_5^i}{2}, U + 2A + \widetilde{z} + 4B\sum_{i=1}^2 \cos\frac{\Lambda_5^i}{2}] \cup [U + 2A + z' - 4B\sum_{i=1}^2 \cos\frac{\Lambda_6^i}{2}, U + 2A + z' + 4B\sum_{i=1}^2 \cos\frac{\Lambda_6^i}{2}]$. *The discrete spectrum of the operator ${}^3\widetilde{H}_t^1$ is consists no more than one point.*

Let $\nu = 3$ and $\Lambda_5 = (\Lambda_5^0, \Lambda_5^0, \Lambda_5^0)$, and $\Lambda_6 = (\Lambda_6^0, \Lambda_6^0, \Lambda_6^0)$.

Theorem 5.12. *The following statements hold:*

1. Let

$$U > \frac{12Bcos\frac{\Lambda_6^0}{2}}{W}, cos\frac{\Lambda_6^0}{2} \geq \cos\frac{\Lambda_5^0}{2}.$$

Then the essential spectrum of the system third triplet-state operator ${}^3\widetilde{H}_t^1$ is the union of three segments $\sigma_{ess}({}^3\widetilde{H}_t^1) = [U + 4A - 12B\cos\frac{\Lambda_6^0}{2} - 12B\cos\frac{\Lambda_5^0}{2}, U + 4A + 12B\cos\frac{\Lambda_6^0}{2} + 12B\cos\frac{\Lambda_5^0}{2}] \cup [\widetilde{\widetilde{z}} + U + 2A - 12B\cos\frac{\Lambda_5^0}{2}, \widetilde{\widetilde{z}} + U + 2A + 12B\cos\frac{\Lambda_5^0}{2}] \bigcup [z'' + U + 2A - 12B\cos\frac{\Lambda_6^0}{2}, z'' + U + 2A + 12B\cos\frac{\Lambda_6^0}{2}]$. *The discrete spectrum of the operator ${}^3\widetilde{H}_t^1$ is consists no more than one point. Here and hereafter, $\widetilde{\widetilde{z}}$ is an eigenvalue of the operator $\widetilde{H}_6$, and z'' is an eigenvalue of the operator $\widetilde{H}_8$.*

2. Let

$$\frac{12Bcos\frac{\Lambda_6^0}{2}}{W} < U \leq \frac{12Bcos\frac{\Lambda_5^0}{2}}{W}.$$

Then the essential spectrum of the system third triplet-state operator ${}^3\widetilde{H}_t^1$ is the union of two segment $\sigma_{ess}({}^3\widetilde{H}_t^1) = [U + 4A - 12B\cos\frac{\Lambda_6^0}{2} - 12B\cos\frac{\Lambda_5^0}{2}, U + 4A + 12B\cos\frac{\Lambda_6^0}{2} + 12B\cos\frac{\Lambda_5^0}{2}] \bigcup [\widetilde{\widetilde{z}} + U + 2A - 12B\cos\frac{\Lambda_5^0}{2}, \widetilde{\widetilde{z}} + U + 2A + 12B\cos\frac{\Lambda_5^0}{2}]$, *and the discrete spectrum of the system third triplet-state operator ${}^3\widetilde{H}_t^1$ is empty.*

3. Let

$$\frac{12Bcos\frac{\Lambda_5^0}{2}}{W} < U \leq \frac{12Bcos\frac{\Lambda_6^0}{2}}{W}.$$

Then the essential spectrum of the system third triplet-state operator ${}^3\widetilde{H}_t^1$ is the union of two segment $\sigma_{ess}({}^3\widetilde{H}_t^1) = [U + 4A - 12B\cos\frac{\Lambda_6^0}{2} - 12B\cos\frac{\Lambda_5^0}{2}, U + 4A + 12B\cos\frac{\Lambda_6^0}{2} + 12B\cos\frac{\Lambda_5^0}{2}] \bigcup [z'' +$

$U+2A-12B\cos\frac{\Lambda_6^0}{2}, z''+U+2A+12B\cos\frac{\Lambda_6^0}{2}]$, *and the discrete spectrum of the system second triplet-state operator* $^2\widetilde{H}_t^1$ *is empty.*

4. Let

$$U \leq \frac{12Bcos\frac{\Lambda_6^0}{2}}{W}, cos\frac{\Lambda_6^0}{2} \leq cos\frac{\Lambda_5^0}{2}.$$

Then the essential spectrum of the system third triplet-state operator $^3\widetilde{H}_t^1$ *is consists of single segment* $\sigma_{ess}(^3\widetilde{H}_t^1) = [U+4A-12B\cos\frac{\Lambda_6^0}{2}-12B\cos\frac{\Lambda_5^0}{2}, U+4A+12B\cos\frac{\Lambda_6^0}{2}+12B\cos\frac{\Lambda_5^0}{2}]$, *and the discrete spectrum of the system second triplet-state operator* $^3\widetilde{H}_t^1$ *is empty.*

Proof. The proof uses representation (19) and the results for the spectra of $\widetilde{H}_6$, $\widetilde{H}_7$, and $\widetilde{H}_8$ in the case $\nu = 3$. □

We have considered the case $\cos\frac{\Lambda_6^0}{2} > 0$, and $\cos\frac{\Lambda_5^0}{2} > 0$, The case $\cos\frac{\Lambda_6^0}{2} < 0$, and $\cos\frac{\Lambda_5^0}{2} < 0$, and other case is investigated similarly. Hence, in third triplet state, the four-electron system with $U > 0$ or $U < 0$ respectively has at most one antibound state or at most one bound state. The essential spectrum of the system is the union of at most three segments.

We see the spectra of these three triplet states are the different.

References

[1] Anderson P W 1961. *Phys. Rev.*, **124** 41–53
[2] Gutzwiller M C 1963 *Phys. Rev. Lett.*, **10** 159–162.
[3] Hubbard J. 1963 *Proc. Roy. Soc. A.*, **276** 238–257
[4] Izyumov Yu A and Skryabin Yu N 1988 *Statistical Mechanics of Magnetically Ordered Systems.* (New York: Consultants Bureau)
[5] Ichinose T 1978 *Trans. Amer. Math.Soc.* **235** 75–113
[6] Ichinose T 1978. *Trans. Amer. Math. Soc.* **237** 223–254
[7] Ichinose T 1982 *Banach Center Publ.* **8** 294–300.
[8] Kanamori J 1963 *Prog. Theor. Phys.* **30** 275–289
[9] Karpenko B V, Dyakin V V and Burdina G L 1986. *Phys. Met. Metallogr.* **61** 702-706
[10] Lieb E 1989 *Phys. Rev. Lett.* **62** 1201–1204
[11] Mattis D 1986 *Rev. Mod. Phys.* **58** 370–379
[12] Naimark M A 1959 *Normed Rings* (Groningen: Noordhoff)
[13] Reed M and Simon B 1978 *Methods of Modern Mathetical Physics*, Vol 1 (New York: Acad. Press)
[14] Reed M and Simon B 1982 *Methods of Modern Mathetical Physics* Vol 4 (New York: Acad. Press)
[15] Shubin S P and Wonsowsky S V 1934 *Proc. Roy. Soc. A.* **145** 159–172
[16] Tashpulatov S M 2014 *Theor. Math. Phys.* **179** 712–728
[17] Tsvelick A M and Wiegman P B 1983 *Adv. Phys.* **32** 453–713
[18] Valkov V V, Ovchinnikov S G and Petrakovskii O P 1988. *Sov. Phys. Solid State.* **30** 3044–3047